7 00

SPACE STATION:
POLICY, PLANNING AND UTILIZATION

Proceedings of the AIAA/NASA Symposium on the Space Station
Arlington, Virginia, July 18-20, 1983

Organized and Operated by the American Institute of Aeronautics and Astronautics (AIAA)

Sponsored by National Aeronautics and Space Administration (NASA)

Edited by
Mireille Gerard and Pamela W. Edwards
AIAA

October 15, 1983
American Institute of Aeronautics and Astronautics
1633 Broadway, New York, NY 10019
212-581-4300

American Institute of Aeronautics and Astronautics
New York, New York

Library of Congress Cataloging in Publication Data

AIAA/NASA Symposium on the Space Station
(1983: Arlington, Va.)
Space station: policy, planning, and utilization

(AIAA aerospace assessment series; v. 10)
1. Space stations—Congresses. I. Gerard, Mireille. II. Edwards, Pamela W. III. American Institute of Aeronautics and Astronautics. IV. United States. National Aeronautics and Space Administration. V. Title. VI. Series.
TL797.A38 1983 629.44'2 83-15699
ISBN 0-915928-80-9

TABLE OF CONTENTS

PREFACE

The AIAA/NASA Symposium on the Space Station was conceived to present to all interested sectors and individuals the current thinking on the designs, technology and potential applications of an Earth-orbiting space station system. The Symposium, which was held in Arlington, Virginia on July 18-20, 1983, provided a forum for presentation, discussion, and debate on the issues, facts, and judgments related to the utility, desirability, and operational feasibility of a space station. This volume is a collection of those presentations and discussions.

The American Institute of Aeronautics and Astronautics (AIAA) is a society of over 34,000 engineers, scientists and students whose knowledge encompasses all elements of the subject described in this book. By cohosting the Symposium and publishing its proceedings, the AIAA has helped to fulfill its primary role as a communicator of new ideas. We hope that this volume will serve as the basis for future technical, policy, and international cooperation discussions on the space station system.

The Symposium Steering Committee

George V. Butler, McDonnell Douglas Astronautics Company

Burton I. Edelson, NASA Headquarters

Robert F. Freitag, NASA Headquarters

Jerry Grey, AIAA

Daniel E. Herman, NASA Headquarters

J. Preston Layton, Consultant

James L. Hieatt, TRW Space and Technology Group

Richard L. Kline, Grumman Aerospace Corporation

Wilfred J. Mellors, European Space Agency

Stanley G. Rosen, Los Angeles Air Force Station

Carl W. L. Watkins, NASA Headquarters

WELCOME

John L. McLucas*
President-Elect
American Institute of
Aeronautics and Astronautics
New York

The AIAA and NASA have joined many times in the past to put together conferences, symposia, and so forth. I believe that we have found these to be mutually profitable.

The fate of the space station concept certainly means much to NASA and its future. There is no doubt that it is one of the most important initiatives that NASA can possibly take. A firm commitment to the space station would no doubt set the course for NASA's major thrust, in a technological sense, for many years to come.

The AIAA lends its support to this symposium in a spirit of open inquiry. We want to look at the pros and the cons of going ahead with a space station at this stage. The viability of the space station concept will depend upon our success in carefully defining the role envisioned for the space station, spelling out the resources it will take to put the program into place, fashioning a realistic timetable for implementing the program, and articulating the benefits that will flow from it.

The extensive roster of subjects to be dealt with during the symposium attest to the numerous ways in which the space station could prove useful, including research and development, military activities, repair and fueling functions, for example, as a transportation hub, and for pharmaceutical manufacturing and materials processing. All of these uses warrant thorough and open-minded investigation. And they are only the benefits that we can foresee standing where we are today. Undoubtedly, once such a structure was put in position and was operating,our experience with its actual use will open up new possibilities that have not been envisioned yet, even by the visionaries in our midst. And it's quite possible that these unanticipated applications will turn out to be the most advantageous of all.

I was intrigued by a report in the July 1983 issue of Science that George Keyworth, President Reagan's science advisor, has called for the United States to consider undertaking an ambitious new space endeavor similar to the Apollo landings. Embarking on the space station program could be the way to answer his initiative. Perhaps Keyworth's comments signal a change in the attitude of the White House toward the space program.

However, it may be premature to jump to any conclusions. I was just looking at the article again, and he says, "If you see the space station as the intermediate platform for extended missions, such as a lunar colony or a Mars mission, its difficult for me to imagine doing that without a manned space station." Whatever the Administration's eventual stance may be, the task of selling the concept of the space station will be a tough one. If it is decided that the space station deserves to become the next major step in the space program, we must be prepared to demonstrate conclusively the return to the American people of such an ambitious undertaking.

On the one hand, I think NASA can count on a large reservoir of good will for its programs and its objectives. On the other hand, the precedent of the lengthy development of the Shuttle will be cited in opposition and will prompt many to look with disfavor on any such similar initiative that NASA might take. To counteract this, we must insure that management of the project is sound and that we do not mislead people by overpromising and overselling the concept.

I note with interest the October 1983 issue of the IEEE Spectrum commemorating the 25th anniversary of NASA. I've had the privilege of previewing a number of articles for that magazine and I find it to be a very well done issue. Tracing the history of NASA and projecting the future is a worthwhile effort, and it certainly behooves each one of us to do our own analysis and review and to decide how our efforts could be put to the common good as we face a new challenge for NASA.

From taped transcript.

*President, COMSAT World Systems Division, Washington, D.C.

KEYNOTE ADDRESS

The Honorable James M. Beggs
Administrator
National Aeronautics and Space Administration
Washington, D.C.

The subject of this symposium is, of course, very timely. We're very pleased to see AIAA supporting a symposium of this type, because it will be of great help to us in focusing our ideas. And, I hope, it will also raise some thoughts in your mind about how we proceed from this point to a successful space station.

Jack Lousma will cover most of the various functions that will be integrated into the space station, so I will not discuss that aspect of the program. Rather, I will restrict my remarks to the whys and wherefores and the timing.

We have been, of course, very pleased with the progress we've made on the Shuttle. It has moved forward very well after a slip of three years. It has demonstrated that the work we have done together over the past 10 years was well conceived and is going to be very fruitful. We are doing very well with the system. Each time it flies, it is demonstrating increased capabilities. And as we move into the time when we will have the TDRS system fully operational, we will have an immense capability to do scientific and engineering work with the Shuttle.

But as that old Russian space pioneer, Konstantin Tsiolkowsky once said, "The Earth is the cradle of mankind, but man cannot stay in the cradle forever."

And while the Shuttle is truly impressive, it allows us only a short time in space. While we can extend that time up to perhaps a month, or maybe even a little longer, we cannot extend it to long-duration, long-endurance space flight. For that purpose, NASA has always had in mind, going back, in fact, to the origins of the Agency, the idea that sooner or later we should have a space station.

A space station would be the node or the point in space in which we would do our long-duration work. It would give us the means to go beyond low Earth orbit and to do all the other things in a manned and unmanned way that we have always dreamed of doing.

The Shuttle is approaching the point when I think it will be routinely operational. We say it's operational now, but we all know that we have some work to do to make it a truly routine operation. But we're making progress in a very dramatic and timely way. We are on schedule in almost everything we are doing -- which is very satisfying.

But with the advent of the Shuttle, as we look around the world, we see that we are being challenged very significantly by a number of nations. The Soviets, of course, have always had an aggressive program. They have always spent significant amounts of resources in pursuing their program. They have done some very impressive things over the last few years; the work they are doing on large boosters, and on their space station with the Salyut series has been very impressive indeed.

But in addition to that, we now see the Europeans, who have long had a desire and an ambition to be more active in space, coming along very rapidly. The advent of the Ariane, which has a space launch capability that is quite remarkable, and the fact that they have announced their intention to go beyond that to enter the field of manned space flight as well, is an indication of the seriousness with which they are approaching the challenge of the future. They intend to be competitive and they intend to be competent, just as we do.

The Japanese also have ambitions in space. They are starting in a modest way, but you can be sure that they will proceed from there and become competitive across the board.

So I think we can anticipate that this competition around the world will be a fact of life in the foreseeable future. The United States could not have expected to have a leadership position forever.

We are still preeminent. But we must proceed step by step on a plan that will provide increasing capabilities as time goes on.

As you know, our planning for the space station has been going on for a little over a year throughout the Agency. We've invited industry, just as we always do, into that planning activity, and we intend to accelerate that in the next year.

From edited transcript.

I believe we will have an approved program in the near future, hopefully within six months to a year.

The purpose of proceeding with this program is, of course, to maintain our leadership. We have held that leadership since we initiated the Apollo program almost 20 years ago. The President's policy, which was reissued and redrawn in 1982, sets that goal of leadership. It will require that we work very hard in the face of the challenge that's being put to us. I know that we can meet that challenge and do the job.

The first 25 years of space have been extremely productive, and I believe that the next 25 years will be even more so. The space station, as we envision it, will be used by all aspects of our program.

It will be used by the scientists, because we can do more science -- and more complex science -- using the space station as a base, and with scientific payloads which can be tended, improved, and repaired as time goes on.

It will be used by the private sector. Jack Lousma will mention the electrophoresis experiments that the McDonnell Douglas and Johnson & Johnson people want to fly on the space station. Certainly, the development of that process will be facilitated by the space station. And I believe the station will make possible the advent of the truly commercial era in space that we have been dreaming of for the last 25 years.

Of course, NASA will use the space station for its own activities and will continue to progress in the areas of technology. In essence, it will be a research center in space, and we hope to attract the international community into this endeavor. We think it is an ideal vehicle for international cooperation.

All of NASA's programs have been, to one degree or another, international, but I think the space station uniquely lends itself to international cooperation. If we can attract that international cooperation, then other nations will be cooperating with us in the resources that they spend, rather than competing with us. If they are spending their money on something that's cooperative, we will avoid duplication of effort. And there's little enough money for this field in the total budget to do the job that needs to be done over the next decade. So, we'd like to spend it on cooperative projects.

I also think that the program will contribute to our national security. As you know, we do not have a commitment on the part of the military that they have a requirement. But I think that as we progress through the planning of this program, we will find that there are some national security implications. This option has always been in our programs, and I think it will be there in the future.

So there are several reasons why we should go forward. The station will enable us to maintain our preeminent position, which is a desire of not only this administration, but generally has been a desire of all administrations, Democratic or Republican, since the beginning of this program.

It will allow us to maintain continuity in the program and to keep the options open for the future. That, I think, is extremely important. We must provide what is the next logical step to the Shuttle, so that we can continue to build on our plan to enhance our capability to do greater and more important things in the future.

A space station will enhance our economic well-being, because it will stimulate technology in a number of areas that are important to the economic condition of the United States. It should open new opportunities for commercial development. I mentioned the ones currently existing. But I think the most important commercial developments are ones that we have not yet dreamed of or conceived of, simply because we have not operated for a long enough period of time to develop and to realize the potential that is there.

It will provide a focus for international cooperation, which means more sophisticated science and a stepping stone for the future. It will also provide an opportunity to enhance our national security.

If we don't build it, what will happen? Well, nothing will happen, at least for a while. I think we can postpone this decision indefinitely if we want to, but if we do, then we can envision a much different world than the one we will see if we go forward.

We will, I believe, see a continual development of international competition, as opposed to international cooperation. I believe that over time we will lose our preeminent position. We will no longer lead in the development of space capability and space exploration, because the Soviets will not stop, the Europeans will not stop, nor will the Japanese stop.

We think it's very important, and we are working with you in this effort. The 25 year partnership we've had has been extraordinarily important to the country. We have shown that this kind of activity can be enormously productive. We will continue to pursue that partnership. In symposia such as this, we will let you know our planning. We hope you will join with us in the planning.

As the Bible says, "Where there is no vision, the people perish." So we hope that we will be able to provide that vision for the future.

I will close by quoting George Will, who, in writing about space, said, "We know next to nothing about virtually everything. It is not necessary to know the origins of the universe. It is only necessary to want to know. Civilization depends not only on any particular knowledge but on the disposition to crave knowledge."

OVERVIEW

Chairman: Philip E. Culbertson

Associate Deputy Administrator
National Aeronautics and Space Administration
Washington, D.C.

We are today in the midst of a debate on the possible merit of this nation undertaking the development of a space station. Fourteen years ago, in the midst of a similar debate on the desirability of developing the Space Shuttle, we held a Symposium much like this one. That meeting encouraged a free exchange of opinions on system concepts, utility, and, most importantly, overall need. It was one in a series of debates that ultimately led to the system flying almost routinely today.

The program of this Symposium states that its objective is "to present in a concise but fully documented manner to all interested sectors and individuals the current thinking on the designs, technology, and applications of an Earth-orbiting space station." I would like to enlarge that objective, however, because when we begin to think about this Symposium over a year ago, we considered it to be an opportunity to review in as much detail as possible all sides of the issue as to the desirability and utility of a space station. Although NASA is enthusiastically pursuing a space station as the next major initiative for the civil space program, we believe that public debate on the pros and cons is an important element of the decision-making process. Therefore, each of the participants was encouraged to candidly provide facts, opinions, and speculations—hopefully differentiating among the three—that would contribute to this decision. I firmly believe that it is from uninhibited exchanges such as this that the national decision-making system performs most effectively.

THE ROLE OF MAN IN SPACE*

Jack R. Lousma**
NASA Johnson Space Center
Houston, Texas

It has been my privilege to be involved in the space program for 17 years, to have flown on the Skylab space station, and to be involved in flying another element of our future program with the space station--the Space Shuttle. It is with that perspective that I describe what I think man's role in space has been and what it will be in the future.

We of course look forward to having a space station, one which will put people in space for long periods of time, tended by the Space Shuttle. I can only say that whoever gets that opportunity has a real adventure in store for them, not only in riding on a Space Shuttle, but in flying on the space station itself.

As I look back on the two flights on which I've flown, the one that resulted in the most personal satisfaction was the Skylab flight. There are several reasons for this: it was a longer flight (it lasted two months); there was an intense commitment to get ready; there was a fair amount of risk to get up there; and we actually became a part of the environment. We felt as though we were in a remote outpost and actually had the opportunity to adapt to being there. In contrast, the Space Shuttle is a quick up-and-down flight. It's a spectacular adventure and the flight plan is packed with dramatic events, but it is over quickly and the opportunity to feel like a space inhabitant is not as prevalent as on a longer space station flight.

These are among the perspectives that I get from having flown in both programs, and I think people who have the opportunity in the future will feel similarly: that both parts of the program are obviously important and will be exciting and rewarding, for not only the nation as a whole but for the space team as well. It's hard to beat the space station flight, and as you look forward to designing one, I think that's a good perspective to keep in mind.

Basically, the thrust of what I want to say here is that man does have a role in space, and it's importance will become even more obvious in the future to the American people, to our government, and to those involved in the aerospace industry.

Man's role in space, of course, is quite well determined already. I want to leave you with that impression. I feel as though I'm preaching to the choir, but I think it's important for us all to know at least that we're singing from the same sheet of music as we project our ideas out into the space community.

After having done this and looking at the tasks that lie before us in the space station, my general opinion is that there is nothing planned that cannot be done by a man in space if we properly prepare him and if we properly prepare the equipment. We must be careful not to be cavalier about what we do nor to take for granted what we have done. Our successes have resulted from proper preparation--of the hardware, the software, the people, the procedures, the flight control, the test and checkout, etc. The whole program has been very, very carefully laid out. For us to become overconfident because we have been so successful would be a mistake; but if we proceed in the same organized fashion that we have in the past, it seems to me that we can expect to be successful again.

I don't want to debate here the issue of manned vs. unmanned spacecraft. I think we ought to make the assumption that there is going to be a very vital and a very active manned program. Clearly, I think we all have to recognize that there is also a very, very important part for the unmanned program to play. There are many things that each program can do well. To debate this issue does not seem appropriate here, so I leave that discussion to other forums.

Man, of course, provides unique capabilities by being in space. We all know that he has a form of intelligence that cannot at this time be duplicated. He has the ability to reason, to make on the spot judgments, to be creative about his work, to iterate, to go back and do it again, and do it better while he's aboard. He has the ability to adapt and to improvise. He has the higher mental function that's required to innovate, to solve problems, to handle problems as they arise, to make decisions, to predict, and to formulate opinions so that he can take a new tack to be successful. His intelligence allows him to use partial information, to make new discoveries, and to handle unscheduled events. These are capabilities of intelligence that are difficult to duplicate and will not be duplicated in the near future by any artificial intelligence.

As far as vision is concerned, we know that unmanned sensors must be used in the nonvisible range. When it comes to operating in the visible range, however, the human eye is hard to duplicate. It is more efficient and more selective, and it can recognize small detail in a broad pattern. It is more sensitive to color differences and different patterns and textures and it can detect small contrast in the visible range at least as well and often better than most sensors can. But we have to admit that sensors of non-visible wavelengths certainly have their application, have been used well in the past, and will continue to be used successfully in the future.

Where manual functions are concerned, it is also difficult to duplicate the hand-eye-brain coordination capability of the human. Human dexter-

*From edited transcript.
**Astronaut.

ity, for example, is difficult to duplicate with robot-type devices. Robot dexterity will be limited to predetermined shapes, predetermined sizes, programmed motions, geometric types of objects, at least in the near future. In a manual dexterity sense, we are moving into broader activity and broader capability by robots. At the moment, however, the manual functions performed by man are considered superior, except where it comes to handling hazardous equipment or to doing repetitive tasks. As we consider a space station we need to think about what jobs a robot can be designed to do, but we should also plan for a person to be there to program the robot, to fix it, to back it up--in other words, to generally tend it.

So if we look at man's unique capabilities, certainly he has many that are not duplicated now and will not be duplicated well into the space station era. Besides, in past space flights, we have found that something always breaks. That seems to be a cardinal rule, and we would be naive if we neglect to recognize it. When it breaks it seldom breaks as we expected that it might or as we had simulated. The human operator has made the difference in those cases. His ability to handle unexpected situations has made the difference between a successful and an unsuccessful manned flight on several occasions.

We have also found in past systems that the hardware and software do not always perform exactly as it had been designed to. It's not ideal. Again, man has filled the gap and made the difference in those cases. Computer space is never as large as we want it to be. We have found again in the Space Shuttle program that man has made up the difference. So if we look at systems developed to date, man has made the difference in many cases where the machine did not have the capability or did not perform totally as designed and where man's presence has made the mission successful.

Basically I want to outline what human capabilities have already been demonstrated in space in zero gravity. I will not consider 1/6th gravity and what has been done on the moon. Certainly a great capability has been demonstrated there, and some of that experience is applicable, but operating in zero gravity is significantly different as far as getting the job done is concerned.

So I want to focus on the capabilities demonstrated in zero gravity, and then I want to look at some of the capabilities that we have yet to develop and demonstrate, and then see what the probability is for success in those undeveloped areas.

As we look at capabilities already demonstrated and those which are planned for use on a space station, repair is the first thought. We also think of EVA for servicing payloads, for pointing instruments with precision to take data, and for doing many kinds of science and applications. We will be using manipulator systems. We must also consider the medical, the physical, and the psychological aspects of being in space for long periods of time. These are areas that we have already addressed, and in which we have already demonstrated a capability. As far as man's role in space is concerned, we can feel confident about our ability to continue to perform successfully in those areas.

A few "space pictures" will demonstrate the capabilities that have already become part of our inventory. No thoughts about EVA would be complete without remembering Ed White on the first EVA. During the Gemini program we found that we could do EVA and that was a very successful one. However, when we tried to do more EVA work in Gemini, we found that we had to improve our preparations--we had some successes, but we also had a number of problems that caused us to rethink our training procedures, hardware development, etc.

By the time we came to the Skylab program, we had learned to track in the water tank zero gravity airplane using actual EVA experience. With improved tracking techniques, carefully designed hardware, and excellent support from the entire space community during the EVA's, we were successful, both in Skylab and STS-6. I'd like to point out, however, the EVA success has been the result of a great deal of hard work by people throughout the space community, by careful planning, and by good training. Those are elements we must continue to apply. We cannot become cavalier or complacent about our ability to perform EVA successfully.

On the sixth Space Shuttle flight, Story Musgrave and Don Peterson effectively demonstrated the successful performance of the Shuttle EMU (the suit and the backpack) and the ability of the crew to move about the cargo bay at will. (See Fig. 1.) Notice that there are handrails and a slide wire available. All of this equipment (a digital display unit, the backpack, and the EMU) was very carefully planned to do this job successfully.

Thus an EVA capability has been developed on the Space Shuttle and we will certainly be able to develop an equally good one on the space station. This is a capability that's been demonstrated. With confidence in the existing equipment, we can look forward to doing as well on the space station.

One of the major efforts on Skylab EVA was spacecraft repair. Not only did we repair things during EVA, but we repaired equipment within the Skylab as well--rate gyros, experiments, tape recorders, etc. The tape recorders failed often (I hope we can make more reliable recorders for the space station). So we did a lot of repair work, both

Fig. 1 Astronauts in the cargo bay of the Shuttle STS-6.

Fig. 2 Repaired solar panel on Skylab.

Fig. 3 Payload servicing on Skylab.

inside and outside the spacecraft. One of the initial problems was that one of the solar panels had come off during launch. The other was hung up. The golden reflective surfaces were exposed because the meteorite shield also fell away during launch. So not only was the capability to make electrical power greatly reduced, but the spaceship became too hot to live in. Both of those problems had to be solved. Some of you were part of the rapid-response team that so professionally and efficiently solved these problems.

The first crew went up and broke the band holding the solar panel close to the skin of the Skylab. Using procedures that were devised on the ground and on the spot, and based on the astronauts' observations when they went out to take a look at the problem, the EVA was improvised with on-board equipment. Through a lot of hard work by ground personnel and the flight crew, the problem was solved by freeing that solar panel to make Skylab at least a one-wing bird that could generate electricity. Figure 2 shows the solar panel deployed. To then solve the thermal problem, a parasol or umbrella was extended out over the spacecraft from the inside to cool it off. Thus, a unique repair capability was demonstrated: to assess the situation in a very short time period, to solve the problem on the ground, to go out of the spacecraft and execute the fix, and to make the spacecraft habitable for the entire planned flight duration, plus some weeks.

The parasol was not designed to last the entire duration of the mission. So on the second flight, a solar shield, or sail, was put out over the parasol. Then the spacecraft became even cooler. This was a fix that was devised in less than 10 days and was successfully deployed to make the entire Skylab series of missions successful.

The original objective of EVA on Skylab was not to repair anything outside at all. It was simply to go out and replace the film on the solar telescopes ever 28 days. The telescopes were located in the hub of the windmill (Fig. 2). There were a number of replacements that were made routinely. But repair was something that was done in real time and demonstrates the real capability that we have. We'll probably get a chance to use this capability again in the future; it demonstrates what we have now and can rely on for future routine and contingency applications in the space station.

We have also fixed a lot of things inside the spacecraft. For example, on STS-3, we had a broken key on one of the keyboards. One of the crew took out a good key from a keyboard that wasn't used too much and replaced the one in the front of the cockpit so we could use it for entry. We also had a lot of in-flight maintenance procedures that had been devised prior to flight. We need to make sure that when we build the space station, we make it so that it can be taken apart and that the proper tools are on board to do the job. We must make a space station that's fixable, not one that is put together in the factory with machines that don't permit us to remove panels to get at equipment. The station must be made in-flight repairable whether we plan to fix it or not, and it must have the capability for on-board maintenance and replacement of failed components.

Figure 3 shows one of our first payload servicing exercises. The Apollo telescope mount on the Skylab required film camera replacement every 28 days, as well as activation at the start of the mission and deactivation at the end. We had an extendable boom (left side of Fig. 3) that was operated by one of the crewmen to send this package out to the guy on the end of the Apollo telescope mount. Owen Garriott is standing at the hub of the windmill to receive that package and to make the replacement. In fact, he is doing payload servicing. This is a capability that we have already demonstrated, and upon which we will have to lean heavily as we move into the space station era.

Precision pointing is another capability we must plan to do well, to find sites or targets on the ground, or to point at celestial objects or places on the sun. We had a battery of eight solar telescopes on Skylab, which could be aimed very precisely. A television screen allowed us to see important phenomena on the sun. These images of, for example, solar flares, could then be used to adjust our instruments and data taken about the event. Thus, already an ability to point instruments very precisely in space has been demonstrated. We'll have the opportunity to do that again in the future on the space station.

We also had instruments with which we could do Earth resources studies. One was a telescope used to find sites on the ground as small as a quarter of a mile square, to track them as the spacecraft moved over that site, to take data, and then, as we moved onward, to search for a new site. We had maps to show us what we should see through our viewfinder. We could, with the telescope at the zoom-out position, find the site on the ground. After pin-pointing the site with the telescope, we turned on the data-taking equipment in our Earth resources package. The data were analyzed on the ground. The ability to point very precisely at locations on the ground is something we have already demonstrated and is a capability that we plan to use in the future.

We expect to be very active in the area of science and applications in the space station. We have also already done a lot of that in past programs. The eight solar telescopes on Skylab operated simultaneously. Every time the sun was up, we were up there taking data, either according to a program preplanned on the ground or, if a real-time event occurred, using the training we had been given over a couple of years to look for important phenomena on the sun. As a result of the data obtained from Skylab about the sun, the textbooks on solar physics have been almost completely rewritten. So, it was a very successful series of scientific experiments.

The white-light coronograph provided a permanent eclipse of the sun. One event we photographed was a large solar explosion, a bubble of gas exploding out through the corona. The scale represents many millions of miles, a very dynamic event that we were able to capture because we were in space, and because we could create a permanent eclipse above the atmosphere. None of the data was attenuated by the atmosphere. It all came in through the telescope and was recorded for scientific analysis later on the ground. We did not have a real-time capability to analyze these data. Real-time data analysis is something that we ought to consider for a space station. The real-time capability is needed to analyze solar data, Earth resources data, etc., so that on board we can go back and, through iteration, enhance the data while we're there, rather than waiting for another flight or another crew to be exchanged.

For stellar astronomy we had a scientific air lock on Skylab to which we mounted several telescopes to look out into deep space. So not only did we do solar physics, we did astronomy as well.

We also demonstrated the capability to do applications-types of experiments with the electron beam welder depicted in Fig. 4. This unit included a chamber in which we could insert cartridges of various metalurgical and crystal formulations, heat them up to a high temperature, and program their cool-down rate. We brought them back to ground for analysis. Major areas of interest in this experiment were: metals melting, sphere forming, exothermic heating, composite casting, growths of single crystals, flammability in zero gravity, and general development of space station technology. This was a very successful set of applications experiments. Thus, we have demonstrated the ability to do what is known as space applications work. This will obviously be very important as we look to the future commercialization of space.

On the STS-6 mission, the crew operated the McDonnell Douglas continuous flow electrofloresis system. And McDonnell Douglas can tell you how successful it was. Results have been very encouraging.

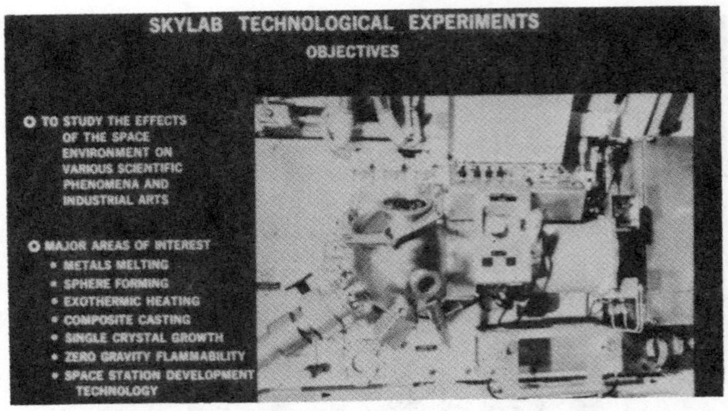

Fig. 4 Electron beam welder on Skylab.

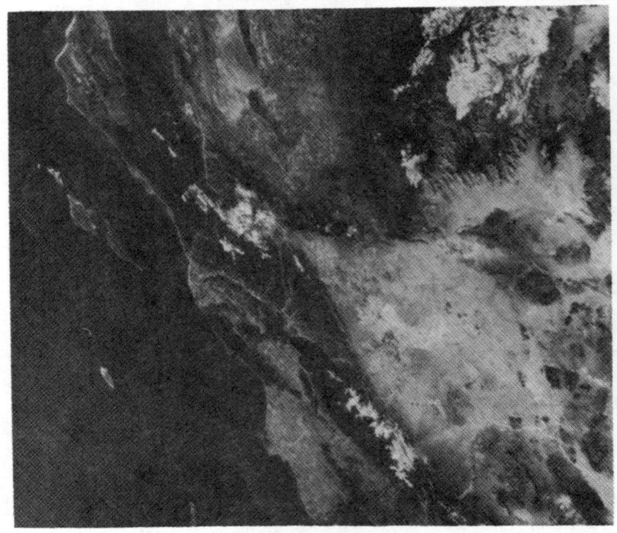

Fig. 5 View of Los Angeles and Vandenberg Air Force Base.

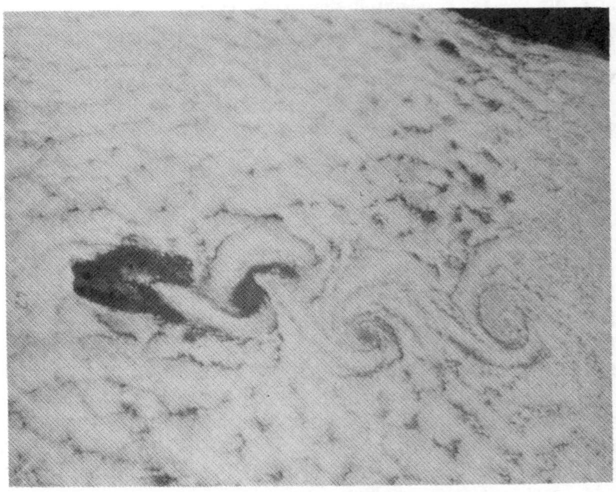

Fig. 6 Meteorologic vortexes off San Diego (from Skylab).

Observations of the Earth will continue to be important in the space station. When people are in space for a long time, they will be trained to look at the Earth with a critical eye. It's remarkable how much you can learn by just observing the Earth and then photographing or taking data on the points of significant interest. For example, on Skylab we saw the northern and southern lights about 18 hours after a solar flare, as we went up over the northern part of our orbit at 50 deg; 45 min later we could see the southern lights caused by the same event. From space we have ability to observe the oceans, to detect fires on the ground, to study worldwide meteorology, and a wide variety of other

Earth related areas. I'm trying to give you an impression of what can be seen with just the naked eye, to say nothing of what can be seen with telescopes or other kinds of devices that operate outside of the visible spectrum.

I have included Fig. 5, not only because it's a striking photograph, but it shows Los Angeles, Vandenberg Air Force Base, and the lake beds at Edwards Air Force Base. The San Andreas fault is clearly visible. It goes north and south between Edwards and Los Angeles and intersects with the Garlock fault, which crosses east and west on the northern area of the lake bed. Observation of the geology of the Earth is a capability we have already demonstrated. We learn to observe important features and events on the Earth's surface, document our observation with photographs or other devices, and return the data for analysis. If appropriate, we can make a real-time report of a transient event.

Another example: internal ocean waves cannot be seen on the surface but can be seen in the sun's reflection on the water. These waves become important not only for basic ocean research and oceanography, but for the understanding of underwater acoustics. The naked eye can also detect and report the location of icebergs. Thus, we have also demonstrated the ability to make important observations of the oceans from space.

Figure 6, showing meteorologic formations generated by vortexes around an island off San Diego, is representative of the ability to study meteorology from space. As we go around the world we recognize weather patterns are more complex than simply a system of fronts that comes across the United States. Other great weather patterns are obvious, and its clear that they're made by some great mechanical processes that are not clearly understood. It is important for us to know more about them. We can contribute to that knowledge from space by gathering data for meteorologists to use. Perhaps in the future, we'll have a meteorologist on board who can, in real time, enhance the collection of this kind of information.

In Skylab, we observed the formation of several hurricanes and tracked them on successive orbits. If you live in Houston, you worry about hurricanes. As a matter of fact, one came right over my house when I was in Skylab, and I had to tell my wife where to find the boards to cover the windows!

The Falkland current was for the first time identified from space on a Skylab flight. It flows off the east coast of South America over to the African continent.

One of man's roles in space is to observe the Earth. The purpose of these Earth-related photos has been to show that we have already demonstrated this capability.

Another capability demonstrated is the ability to use manipulator systems. The remote manipulator system (RMS) is certainly a tribute to Canadian engineering, and we're happy to see our friends from the National Research Council and SPAR Aerospace participating in this symposium. Certainly they can be proud of their RMS (Fig. 7). Certain adaptations of it will stand us in good stead on

Fig. 7 Remote manipulator system on the Shuttle.

Fig. 8 Manned maneuvering unit on Skylab.

the future space station. We've demonstrated an ability to use a manipulator system, and we've improved our knowledge of that on several flights already.

Where living in space is concerned, of course, we want to know how long we can survive and do useful work there. Medically speaking, on Skylab we determined that we can stay in space for at least three months. We need to obtain longer-duration data rather than simply extrapolating our present data out to six, eight, or ten months. We can use some of the Russian long-duration medical data as a basis of comparison, but I think we should collect our own information by scheduling a few people for extended space flights.

We have found that the cardiovascular and pulmonary systems and other bodily functions seem to adapt to being in the space environment. Different parts of the body adapt over a differing period of time. Some of those adaptations take up to several weeks. In any case, we have found that the body is an amazing adaptive device, and we can expect to move forward medically with confidence to longer flights from the data we've gathered so far. However, we need to carefully examine the medical aspects of long space station flights.

On Skylab, we had an ergometer and other exercise devices to definitively measure our cardiovascular health and to stay in good physical condition.

Since physical well-being especially is important on long flights, we had good food on the Skylab. I think it's important to think of that in the future. Space shuttle flights and space station flights are clearly different. A Space Shuttle mission is analogous to a camping trip. You know it will be short; you will be home in a few days; and you can live with most anything. In Skylab, we planned to go for a long period of time. It was analogous to a lengthy expedition. How you eat and sleep, your sense of well-being, sanitation facilities, etc., become more important. They've got to be good and work right, because they directly affect the amount of work you can produce and your overall health and attitude while there.

Another feature that enhances the sense of well-being is the ability to keep clean and to carry out daily hygiene functions quickly and with no fuss or muss. That was the kind of the system we had on Skylab. The hygiene facility was a great improvement from what was on Apollo. Such facilities were also part of a medical experiment to determine what happens to the body while in space. Most of the waste products were processed and returned to Earth for analysis.

We also had a shower on Skylab. We didn't use it very much because it was difficult and time-consuming to operate. If we have one on the future space station, it's hard to predict how often it will be used. The Skylab atmosphere was very dry, and we never got sweaty unless we were exercising. We did use the shower a little bit in order to check it out, however. In any case, we demonstrated on Skylab that we can live comfortably, and work productively for extended durations if the design is right.

So far, I have discussed capabilities that we have demonstrated in the past, those that we can rely on to be similar to what we can expect in the future. We can look forward with confidence in all of these areas as we go on into the space station program. I simply want to point out that given the properly designed equipment, training procedures, mobility devices, and stabilization, man has already demonstrated he can accomplish almost any zero-gravity task, and that he has already proved he can perform productively in space for long periods of time.

In moving on, I want to briefly touch on capabilities we have yet to demonstrate. As we inventory other types of capabilities that we plan to develop in the future on the space station, I think we can look forward with confidence to being successful in them, too.

To add to our repertoire of capabilities, we plan to use a manned maneuvering unit (MMU). We plan to do payload servicing, free-flyer servicing, repair, proximity operations, and rendezvous. We plan to do some propellant and cryogenic handling as well as space construction. We're going to develop a teleoperator maneuvering system (TMS) and an orbital transfer vehicle (OTV). These are capabilities that we haven't totally demonstrated in the past, but I think we can demonstrate them in the near future. Many of them are already in the works. The message of this part of my presentation is to simply point out that we will be successful in demonstrating these new capabilities for the space station if we do not compromise, forget, or become complacent about the principles we have learned in the past.

We have already flown a maneuvering unit in space. The one shown in Fig. 8 was flown in the Skylab. It flew very smoothly with several control modes. It was serviced with nitrogen as the propellant, and we could recharge the batteries. The current maneuvering unit is a follow-on to this one.

We're training with the new MMU in the water tank. We plan to fly the new MMU on STS-11 for its first checkout, and then to use it operationally to fly over to the Solar Maximum Satellite on STS-13. This unit is coming along very well, and I think we can look forward with confidence to the use of the unit. It looks quite different than the one we flew in Skylab but has many of the same features incorporated in it, and the Skylab model provided much of the information needed to develop the new Shuttle MMU. In a sense, some of the MMU capability has already been demonstrated, and we can look forward to additional verification with optimism.

To repair the Solar Maximum Satellite, which is a free-flying payload, the crewman will fly over to stabilize it. Then the Space Shuttle will move close to it, grab it with the RMS, and place it in the payload bay to replace the failed components within it. In order to do this we must demonstrate our ability to make a rendezvous with the Space Shuttle. We will be doing that on STS-11 and then again on STS-13. Proximity operations, the ability to fly in close formation and to grapple free-flying payloads with the RMS, became a reality on STS-7. It appears that proximity operations can be conducted routinely, and that as we move into the space station program, we can expect both rendezvous and proximity operations techniques to be proved.

We'll have a space telescope eventually. We have been working diligently on how we might service it, and this also appears feasible.

We haven't done much space construction. We will be required to work at that, but again I think it's feasible. Based on our understanding of what we've been able to accomplish in the past, it's clear that if we make our systems properly we can expect to be successful in space construction. We need to make sure that the various components fit before they leave the ground--so that when you put them together, they function. We must provide for that in the design, test, and checkout of any project. I believe that space construction will not only be necessary but will be demonstrated very successfully. We might, in addition to using the MMU and other mechanical means, use the RMS to do space construction.

The teleoperator maneuvering system (TMS) is not a new concept. We already have a start on a TMS-type vehicle. There were plans for a Skylab reboost mission a few years ago. The teleoperator was a booster vehicle that could be docked to the Skylab. The size of a truck, it was to be carried in the payload bay of the Space Shuttle. It was remotely controlled from inside the cockpit and had a television system on it. We were going to effect a rendezvous to about 500 ft, deploy the teleoperating maneuvering system, maneuver it remotely, and hook it to the Skylab. Its booster system would then have been used to either put the Skylab higher or to a known place in the water. This was a project that we worked on for about a year, and it was, in our opinion, very feasible. Had the timetable of the Space Shuttle and the Skylab coincided better, the mission would have been flown.

So in the future, teleoperator systems, whatever shape or form they take and whether they have a man in them or are operated remotely, have their groundwork laid. We can be confident they can be successful.

When we consider an orbital transfer vehicle (OTV), we're thinking of receiving a vehicle and a satellite from the Shuttle, servicing and assembling them, checking them out, and then launching them. We've done the launch part already. We've already launched high-energy stages from the Orbiter on several Shuttle flights. I think we have the technology on hand and, if we plan and design properly, can expect to be successful in all other areas of OTV operations.

An OTV will probably be unmanned at first, but eventually it may be manned. I think OTV and TMS operations are feasible and I am optimistic about their outcome, but these areas will be among the most challenging facing us as we go forward with a space station program.

Finally, I believe America will eventually have its space station. We don't know what it'll look like. It will incorporate a wide variety of capabilities. We have talked about most of them with respect to our ability to conduct manned operations. It is my experience that if a crewman has proper training and equipment, is provided with a means to get to his workstation, and adequate retraint when he arrives, he can work in space about as successfully as he can in his own garage or laboratory. My message to you is that we will successfully demonstrate all these new capabilities as well as take advantage of already proven capabilities, all of which are required for a successful space station program, if we diligently adhere to the principles that we have acquired from the past. If we do not become overconfident, complacent, or cavalier about the way in which we plan our future activities and design the equipment to support them, then man's role in space will certainly be even more productive in the future than it has been in the past.

SPACE STATIONS:
A HISTORICAL PERSPECTIVE

John M. Logsdon*
National Air and Space Museum
Washington, D.C. 20560

Abstract

This paper discusses the historical evolution of the space station concept, with particular attention to NASA plans in the 1960-1980 period. Emphasis is given to the changing justification presented for station development during that period and to the political context within which station proposals were evaluated.

Introduction

Almost from the first time humans thought about leaving the surface of this planet, one theme has been the creation of some form of human outpost in space. In fiction, and during this century in increasingly specific engineering detail, the space station concept has been extensively discussed. In one of the two major space-faring nations, the Soviet Union, a fairly rudimentary but still very capable space station program, centered on the Salyut spacecraft, has been ongoing since 1971. In the other space power, the United States, the desirability of developing some kind of permanent manned presence in space is now under consideration, as it has been on several occasions over the past twenty-five years. This report contains a historical overview of those past occasions.

Space Stations of the Imagination 1/

Perhaps the first detailed plan for a space station was contained in an 1869 Atlantic Monthly article by Edward Everett Hale, "The Brick Moon." Hale proposed a 60-meter-diameter satellite, made of bricks and housing a crew of 37. From a 6000 km orbit, the brick moon would serve as a navigational aid for ships; communications were by Morse code, with signals created by the station's occupants jumping up and down on the satellite's exterior surface!

Space pioneers Konstantin Tsiolkovsky and Herman Oberth both gave some attention to building orbital stations. Tsiolkovsky discussed the idea as early as 1911, and in 1923 wrote of staying "at a distance of 2000-3000 versts (a Russian unit of distance equal to 0.6629 miles) from the Earth, as its Moon. Little by little appear colonies with supplements, materials, machines, and structures brought from Earth." In his 1923 book The Rocket into Interplanetary Space, Oberth first described an orbiting manned satellite as a "space station," and proposed that it could be used as an Earth observation site, world communications link, weather satellite, or as an orbital refueling station for outward-bound space vehicles.

Others in Europe wrote of stations during the twenties and thirties, and their ideas were quite familiar to the team of German rocket scientists led by Wernher von Braun when it began operation in the United States after World War II. Some form of large space platform was a central part of von Braun's thinking throughout the 1950s, and his ideas got wide publicity in a Collier's magazine special section of March 22, 1952 titled "Man Will Conquer Space Soon." Von Braun's contribution to the magazine was an article called "Crossing the Last Frontier," in which he claimed that "scientists and engineers now know how to build a station in space that would circle the earth, 1,075 miles up . . . . If we do it, we can not only preserve the peace but we can take a long step toward uniting mankind." Von Braun's plan called for a triple-decked, 250-foot wide, wheel-shaped station in polar orbit which would be a "superb observation post" and from which "a trip to the moon itself will be just a step." The main station would be accompanied by a free-flying unmanned but man-tended astronomical observatory. 2/

Von Braun's detailed description was only one of the many concepts developed in the years after World War II but prior to the 1957 launch of Sputnik and the formal beginning of the Space Age; even before the United States had an official space program, most of the possible uses of a space station had been identified by visionaries who dreamed of space travel. The past twenty-five years, in this country at least, have been spent in debating whether to translate some form of that vision into reality.

Planning a Post-Mercury Space Program 3/

Even before it opened its doors for business on October 1, 1958, NASA had been assigned the responsibility for manned space flight and had developed Project Mercury as its initial manned activity. The NASA leadership set as a high priority task developing a long-range plan for the agency's first decade. A space station was a leading candidate for a post-Mercury goal. The House Space Committee in early 1959 concluded that stations were the logical follow-on to Mercury, and von Braun (then still working for the Army) presented a similar view in his briefings to NASA. At this time, the German rocket team had developed an elaborate scheme, called Project Horizon, for Army utilization of space, including military outposts on the lunar surface.

In the first half of 1959, NASA created a Research Steering Committee on Manned Space Flight, chaired by Harry Goett. At the first meeting of this committee members placed a space station ahead of a lunar expedition in a list of logical post-Mercury steps. In subsequent meetings, the debate centered on the research values, especially with

*Chair in Space History, Member AIAA

respect to biomedical studies, of a station versus the excitement of a lunar landing goal. While some members of the committee argued that "the ultimate objective of space exploration is manned travel to and from other planets," the Langley representative argued for an interim step, since "in true space flight man and the vehicle are going to be subjected to the space environment for extended periods of time and there will undoubtedly be space rendezvous requirements. All of these aspects need extensive study. . . . the best means would be with a true orbiting space laboratory that is manned and that can have a crew and equipment change." 4/ Ultimately, the Goett committee recommended that a lunar landing be established as NASA's long-range goal, on the grounds that it was a true "end-objective" requiring no justification in terms of some larger goals to which it contributed.

At an August 1960 industry briefing on NASA's future plans, George Low presented a scheme in which a manned lunar landing and creation of a space station were given equal treatment as long range goals of the NASA program; Low told the conference that "in this decade, therefore, our present planning calls for the development and demonstration of an advanced manned spacecraft with sufficient flexibility to be capable of both circumlunar flight and useful earth orbital missions. In the long range, this spacecraft should lead toward manned landings on the moon and planets, and toward a permanent manned space station." 5/ Low also announced the name of the advanced spacecraft program, aimed both at the moon and at space stations; it was to be called "Project Apollo."

All of the debate was effectively terminated by John F. Kennedy's May 1961 announcement that "I believe we should go to the moon." As NASA planners accepted Kennedy's challenge and chose lunar-orbit rendezvous as the approach to accomplishing a lunar landing mission before 1970, any change of developing a space station in the 1960s disappeared. The rest of the decade was spent studying the concept, with the anticipation that it would be the logical post-Apollo program.

Space Station Plans During the 1960s

Throughout the sixties, there were almost always funds available for engineering and design studies of future concepts. These studies were conducted both within NASA and by the various aerospace contractors (particularly those without a major role in Apollo), and they resulted in the examination of a wide variety of concepts, ranging from inflatable balloon-like structures, through the use of refurbished rocket stages, to very large stations requiring the use of Saturn V boosters to put them in orbit. Three NASA field centers, the Manned Spacecraft Center in Texas, the Marshall Space Flight Center in Alabama, and the Langley Research Center in Virginia, managed these in-house and contractor studies, and they were coordinated by the Advanced Missions Office of the Office of Manned Space Flight at NASA headquarters in Washington. 6/

Advanced Planning, 1962-1968

While the manned flight centers at Houston and Huntsville were focussing most of their energies on getting Apollo started in the early 1960s, the Langley Research Center was giving substantial attention to the theoretical and engineering aspects of space station design. These efforts dated from at least mid-1959, and by 1962 enough work had been done to form the basis for a space station symposium. 7/ Langley researchers noted that "a large manned orbiting space station may have many uses or objectives." Among these objects they listed:

1. Learning to live in space
   a. artificial-gravity experiments
   b. zero-gravity experiments
   c. systems research and development
2. Applications research
   a. communications experiments
   b. earth observation
3. Launch platform experiments
4. Scientific research

With respect to launch platform experiments, Langley suggested that "the space station with its crew of trained astronauts and technicians should be a suitable facility for learning some of the fundamental operations necessary for launching space missions from orbit. The new technologies required for rendezvous, assembly, orbital countdown, replacement of defective parts, and orbital launch can be determined." 8/

Among the various space station studies carried out by Langley contractors during the first half of the 1960s, perhaps the most detailed was that of a Manned Orbital Research Laboratory (MORL) conducted by Douglas Aircraft from 1963 to 1966. Douglas had had some prior interest in space stations; in 1960 it had built a full-scale mockup of a four-person astronomical space observatory as the central theme of an "ideal home exhibition" held in London. This station was to be constructed inside the fuel tank of a second-stage booster, a Douglas idea which ultimately found use in the Skylab program over a decade later. 9/

In this study, a baseline technical concept for a Manned Orbital Research Laboratory (MORL) was established first, then the "utilization potential" of such a station examined, i.e., design preceded requirements. When the original design was compared to various requirements, it was inadequate, and a larger station in a different orbit evolved as the final result of the study effort. The study found that the highest utilization potential came from "key engineering and scientific research studies augmented by specific experiments directed toward potential Earth-centered applications."

The MORL study predicted the need for "hundreds of thousands of manhours" in orbit to carry out all useful applications; this implied a long-range requirement for "near permanent operations and support of probably several space stations." The study also noted, foreshadowing a future issue, that "the limiting factor on the number of such stations, and the crew size of each station, appears to be the cost of logistic support." The final MORL concept, although basically a zero-gravity station, had an on-board centrifuge for reentry simulation, testing of physical condition, and the physical therapy if zero-gravity conditions were debilitating for the crew. 10/

Management Attention to a Space Station

By mid-1966, with an Apollo Applications Program being defined and Apollo itself pushing toward an initial lunar landing in 1968 or 1969, NASA top-level management recognized that it was "timely that we update our studies of a permanent manned station." This agency-wide study was to be divided into two parts; one effort would focus on system design, the other on the "need for, the requirements of, and the constraints on a space station" to support a number of objectives:

1. Astronomy with large optical and radio telescopes
2. Geographical studies with emphasis on earth resources
3. Meteorological sensor development
4. Biological research
5. Aeromedical research and development
6. General laboratory research and development in advanced technology
7. Flight operations development directed toward more efficient resupply, service, and orbital operations
8. Long duration flight leading to a future missions capability of manned flight to the planets.

The study charter emphasized that "it is still a question whether a permanent space station is the best approach to achieving the envisioned mission objectives," and called for a review of both advantages and disadvantages of a station as the means for accomplishing them. 11/

The requirements segment of the 1966 space station study was directed by Charles Donlan; the design segment, by E.Z. Gray. It was the Donlan group in which the conflicts between various experimental objectives and design requirements were assessed. The Donlan study found that "the most difficult problem to resolve is the matter of artificial gravity. If artificial gravity becomes a firm requirement for crew comfort," it noted, "its implementation can have a very large impact on space station design" since "zero gravity is a mandatory requirement for major portions of the experiment programs." Other conflicts included those between astronomical observations and earth observations, which are "obviously in conflict in both direction and stabilization modes." The conclusion of this assessment was that "the manned space station concept emerges as an unparallelled opportunity for learning how to exploit the features of space," and that "the prime justification for a manned station (is) associated with the potential the station offers for undertaking broad-based research and development programs in science and technology addressed to all the space objectives of the United States." The group considered both a single station in which there would be an attempt to integrate various conflicting experiments by sequential scheduling, and a concept of splitting the station into several components, each optimized for a particular set of activities. The study concluded that "a minimum station that would appear to satisfy much of the program needs would operate in 200-mile orbit at an inclination of 50°, be capable of operating for a five-year period, continuously or intermittently, and large enough to house a staff of from 8 to 12 people." 12/

Attempts to Gain Approval for Space Station Program

Based on the 1966 space station study and the lead time required for get a space station program defined and implemented, NASA leadership in the fall of 1966 decided to go ahead with what the Donlan Committee had defined as a minimum space station program, and requested $100 million in the FY 1967 budget for detailed definition studies. The Bureau of the Budget totally rejected this request, allocating no funds specifically for space station work,13/ and throughout 1967 and 1968 preliminary studies continued inside NASA at a low level of funding.

An Unsuccessful Space Station Initiative

When the Nixon Administration took office on January 20, 1969, space was not high on its list of concerns. Thomas Paine was Acting Administrator as the White House tried to convince others to take the NASA job.

It is standard procedure for a new administration to conduct an early review of the outgoing government's just-submitted budget. President Nixon and Bureau of the Budget Director Robert Mayo asked for such a reappraisal with the guidance to agencies that they should attempt to reduce the budget submitted by a Democratic President. Paine and his closest advisers decided to take an opposite tack.

In an unusual move for an agency leader in a new administration, much less one serving in an acting capacity, Paine appealed directly to the President in support of the manned space flight program. He argued that "positive and timely action must be taken by your Administration now to prevent the nation's programs in manned space flight from slowing to a halt in 1972" and suggested that "the nation should. . . focus our manned space flight program for the next decade on the development and operation of a permanent space station -- a National Research Center in earth orbit -- accessible at reasonable costs to experts in many disciplines who can conduct investigations and operations in space which cannot be effectively carried out on earth." Paine told the President that "the case that a space station should be a major future U.S. goal is now strong enough to justify at least a general statement on your part that this will be one of our goals." Paine argued that

> There are many potential valuable uses of such a space station, and new ones will be found as experts in many fields become familiar with the possibilities and are able to visit and actually use it. However, we believe strongly that the justification for proceeding now with this major project as a national goal does not, and should not be made to depend on the specific contributions that can be foreseen today in particular scientific fields like astronomy or high energy physics, in particular economic applications, such as earth resources surveys, or in specific defense needs. Rather, the justification for the space station is that it is clearly the next major evolutionary step in man's experimentation, conquest, and use of space. The development of man's capability to live and work economically

and effectively in space for long periods of time is an essential prerequisite not only for operations in earth orbit, but for long stay times on the moon and, in the distant future, manned travel to the planets. It is for these reasons that I believe that space station development should become one of your Administration's principal working goals for the nation over the next decade. 14/

NASA was unable to get much support for an early Presidential commitment to a space station. The Bureau of the Budget was opposed, and the President's Science Advisory Committee saw "no great urgency," commenting that "from a programmatic standpoint, the arguments in favor of early action appear very weak." 15/

The space station did gain some support from the Department of State, which saw "a close relationship between our space program and foreign policy objectives. Thus, an ongoing, challenging and successful space program is important from the viewpoint of these objectives -- particularly one designed and funded to afford increasing opportunities for international cooperation." 16/

The Department of Defense position was the DOD "does not have or anticipate projects which require a space station as defined by NASA. DOD has great interest in the development of a lower cost transportation system suitable for their uses as well as for NASA's." 17/

By the end of March 1969, Thomas Paine's attempts to gain early approval of station development from the new Nixon Administration had failed. NASA had built up substantial commitment to the space station concept through the 1960s, but when it came time for the country to decide, through the policymaking process, whether the station was a "good buy," the response was negative; the space shuttle was a preferred alternative. The reasons for this negative assessment were already clear for NASA to see by March 1969, but it took over a year for NASA's leadership to acknowledge the situation and to steer the agency away from the station and behind the shuttle.

Space Station Studies Go to Phase B (and Then Nowhere)

While at the policy level NASA's hope for a quick Nixon commitment to the space station was being dampened, at the engineering level the momentum behind the program was building. After conducting preliminary Phase A studies during 1967 and 1968, NASA was prepared in early 1969 to involve the aerospace industry in defining the program through two Phase B studies. NASA's hopes were that these program definition studies would provide the technical basis for a start on space station development within a year or two. These studies were initiated in September 1969, and extended over most of the next two years. But events at the policy level made it increasingly unlikely that the space station program would ever proceed beyond the Phase B stage, at least in the 1970s.

Defining the Preferred Concept and Its Rationale

One problem, perhaps the key one, was that NASA found it quite difficult to tell both prospective contractors and the political leadership what kind of station, for what purposes, it wanted to develop. Thomas Paine, bullish after Apollo 8, was pushing for a more exciting space station concept than that emerging from NASA's in-house planning process; Paine wanted to see what the various center directors thought, and for that purpose convened a January 27, 1969 meeting.

That meeting dramatized the lack of agreement within NASA on basic parameters of the space station program. Abe Silverstein, Director of the Lewis Research Center, commented that "NASA is asking (in mushy language) for something we should know ourselves before going out. . . . This program must be saleable in terms people can understand. . . .Low cost in orbit and space station capability cannot be achieved by funding a series of 'blue book' industry studies; that only loses us a year and several million dollars."

Wernher von Braun of MSFC struck a theme that Paine found quite attractive:

> NASA should now tell the contractors what we want in the long run, what we forsee as the ultimate--the long-range--the dream--station program. NASA should spell out the sciences, technology, applications, missions and research desired. Then NASA should define a 1975 station as a core facility in orbit from which the ultimate "space campus" or "space base" can grow in an efficient orderly evolution through 1985.

Finally, MSC Director Robert Gilruth told Paine: "This work statement doesn't set NASA's sights high enough for the future. We should now be looking at a step more comparable in challenge to that of Apollo after Mercury. The space station size should be modular and based on our Saturn V lift capability into 200-mile orbit. Three launches should give us one million pounds in orbit, including spent stages. That is the number we should be planning for the core size." 18/

New guidelines for the space station were developed in the weeks following the January 27 meeting under the direction of Charles Matthews, Deputy Associate Administrator of Manned Space Flight at NASA Headquarters, with participation by staffers from OMSF, MSC, MSFC, and LRC; Matthews was to be in charge of the space station program during the Phase B period. Center representatives were still in significant disagreement. The MSC representative noted that "crew sizes smaller than 20 men represent sorties into space as opposed to what Gilruth would like to see as an agency goal, namely, establishing an earth-like environment in space." The MSFC representatives "didn't agree that the 1975 module had to be designed for artificial 'g' or had to be a part of the 'ultimate' Space Station," and felt that OMSF was "getting away from what Dr. von Braun wants... (6 to 9 men, etc.)" One of the headquarters participants "didn't agree at all that there is going to be a 100-man type Space Station," particularly "if very low cost transportation is feasible and available." 19/

Out of this lack of consensus within NASA came a rapid change from the January concept of a space station program; Paine had heard, and agreed with, the advice of his center directors that NASA should be bolder in its plans. In February, Aviation Week reported that "all previous concepts have been retired from active competition in favor of a large station," with the focus on "a 100-man earth-orbiting station with a multiplicity of capabilities" and the launch of the first module of the large space station, with perhaps as many as 12 men, by 1975." Top NASA officials were reported to have rejected earlier space station plans as "too conservative." 20/

Phase B Competition

NASA issued a statement of Work for the Phase B Station Program Definition on April 19, 1969. The statement described the space station as "a centralized and general purpose laboratory in Earth orbit for the conduct and support of scientific and technological experiments, for beneficial applications, and for the further development of space exploration capability" and noted that the work requested would include "the Space Base but will focus on the mid-1970s Space Station as the initial but evolutionary step toward the Space Base."

The initial space station was to have a crew of 12, and normally to operate in a zero gravity mode, but there would be during the early weeks of its mission an assessment of the effects of artificial gravity; a counterweight would be tethered to the station and the configuration spun to provide the gravitational effect. The station was to be 33 feet in diameter and was normally to operate in a 270 nautical mile, 55^0 orbit, but also be capable of operating in polar and slightly retrograde orbits. 21/

Shortly after the original proposals in response to this statement of work were received by NASA, a new requirement was added. Not only was the space station to be designed so it could be the core around which a space base could be developed; the station module would also be the core of a spacecraft designed for a manned trip to Mars. This requirement was a reflection of the high hopes for all of NASA's future manned programs which were pervasive in the immediate aftermath of the first lunar landing. NASA had developed by mid-1969 an extremely ambitious "integrated plan" which called for early development of the space station, a space base, a lunar orbit base, a lunar surface base, and a space shuttle and for the initial manned expedition to Mars in the 1980s. 22/

On July 22, NASA awarded Phase B contracts of $2.9 million each to North American Rockwell and McDonnell Douglas. The studies were to run for 11 months beginning in September; MSC would manage the North American effort, and MSFC, the McDonnell Douglas study.

One source of continuing tension during the course of the Phase B studies was the difficulty of integrating station design and the candidate experiments proposed for the station. These studies were compiled into a thick document known universally as the "Blue Book." One participant in the study later noted that "the candidate experiments compiled in the NASA Blue Book are too costly to be considered as a whole, are somewhat duplicated..., have not been verified as the true experiment goals...." 23/

Study Effort Redirected

Throughout the first half of 1970, the space station studies gathered momentum, aimed toward July completion and a subsequent NASA request to begin station development in late 1971. However, in effect the decision that there would be no space station had already been made.

Thus, on July 29, 1970, Charles Matthews ordered MSC and MSFC to terminate on-going Phase B activity and to redefine the effort in a fundamental way. NASA leadership had become convinced that the Saturn V program, which had been in terminal condition for almost two years, was finally dead, i.e., there would be no booster capable of launching a 33-foot station. The only launch vehicle available for use in putting the space station into orbit would be the space shuttle, with its 15-foot by 60-foot payload bay. What had started out as the supply vehicle for the station was now its key to survival.

It took some doing to skew the study effort toward components of 14-foot diameter; one study contractor commented that "people who were eager to fly in a 33-foot station found the prospect of long stays in the 14-foot station not very attractive." But NASA did ultimately issue Phase B extension contracts for a modular space station study effort to extend through most of 1971, and North American Rockwell and McDonnell Douglas went to work on the new concept.

Phase B Studies "Fade Away"

By the time these studies began, however, the likelihood that they would lead to an early commitment to station development was already vanishingly small. NASA had suffered a number of shocks in late 1969 and through 1970 in its attempts to get an ambitious post-Apollo program approved, and by the summer of 1970 it had become quite clear to NASA leaders that only one big program had any chance of Presidential and Congressional approval, and that it was not the space station program. From its start as the "advanced logistics system" for the station and space base, the space shuttle had garnered the interest of the Air Force and many within NASA, and in mid-1970 the agency leadership decided to make the shuttle its top priority program.

In September 1970, NASA was simultaneously:

- conducting public meetings to inform potential users of the characteristics and potentials of a 33-foot space station

- developing plans for the study of a modular, shuttle-launched station because the 33-foot concept was dead, and

- preparing a budget request in which space shuttle development was given priority, and the space station dropped from separate program status to an advanced mission study.

Station studies continued through 1970, 1971, and 1972, with the final in-house studies being focused on a single "research applications module" carried into orbit by a shuttle.* This was all that remained of what, only a few years earlier, had been plans for truly large facilities in earth orbit. As a final indication of this reality, on November 29, 1972 the Space Station Task Force was abolished, then immediately reincarnated as the Sortie Lab Task Force. NASA was able to gain approval for shuttle development in early 1972, and that task occupied the agency's energies throughout the decade. Until the Shuttle was ready, the dream of permanent human facilities in space would once again have to wait.

During the 1969-1971 period, the political leadership of the country applied its concept of national interest and national priorities to the space program; through that process, the technological aspirations of NASA were put under firm political control. What happened to NASA's space station plans is best viewed, not in terms of NASA "winning" or "losing," but in terms of what happens when an agency's aspirations are significantly at variance with what political leaders judge to be both in the long-term interests of the nation and politically feasible.

Space Station Planning During the 1970s

During 1973, NASA carried out a "quasi-space station" effort, Skylab. Skylab was not a true space station, as defined in this paper, because the modified Saturn upper stage used to house the orbital facility was not equipped for resupply of key expendable items, and thus the facility could not be used over a multi-year period. The orbital workshop was launched in May 1973, and during the year three three-person crews occupied Skylab for periods of 28, 59, and finally 87 days. The results of the Skylab program provided a major stimulus to NASA's once again beginning to examine the space station concept in the mid-1970s. 24/

Other influences in this direction included the need to begin to identify potential "post-shuttle" programs and new requirements for manned space operations emerging from a number of study efforts being carried out by NASA in the 1974-1975 time frame. In order to build a plausible rationale for once again proposing a space station as an element of NASA's program, it would be necessary to identify some high priority missions which could not be accomplished using the space shuttle, with its 7-30 day orbital staytime, the Spacelab facility for manned experimental activities, and the shuttle's significant capability for lifting large and/or heavy cargoes to low-earth orbit. Studies which established requirements for large structures in both low-earth and geosynchronous orbit, structures which could only be constructed in space, seemed to provide the needed rationale, and space construction became a major theme in space station studies during the 1975-1980 period.

*This was the origin of the Spacelab program.

Development of the "Construction Shack" Concept

The first NASA foray into a new station effort was a 1975 study of "Manned Orbital Systems Concepts" carried out by McDonnell Douglas Astronautics under the technical direction of the Marshall Space Flight Center. This study "examined the requirements for. . . a cost-effective orbital facility concept capable of supporting extended manned operations in Earth orbit beyond those visualized for the 7- to 30-day Shuttle/Spacelab system." Study guidelines included use of available hardware developed for the Skylab, Spacelab, and Shuttle programs "insofar as practical" and an initial operational capability in late 1984.

The context for the MOSC study included a growing concern about the Earth's resource limitations, population growth, and environmental stresses, driven by the widely publicized "limits to growth" debate of the early 1970s. The study noted that "the planning and development of future space programs cannot be done in isolation from the many critical problems facing the peoples of the world during the coming decades" and that "there will continue to be many conflicting and competing demands for resources in the years ahead." This context skewed the emphasis in establishing missions for the facility to "the research and applications areas that are directly related to current world needs."

Though oriented more directly than past station concepts to high priority global problems, the MOSC study still emphasized the "science and applications research facility" rationale; although such activities as assembly of large structures and operating space manufacturing facilities were examined during the study, the emphasis was on a facility which would "enable the scientific community to pursue programs directly related to the improvement of life on Earth." The final MOSC configuration called for a four-man modularized facility; the manned module would be based on the Spacelab design, and Spacelab pallets would be used to support unpressurized payloads. Total program costs for development and operation of the initial MOSC facility were estimated to be $1.2 billion. 25/

Rather than attempt to gain approval to take the MOSC effort to a Phase B stage, NASA in the fall of 1975 decided to conduct further studies in which the emphasis was shifted from research in orbit to space construction. In explaining its study plans, NASA noted:

> Earlier space station studies emphasized the "Laboratory in Orbit" concept. Emphasis is now being placed on a Space Station as an "Operational Base" which not only involves a laboratory but also such uses as

(a) an assembly, maintenance, and logistics base for conducting manned operations involving antennas, mirrors, solar collectors, transmitters; (b) for conducting launch and retrieval operations for orbit-to-orbit and Earth-departure vehicles which may require assembly or propellant transfer in orbit; (c) for conducting retrieval, maintenance and redeployment operations for automated satellites; (d) for managing clusters of spacecraft and space systems as a central base for support for common services...26/

There were a number of reasons for NASA's switch in emphasis in space station justification. There was no evidence that the scientific community was any more supportive of a manned orbital laboratory concept in 1975 than it had been in 1970; prior attempts to justify a space station on its use as a space-based R&D facility had not been successful. More positively, the mid-70s saw a number of studies of the potentials of space operations for addressing problems on earth. These included the NASA "Outlook for Space" study, Gerard O'Neill's proposals for developing very large human habitats in space, and particularly the bold proposal by Peter Glaser for creating large solar power satellites (SPS) in geosynchronous orbits.

Developing a SPS would require extensive on-orbit manned operations and the capability for assembling very large structures in space. Similar construction requirements were derived from less grand schemes involving large antennas in space for communications use and scientific investigations.

By the end of 1975, NASA had developed a plausible argument that space construction might be a major requirement of its programs during the 1980s, and wanted to explore the role of manned orbital facilities in carrying out these construction efforts. In December 1975 the agency issues a request for proposals for a "Space Station Systems Analysis Study"(SSSAS); the study effort was to be focused around the use of a space station to "serve a wide range of operational base and space laboratory activities," such as using the station "as a test facility and construction base to support manufacturing, fabrication and assembly of various sizes of space structures."27/

Two parallel SSSAS studies were conducted, one by McDonnell Douglas and one by Grumman. One finding of the studies was that scientific efforts could "go along for the ride" on space stations capable of supporting construction, materials processing, and power generation objectives. One aerospace publication reported that:

> The space base concept is one whose time seems to be coming rather quickly. Until recently, space stations have been thought of mainly as..."the traditional laboratory in the sky." Some observers were surprised when construction, materials processing and power were given roughly equal status with science...Now, the balance has shifted further...space construction work is the "prime focus" of the studies. 28/

When NASA began its SSSAS effort in late 1975, its hope had been to use the Phase A study results as the basis for a Phase B space station "new start" in fiscal 1979, i.e., sometime after October 1978. However, NASA was unable to get the approval of the Office of Management and Budget (OMB) to proceed on a schedule which would have made such a new start possible. For one thing, much of NASA's justification was based on using a space station during the development of a solar power satellite, and OMB was strongly opposed to the SPS concept and had taken it away from NASA planners. Both the outgoing Ford Administration and the incoming Carter Administration rejected a NASA request for $15 million for space station studies in the FY 1978 budget. 29/ (In FY 1977, the budget requested for all advanced systems studies had been $13 million.) This rejection, moreover, was quite emphatic; NASA was told that OMB was very unlikely to consider any substantial funding for the space station until the space shuttle program was clearly a success. At this time, the shuttle program was experiencing technical difficulties and cost and schedule overruns, and the White House was giving serious consideration to major changes (including possible termination) in the program. This was hardly a congenial context for NASA to propose a major follow-on program. Recognizing that NASA was not going to be able to start on a major space station effort anytime soom, by the spring of 1977 some NASA officials were suggesting that "the (shuttle) orbiter is a significant space station in itself," and were looking towards ways to enhance shuttle capability to perform many of the missions that the SSSAS studies had assigned to a space station. 30/

Johnson Invents the Space Operations Center

Rather than being the year in which significant momentum behind a space station program was developed, 1978 turned out to be a year in which there was essentially no space station activity per se. The SSSAS studies had identified, as important steps in extending the capabilities of the space shuttle, the development of an in-orbit power supply and of shuttle-tended unmanned orbital platforms for various science and applications payloads. Both Johnson Space Center and Marshall Space Flight Center were studying orbital power supplies during 1978; the Johnson Space Center concept was called a power extension platform (PEP), while Marshall Space Flight Center was examining a 25 KW power platform. Marshall also initiated studies of an unmanned Science and Applications Space Platform (SASP), and most of the MSFC study activities during the 1978-1980 period were devoted to these two program concepts.

While Marshall's emphasis was on an evolutionary approach to space platforms, the leadership of the Johnson Space Center had, by early 1979, decided that the center's efforts should refocus on a major space station effort. A major factor influencing JSC thinking was "a need for a real goal to maintain the dedication of present participants in the space program and the interest and enthusiasm of young people in space technology in order to motivate their pursuing engineering and science careers." 31/

Based on these considerations, JSC during 1979 conducted an in-house study of a concept identified as a Space Operations Center (SOC). This study was based on two assumptions: "that the next 10 to 20 years will include requirements for large, complex space systems" and "that geosynchronous orbit is clearly a primary operational area in space in the coming decades." If these assumptions were valid, JSC argued, then "the space construction and servicing of these future systems will be more effective with a permanent, manned operations center in space."

The primary objectives of the SOC were identified as:

- The construction, checkout, and transfer to operational orbit of large, complex space systems.

- On-orbit assembly, launch, recovery, and servicing of manned and unmanned spacecraft.

-Further development of the capability for permanent manned operations in space with reduced dependence on Earth for control and resupply.

The SOC study noted that this list of objectives

> noticeably does not include onboard science and applications objectives, although the free-flying satellites which would be serviced would include mostly those of this genre. The primary implication of this omission is that experiment and applications requirements will not be design drivers; the SOC will be "optimized" to support the operational functions of the objectives. However, experiments or applications which can tolerate the operational parameters of the SOC can be operated onboard, or an entire dedicated module could be attached to an available berthing port.

The study developed a concept of a self-contained, continuously-manned orbital facility built from several shuttle-launched modules. The initial SOC crew would be 4-8 people. In addition to a core facility, the full capability SOC would require a construction facility and a flight support facility. The costs of this fully-capable SOC were estimated at $2.7 billion, with the total facility in place 9-10 years after program initiation. 32/ During 1980, there were several contractor studies, most notably by Boeing, exploring the SOC concept.

Conclusion

It should be evident from the whole of this paper that there can be no obvious cut-off point for an account of the development of the space station concept; this account arbitrarily breaks the story at the end of 1980. In recent years NASA has become engaged in an agency-wide effort to define a space station program which will represent an attractive investment opportunity to the nation's leaders. This effort is drawing upon over two decades of NASA space station planning in the context of the success of the space shuttle, increasing commercial and national security interest in space activities, and NASA's need to identify its post-shuttle directions. Perhaps this time around, the space station will indeed be accepted as the "logical next step" in evolving U.S. capabilities to operate in space. If this is the case, then the years of effort described herein will have been justified.

This paper is condensed from a larger study performed for the Johnson Space Center under NASA Contract NAS9-16461.

References

1. Much of this section is based on papers by Frederick I. Ordway, III, "The History, Evolution, and Benefits of the Space Station Concept," presented to the XIII International Congress of the History of Science, August 1971; and Leonard David, "Space Stations of the Imagination," AIAA Student Journal, Vol. 20, No. 4, Winter 1982/1983.

2. Wernher von Braun, "Crossing the Last Frontier," Collier's, March 22, 1952, pp. 25-29, 72-74.

3. This history is recounted in John M. Logsdon, The Decision to Go to the Moon: Project Apollo and the National Interest (Cambridge, Mass: MIT Press, 1970), chapter 2; and W. David Compton and Charles D. Benson, Living and Working in Space: The History of Skylab (Washington: NASA SP-4208, 1982), chapter 1.

4. Bruce Loftin, as quoted in the Minutes of the Research Steering Committee on Manned Flight, meeting of May 25-26, 1959.

5. George Low, "Manned Space Flight," in NASA, NASA-Industry Program Plans Conference, July 1960, p. 80.

6. Studies during the 1960s at LRC, MSC, and MSFC are summarized in Langley Research Center, Compilation of Papers Presented at the Space Station Technology Symposium, February 11-13, 1969.

7. The early Langley studies are summarized in Langley Research Center, A Report on the Research and Technological Problems of Manned Rotating Spacecraft, NASA Technical Note D-1054, August 1962.

8. Ibid.

9. George V. Butler, "Space Stations, 1959-To?," in B.J. Bluth and S.R. McNeal, Update on Space, Vol. I (Granada Hills, CA: National Behavior Systems, 1981), p. 8.

10. Douglas Missile and Space Systems Division, Douglas Aircraft Co., "Report on the Development of the Manned Orbital Research Laboratory (MORL) System Utilization Potential," Report SM-48822, January 1966.

11. Memorandum from Deputy Administrator, "Preliminary Study of a Manned Space Station," July 19, 1966.

12. The results of the study were presented in two briefings given by Charles Donlan on September 29, 1966 and several times in November 1966.

13. Space Business Daily, March 22, 1967, p. 122.

14. Memorandum from Thomas Paine to the President, "Problems and Opportunties in Manned Space Flight," February 26, 1969.

15. Briefing Memorandum prepared by Bureau of the Budget, Economics, Science, and Technology Division, "President's Task Group on Space...Meeting No. 1," March 6, 1969.

16. R.F. Packard, "Department of State Observations," draft, March 13, 1969.

17. Milton Rosen, "Memorandum for the Record," March 13, 1969.

18. Thomas Paine's Notes from Meeting on Space Stations, January 27, 1969.

19. Robert Lohman, "Notes from Mr. Matthews' Meeting, Friday, January 31, 1969, on Space Station Issues," February 4, 1969.

20. William Normyle, "NASA Aims at 100-Man Station," Aviation Week and Space Technology, February 24, 1969, p. 16.

21. NASA, "Statement of Work: Space Station Program Definition (Phase B)," April 14, 1969.

22. NASA's attempts to gain support for this large program are described in John Logsdon, "The Policy Process and Large Scale Space Efforts," Space Humanization Series, Vol. 1, No. 1, (1979), pp. 65-80.

23. Jack C. Heberling, "The Management Approach to the NASA Space Station Definition of Studies of the Manned Spacecraft Center," NASA Technical Memorandum X-58090, June 1972, p. 30.

24. Compton and Benson, op. cit., provide a history of the Skylab program.

25. McDonnell Douglas Astronautics, Manned Orbital System Concepts Study, Book 1 - Executive Summary, September 30, 1975, pp. iii, 36, 1-2, 30.

26. U.S. Senate, Committee on Aeronautical and Space Sciences, NASA Authorization for FY 1977, Hearings, p. 1046.

27. NASA Press Release, "NASA Seeks Proposals for Space Station Studies," December 11, 1975.

28. "Operational Base Concepts Gain in Space Station Studies," Aerospace Daily, September 13, 1976, p. 54.

29. Aviation Week and Space Technology, April 18, 1977, p. 42.

30. Ibid.

31. NASA, Lyndon B. Johnson Space Center, Space Operations Center: A Concept Analysis, November 29, 1979, pp. 1-1, 1-2.

32. Ibid, pp. 1-8, 1-12, 1-13, 1-19, 1-24.

INTEGRATED REQUIREMENTS FOR A SPACE STATION

E. Brian Pritchard*
NASA Langley Research Center
Hampton, Virginia

Abstract

As the first step in NASA's planning for a potential space station system, a major effort was undertaken to define realistic missions that would be enabled by or have materially benefitted from the permanent presence of man in space. Studies were conducted by NASA advisory boards, in-house panels, and industry under contract to NASA. Additional studies were undertaken by the international community. The results of these studies have been integrated into a single time-phased mission set that provides the foundation for current studies of space station system architecture. To meet the needs of missions in the science and applications, commercial, and technology development areas, the space station must be a research laboratory, an observatory, a service center, a communications and data processing mode, a transportation mode, a storage facility, and a construction/assembly base.

Introduction

An informed decision to proceed with development of a US space station must rest on a clearly defined national need for such a system. To this end, NASA funded the studies, Space Station Needs, Attributes, and Architectural Options. In addition, a Space Station Mission Requirements Working Group was established to direct the industry studies and to integrate in-house activities, and special studies such as studies of the Space Science Board and the Space Applications Board. This group was supported by three mission area panels: science and applications, commercial, and technology development. The Working Group also maintained liaison with the international community performing similar studies.

Upon completion of the industry and in-house studies, the Space Station Mission Requirements Workshop was held at the NASA Langley Research Center, May 2-13, 1983. The purpose of this workshop was to review study results and to synthesize these results into a single set of time-phased missions and their associated requirements. In developing this mission set, the following fundamental ground rules were established:

> The Mission Set must be valid, demonstrated either through a funding commitment as in the commercial case or through endorsement by the scientific community and inclusion in the NASA OSSA long-range plan as in the science and applications case. Also, the mission set must be realistic -- programmatically, scientifically and technologically.

It is also important to note that we did not restrict the mission set to those missions that could be accomplished only onboard a manned space station. In fact, the definition of a space station as used here is: the space station system includes a manned facility, co-orbiting free flyers and platforms, remote free flyers and platforms, an orbital maneuvering system, a reusable orbital transfer vehicle, and storage and servicing facilities. Although this total system capability is not expected to be available in the initial space station system, it is required by the mid-1990s.

The purpose of this paper is to present the results of the study activities discussed above in terms of time-phased mission sets and their associated requirements that must be accommodated by the space station. The mission requirements development is a continuing process. Thus, the missions presented here should not be considered a final mission set to which the space station must be designed. Missions and their requirements will continue to evolve in the months and years ahead, particularly as experiments performed onboard the operational Shuttle fleet provide new insights into science, technology, and commercial ventures for the future.

Time-Phased Mission Set

Science Applications Missions

The time-phased mission set developed at the Mission Synthesis Workshop, as modified by subsequent analysis, is presented in Figs. 1-6. The science and applications missions (Figs. 1 and 2) are divided into the current NASA disciplines. The code number relates the mission to the NASA data base maintained at the Langley Research Center. This data base includes all pertinent data for the missions currently available. The data base is maintained by the NASA Mission Requirements Working Group and is updated periodically as additional information becomes available. The 21 missions included under astrophysics range from payloads attached to the space station to free flyers at high-inclination and/or high-energy orbits. The length of the bar denotes the duration of the mission and desired inclination and accommodation. Included in the mission set are free-flying missions that will be launched prior to the space station era, such as the Space Telescope and the Gamma Ray Observatory, but which will be serviced by the space station. Others such as OPEN and the Very Long Baseline Interferometer will not interact with a space station at 28.5 deg inclination but are included for completeness of the total mission set. The triangle in 1991 for the Solar Dynamics Observatory denotes launch with no potential interaction with the space station. A number of missions are shown as attached either to the space station or to a space platform. Decisions on the location of these missions will depend on a number of factors. The dominant factors are the ability to point instruments accurately with acceptable jitter, the degree of contamination of

*Special Assistant to the Director for Space.

the environment around the space station, the frequency of service required by the payload, and the cost effectiveness of one accommodation mode vs the other.

In earth science and applications five missions have been defined. The Lidar facility is envisioned as a research facility for development of Lidar technology and techniques as well as scientific studies of the tropical atmosphere. Once the development is complete, Lidar instruments would be placed on the Earth Science Research Platform located in a polar or near polar orbit. The Earth Science Research Platform is an evolutionary interdisciplinary facility for study of the Earth, the oceans, the atmosphere, and biogeochemical cycles. The GOES Follow-on is a mission to geosynchronous orbit for development and demonstration of advanced instrumentation in meteorology and climatology.

Solar system exploration missions are shown as launches and may or may not interact with the space station (Fig. 2). If a reusable orbital transfer vehicle (OTV) is available by the mid-1990s, it could be used as a launch stage for some missions. The only mission shown that would be enabled by a space station is the Mars Sample Return Mission which requires on-orbit assembly and could use the space station for sample analysis on return.

Life sciences missions have two focii: 1) studies of long-duration weightlessness effects on humans, animals, and plants in an onboard lab facility and 2) the development of a fully closed life support system. Initial activities in the Life Sciences Lab are devoted to research on plants humans, and small animals. Later in the decade (1995) an animal and plant vivarium is added.

The Materials Processing Sciences R&D Facility was identified as a major need by both the Science and Applications and Commercial Panels. It is shown in both Figs. 2 and 3 for this reason.

The only activity envisioned in the communications area under science and applications is an experimental geosynchronous platform that would support a number of communications payloads. This platform could also be a key element in the development of techniques for geoplatform servicing by the combination of the OTV and a "smart" orbital maneuvering vehicle (OMV).

Commercial Missions

The commercial mission set is presented in Figs. 3 and 4. The MPS Processing Lab 1 is required at space station IOC. A volume of 60 m^3 is required with an average power requirement of 8 kW. It is anticipated at this time that this unit would be provided by the government and used by government, academic, and industry researchers. MPS lab 2 would probably be provided by the commercial sector. The volume and power requirements for this lab are 60 m^3 and 15 kW, respectively.

A total of 10 commercial production units are shown for the decade of the 1990s. The electrophoresis operations in space (EOS) unit would be provided by McDonnell Douglas/Johnson & Johnson and produce a variety of pharmaceutical products. The electroepitaxial crystal growth (ECG) unit provided by Microgravity Research Associates would produce high-purity 5 cm diam crystals of gallium arsenide. These two activities are currently in the research phase under joint endeavor agreements with NASA. Additional units with high potential for implementation over the remainder of the decade are:

1) Isoelectric focusing (IEF) -- biological products directional.

2) Directional solidification crystal growth (DSCG): gallium arsenide, HgCdTe, and other crystals.

3) Vapor crystal growth (VCG): HgCdTe and other crystals.

4) Optical fiber: high-quality optical fibers.

5) Solution crystal growth: crystals with fast switching electronic characteristics.

6) Iridium crucible: high-purity iridium crucibles.

7) Biological processes: proprietary process for production of biological materials.

8) Merged technology - catalyst: proprietary process.

We do not expect that all of the above will prove to be viable commercial endeavors. Some will, but others will be replaced by new products and processes that will be developed either on Shuttle missions or in the MPS R&D labs.

Analysis of the commercial potential for Earth and ocean observations indicate that the return on investment would be adequate for commercial entities to develop and operate the instruments but not adequate for provision of the spacecraft as well. Thus, it is assumed here that instrumentation would be provided by industry and accommodated on a NASA space platform such as the Earth Science Research Platform. Instruments with the greatest potential are the stereo multilinear array, the stereo synthetic apperture radar, and a coastal zone color scanner-type instrument.

The commercial space communications field is the first success story in the industrialization of space. Projections of future missions and needs vary widely. We have attempted here to strike a middle ground with an average of 15 satellite deliveries per year divided into the mass categories accommodated by the four upper stages indicated in Fig. 4. It is assumed here that a cost-effective reusable OTV introduced in 1994 will begin to provide transportation to geosynchronous orbit and by 1997 will have captured the full 15 missions per year. It appears that a communications test lab attached to the space station is required. This is basically an on-orbit antenna test range. Significant technology development is required to move to large antennas for land mobile communication and for laser communications. These activities, while identified by the Commercial Panel are included under technology development missions for requirements bookkeeping purposes.

With the advent of the OTV, geosynchronous satellite servicing becomes feasible. Thus, we show servicing missions beginning in 1995 and grow-

CODE	NAME	YEAR (91–00)
	ASTROPHYSICS	
SAA 0001	SPECTRA OF COSMIC RAY NUCLEI (SCRN)	ATT 28.5° SP/SS: 91
SAA 0002	SPACE PLASMA PHYSICS (ACTIVE)	ATT 28.5° SS: 92–93
SAA 0002A	SPACE PLASMA PHYSICS (PASSIVE)	ATT 28.5° SP: 92–93
SAA 0006	STARLAB	ATT 28.5° SP/SS: 91–93
SAA 0003	SOLAR OPTICAL TELESCOPE (SOT)	ATTACHED 28.5° SP/SS: 92–96
SAA 0009	PINHOLE/OCCULTER FACILITY	ATT 28.5° SP/SS: 97–98
SAA 0011	ADVANCED SOLAR OBSERVATORY (ASO)	ATT. 28.5° SP/SS: 99–00
SAA 0004	SHUTTLE IR TELESCOPE FAC. (SIRTF)	ATT: 93; ATT 28.5° SP/SS: 95
SAA 0005	TRANSITION RADIATION & ION CAL (TRIC)	ATT 28.5° SP/SS: 94–95
SAA 0007	HIGH THROUGHPUT MISSION (HTM)	ATTACHED 28.5° SP/SS: 96–99
SAA 0008	HIGH ENERGY ISOTOPE	ATT 28.5° SP/SS: 97–99
SAA 0012	SPACE TELESCOPE (ST)	FREE FLYER 28.5°: 91–00
SAA 0013	GAMMA RAY OBSERVATORY (GRO)	FF 28.5°: 91–93
SAA 0014	X-RAY TIMING (XTE)	F.F. 28.5°: 91–92
SAA 0019	FAR UV SPECTROSCOPY EXP. (FUSE)	F.F. 28.5°: 93–96
SAA 0010	CORONA DIAGNOSTIC MISSION (CDM)	F.F. 28.5°: 99–00
SAA 0015	OPEN	F.F. 0° & 80°: 92–95
SAA 0016	SOLAR MAX MISSION	F.F. 28.5°: 91–93
SAA 0017	ADV X-RAY ASTROPHYSICS FÁC (AXAF)	FREE-FLYER 28.5°: 91–00
SAA 0018	VERY LONG BASELINE INTERFERO. (OVLBI)	FREE-FLYER 57°: 95–97
SAA 0020	LARGE DEPLOYABLE REFLECTOR (LDR)	F.F. 28.5°: 98–00
SAA 0021	SIRTF SUNSYNCH	F.F. 97°: 98–00
SAA 0022	SOLAR DYNAMICS OBSERVATORY	△ 91
	EARTH SCIENCE & APPLICATIONS	
SAA 0201	LIDAR FACILITY	ATT 28.5° SS/SP: 91–93
SAA 0202	EARTH SCIENCE RESEARCH	FREE FLYER 90°: 91–00
SAA 0203	TOPEX	FREE FLYER 63.4°: 91–94
SAA 0204	GEOPOTENT RESEARCH MISSION (GRM)	FF 90°: 91
SAA 0205	GOES FOLLOW-ON	△ 96

Fig. 1 Time-phased mission set, science and applications (as of June 1983).

CODE	NAME	YEAR (91–00)
	SOLAR SYSTEM EXPLORATION	
SAA 0101	MARS GEOCHEM/CLIMATOL ORBITER (MGCO)	△ 1990
SAA 0102	LUNAR GEOCHEM ORBITER	△ 93
SAA 0103	COMET HMP RENDEZVOUS	△ 1990
SAA 0104	VENUS ATMOSPHERE PROBE	△ 94
SAA 0105	TITAN PROBE	△ 93
SAA 0106	SATURN PROBE	△ 98
SAA 0107	MAIN BELT ASTEROID RENDEZVOUS	△△ 96
SAA 0108	SATURN ORBITER	△ 98
SAA 0109	NEAR EARTH ASTEROID RENDEZVOUS	△ 99
SAA 0110	MARS SAMPLE RETURN	△ 99
	LIFE SCIENCES	
SAA 0307	LIFE SCIENCES LAB	PRESSURIZED MODULE: 91–00
SAA 0306	CELSS PALLET	ATTACHED S.S.: 93–98
SAA 0305	DEDICATED CELSS MOD.	PRES. MOD.: 99–00
	MATERIALS SCIENCE	
SAA 0401	MPS R&D FACILITY	PRESSURIZED MODULE: 91–95; REQMT'S INCLUDED IN COM 1201
	COMMUNICATIONS	
SAA 0501	EXPERIMENTAL GEO. PLATFORM	△ 94

Fig. 2 Time-phased mission set, science and applications (as of June 1983).

ing over the remainder of the decade. COM 1121 involves storing a spare direct broadcast satellite (DBS) at the space station. Upon failure of an operational DBS the spare could be rapidly reconfigured to match the failed satellite characteristics and transported to geosynchronous orbit by the OTV. The missions listed under industrial services are not missions as such. They are potential services which industry could provide to the government and to other industries at a profit.

Technology Development Missions

The technology development missions set shown in Figs. 5 and 6 was primarily developed by the NASA Technology Development Missions Panel. The missions are grouped according to the NASA OAST technology disciplines. In general, the technology development missions require attachment to the space station and a large amount of crew EVA. Key missions in the materials and structures discipline

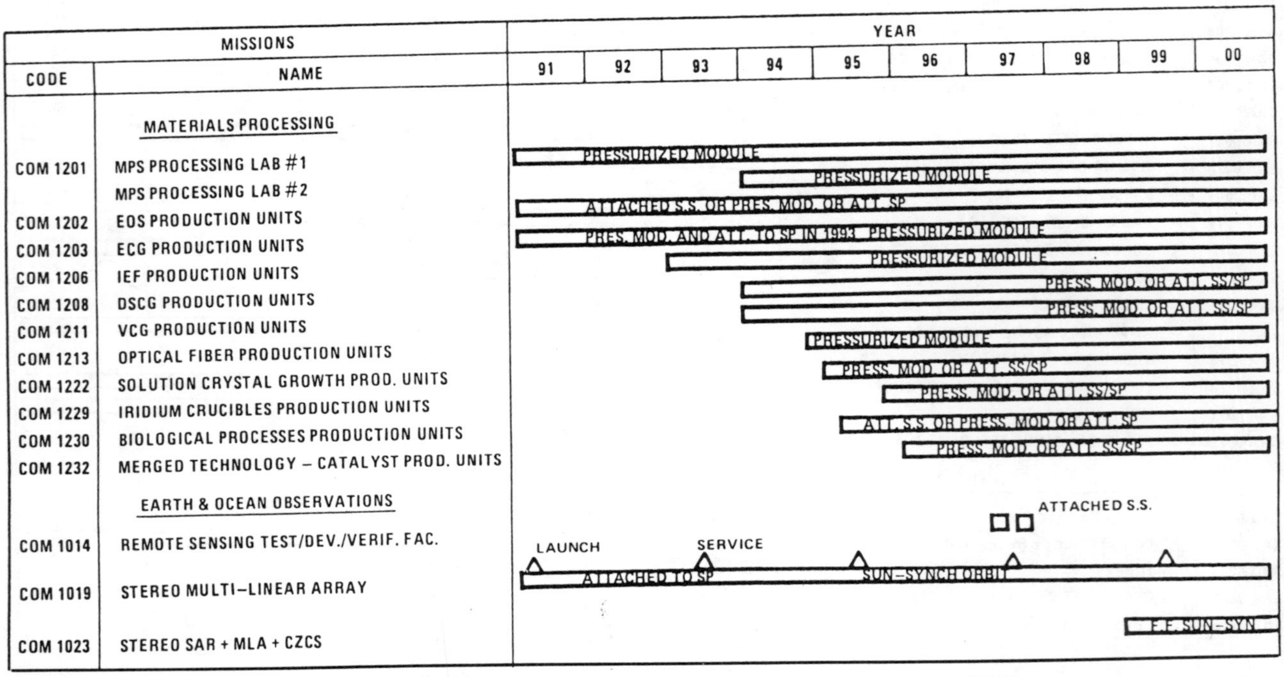

Fig. 3 Time-phased mission set, commercial (as of June 1983).

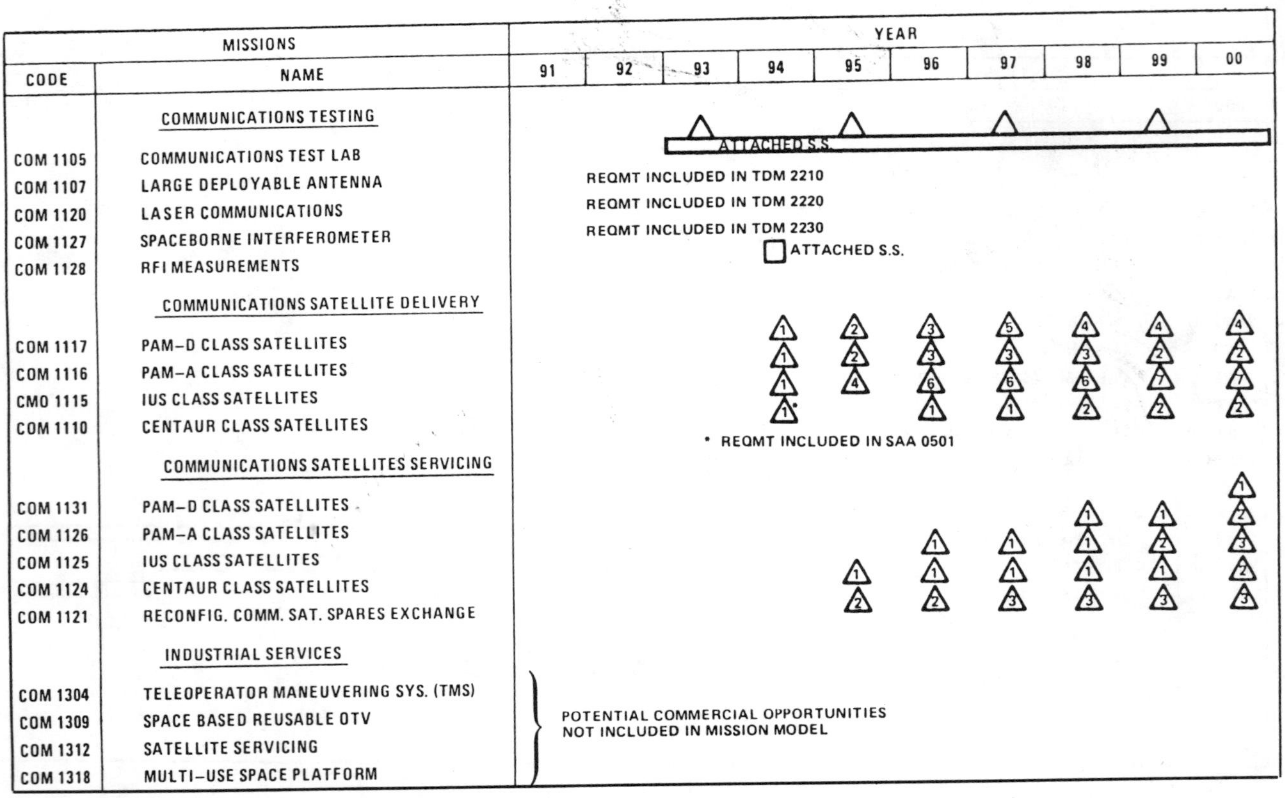

Fig. 4 Time-phased mission set, commercial (as of June 1983).

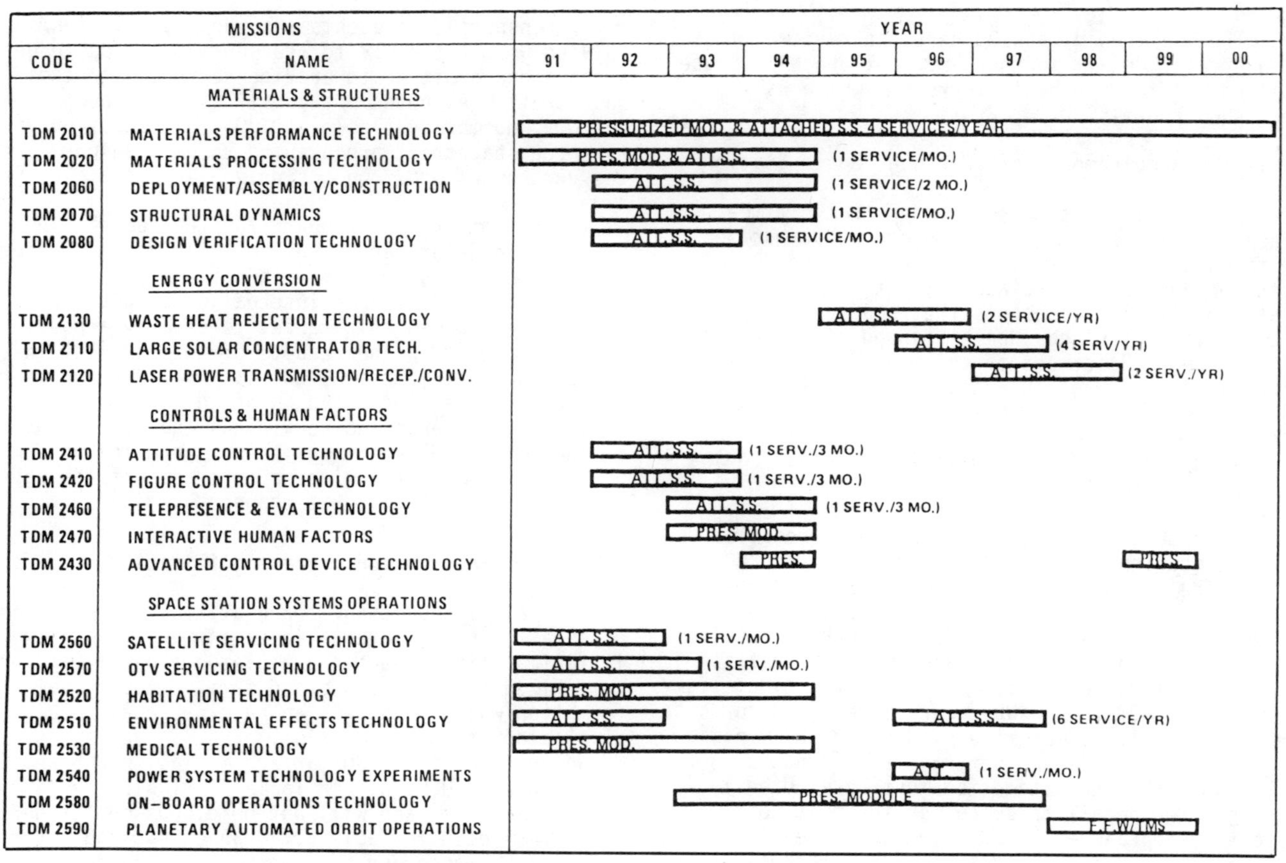

Fig. 5 Time-phased mission set, technology development (as of June 1983).

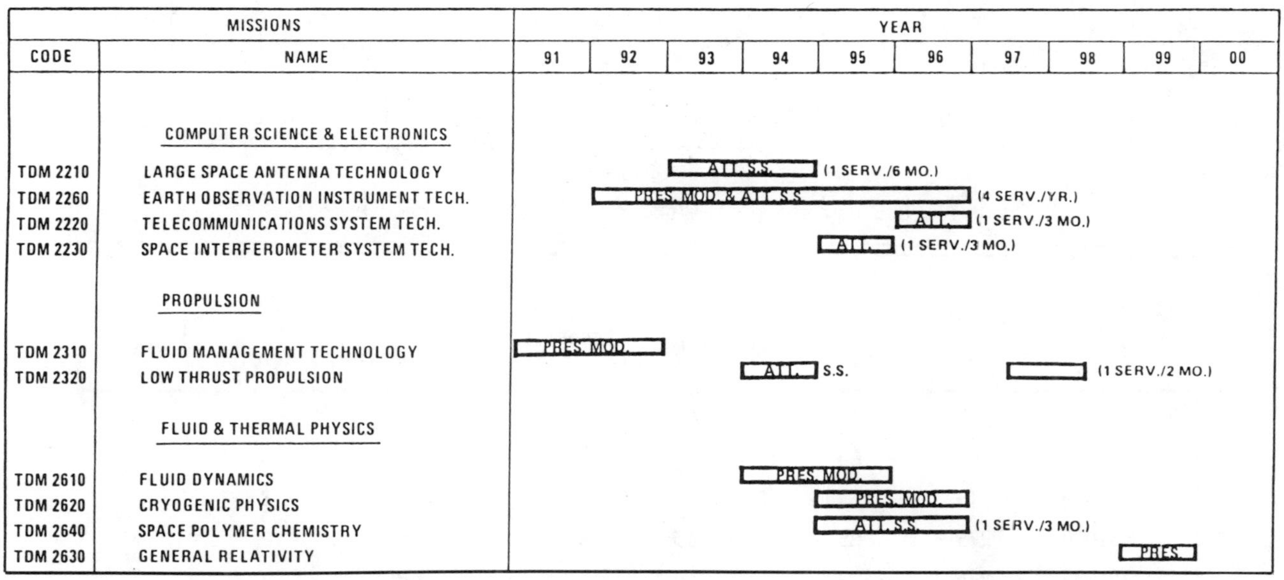

Fig. 6 Time-phased mission set, technology development (as of June 1983).

are those for development of space construction technology that are enabling for the development of future large antennas, both for commercial communications missions and for future science missions. For example, the large deployable reflector astrophysics mission also requires this technology.

Although satellite and OTV servicing technology development will be initiated on the Shuttle, this development will continue on the space station for several years. As shown in Figs. 5 and 6, most of the technology missions are undertaken over the first half of the decade with a substantial decrease in activity over the latter part of the decade. We may be overly optimistic in our time phasing, but the technology to be developed early is required for missions in the late 1990s. We may also be short sighted in terms of our ability to predict the technology development needs for those missions in 2000 and beyond.

The missions discussed here are reviewed in much greater detail in the Report of the NASA Space Station Mission Synthesis Workshop, which will be published shortly.

Mission Requirements of 28.5 deg Space Station

The requirements associated with the time-phased mission set of Figs. 1-6 can be summarized in terms of those associated with the manned space station and those associated with the co-orbiting platform. Table 1 presents the requirements which the mission set places on the manned space station at 28.5 deg. Placement of the manned space station at an inclination of 28.5 deg is driven primarily by two factors. The majority of U.S. missions considered can be accommodated at 28.5 deg and this is the inclination to which maximum payload can be delivered by the Shuttle.

As noted in Table 1, high power is required from the beginning of the station era -- 55 kW in 1991. The primary driver in power is the commercial materials processing area. In 1991, the two production units and the MPS lab require a total of 43 kW. IVA crew time associated with materials processing is also large. The current estimate is that about 1400 crew hours/year is required for each production unit to maximize production and minimize down time. Obviously, the tradeoff between crew time and increased automation must be studied in much greater detail to establish the most cost-effective approach to production unit operation. EVA hours are primarily driven by the technology development missions and reach a maximum in 1993. Note that this is actual work time and does not include the time associated with getting into and out of the suit or prebreathing (if necessary). The number of pressurized modules is largely due to commercial production units.

Free-flyer servicings are those servicings associated with science mission free flyers at 28.5 deg inclination.

Transportation events are those activities involving use of the OTV to transport payloads to geosynchronous orbit or to provide servicing to those spacecraft. As noted previously, it is assumed that the OTV becomes available in 1994. In addition, up to three spacecraft are assumed to be transported to geosynchronous orbit on a single OTV flight.

The summary requirements for the co-orbiting 28.5 deg platform are given in Table 2. These requirements are associated with U.S. science payloads. Accommodation of international instruments, while not shown in this table, is clearly a possibility as noted in the results of the parallel international utilization studies. The number of servicings per year indicated is based on the need for payload interchange and servicings and require use of the OMV to either bring the platform to the proximity of the manned station or to transport the payload to the platform for exchange.

International Space Utilization

Parallel studies of space utilization have been undertaken by the international community. These initial studies have been completed recently and are summarized here only in broad terms. Additional information was provided in an international panel discussion at this symposium. ESA, Japanese, and Canadian results are summarized in Tables 3-5 and compared with U.S. use in Table 6. Tables 3-5 are self-explanatory and will not be discussed further here. As noted in Table 6, a manned R&D lab for materials science and space processing is required from the viewpoint of all countries. Fundamentally, all the countries note a substantial role for a manned space station in both science and applications and technology development. Operational support to unmanned facilities is also viewed as a major space station role.

Although U.S. requirements for the low-inclination platform are relatively modest, as noted in

Table 1 28.5 deg space station, summary of U.S. requirements.

Year	Mass Kg	Pressurized Volume M^3	Power kW	IVA hrs	EVA hrs	# Pressurized Modules	# Attached Payloads / # Ports	# FF Servicing	# Transportation Events
1991	35900	195	55	12300	830	4(2)[2]	6/2	3	-----
1992	49900	195	60	12900	1450	4(2)	13/3	3	-----
1993[1]	55000	215	63	16600	1520	5(3)	12/4	3	-----
1994	61000	275	88	24800	1330	8(5)	11/4	2	4
1995	101000	275	101	25200	230	11(8)	7/4	2	6
1996[1]	107000	310	106	29500	160	14(10)	10/4	2	12
1997	101000	305	121	30200	150	14(10)	8/4	1	10
1998	99000	305	112	27400	140	14(10)	5/4	2	13
1999[1]	96000	370	130	27500	110	15(10)	2/2	2	14
2000	94000	365	129	26900	70	15(10)	2/2	2	14

Notes: (1) EEG production facility additional units brought on line in 1993, 96, and 99 are assumed to be co-orbiting free flyers.

(2) Number in parenthesis is the number of commercial production units provided by industry.

Table 2 28.5 deg platform, summary of requirements.

YEAR	POWER kW	RECORDING DATA RATE MBPS	# ATTACHED PAYLOADS	PAYLOAD MASS KG	# SERVICINGS
1991	2.5	16	2	6400	2
1992	3.1	66	3	11500	3
1993	4.7	67	4	18500	2
1994	1.9	50	2	12400	3
1995	3.2	51	3	19400	2
1996	3.3	51	2	16600	3
1997	2.6	2	3	16600	3
1998	2.6	2	3	16600	2
1999	4.9	44	3	25500	3
2000	2.6	42	1	12500	2

Table 3 ESA space utilization (128 payloads).

- o MATERIAL SCIENCE AND SPACE PROCESSING
  - AUTOMATED PROCESSING FACILITY (20 KW)
  - MANNED R&D FACILITY (10 KW)
- o LIFE SCIENCES
  - FREE FLYER (EURECA)
  - MANNED R&D FACILITIES (SHORT AND LONG MODULES)
- o SPACE SCIENCES AND APPLICATIONS
  - EARTH OBSERVATIONS FREE FLYERS AND PLATFORM (57° TO POLAR)
  - ASTRONOMICAL OBSERVATIONS FREE FLYERS AND PLATFORM (LOW AND HIGH INCL.)
- o TECHNOLOGY AND OPERATIONS
  - ASSEMBLY/CONSTRUCTION
  - SATELLITE SERVICING
- o NEW FIELDS
  - CALIBRATION LAB
  - MODULAR OTV
  - SOLAR ENERGY TRANSMISSION
  - SPACE PROPULSION SYSTEM DEVEL.
  - SPACE DEBRIS REMOVAL

Table 2, the international community also sees a need for such a platform for astronomical observations. Japan sees a need for a manned facility at high inclination primarily to support real-time oceanographic applications. Although the U.S. is the only country calling for an attached MPS processing facility, the German Columbus study focuses on a Spacelab derivative system that can operate either in an attached or co-orbiting mode.

Special Requirements

Each mission area has its own special requirements that will impact the space station design. Science and applications missions that are instrument dependent, whether Earth observations or astronomical, require stringent control of the contamination environment around the station. These requirements are defined in more detail for representative instruments in the Mission Synthesis Workshop Report. These missions also require highly accurate pointing and stability.

Commercial missions may require both physical and communications security for proprietary processes and/or experiments. Rapid replacement of DBS satellites in geosynchronous orbit requires the capability for on-orbit storage of complete spacecraft. Finally, the materials processing facilities require storage of both raw materials and spare equipment.

Technology development missions are unique in their requirement for large amounts of EVA time. A corolary to this requirement is the need for improved suits, tools, and techniques to maximize man's productivity in extravehicular activities.

Table 4 Japanese space utilization.

- o MATERIAL SCIENCE AND SPACE PROCESSING
  - MANNED R&D FACILITY
  - PROCESSING FACILITY
- o LIFE SCIENCES
  - MANNED R&D FACILITY (BIOLOGY, SPACE MEDICINE, CELSS)
- o SPACE SCIENCES AND APPLICATIONS
  - EARTH OBSERVATIONS MANNED AND UNMANNED FACILITIES (NEAR POLAR)
  - ASTRONOMICAL OBSERVATIONS FREE FLYERS/PLATFORM (LOW AND HIGH INCL.)
- o TECHNOLOGY AND OPERATIONS
  - FREE FLYER FOR SPACE PLASMA, ADVANCED PROPULSION, MICROWAVE ENERGY TRANS., SOLAR ARRAYS AND CONCENTRATORS
  - ATTACHED MODULE FOR LARGE STRUCTURES, LONG DURATION EXPOSURE, ZERO GRAVITY LIQUID HANDLING
- o COMMUNICATION TECHNOLOGY
  - LARGE ANTENNAS CONSTRUCTION
  - SATELLITE ASSEMBLY AND TEST
  - MAINTENANCE

Table 5 Canadian space utilization (37 payloads).

- o MATERIAL SCIENCE AND SPACE PROCESSING
  - MANNED R&D FACILITY
  - AUTOMATED PROCESSING FACILITY
- o LIFE SCIENCES
  - MANNED R&D FACILITY
- o SPACE SCIENCES AND APPLICATIONS
  - EARTH OBSERVATIONS FREE FLYERS/PLATFORM (SUN SYNC., 4-7 KW, 120-240 MBPS)
  - ASTRONOMICAL OBSERVATIONS FREE FLYERS/PLATFORM (LOW AND HIGH INCL.)
- o TECHNOLOGY AND OPERATIONS
  - LARGE STRUCTURES CONSTRUCTION/ASSEMBLY
  - SOLAR CELLS/ARRAYS
  - SENSOR DEVEL.
  - OTV SRM
  - MAINTENANCE, SERVICING, REPAIR

Space Station Functional Characteristics

To meet the requirements associated with the time-phased mission set presented in Figs. 1-6, the space station must have the characteristics described in Table 7. Initially, the manned element placed at 28.5 deg must function as a research laboratory, an astrophysical observatory, a service center (which requires basing and support of an OMV), and a communications and data processing mode (for onboard and co-orbiting experiments). In addition, both low- and high-inclination platforms are required.

By the middle of the 1990s, the manned element must additionally serve as a transportation mode with a space-based OTV, provide storage for complete spacecraft and serve as a construction and assembly depot for construction of spacecraft larger than those which can be accommodated in the Shuttle cargo bay.

By the year 2000, we will need enhanced capability in all areas as the traffic induced by the existence of the Space Station with the characteristics described here places increased demands on space station services. Modest replication of the manned element (one- to four-man crew) in near polar orbit should also be expected.

User Concerns and Challenges

Many elements of the user community have concerns about the space station. The three most prevalent concerns are:

1) How will NASA insure "user friendliness?" This includes rapid and guaranteed access; adequate

resources (including storage); minimum "red tape"; security for proprietary experiments, process, and data; contamination and pointing control; and reasonable user charge policy.

2) How cost effective are satellite/platform servicing and space-based OTV?

3) Will space station development reduce funding available to traditional space science and applications disciplines?

As a result of these concerns, a list of challenges to NASA as it proceeds with its space station planning activity can be developed. The list below was developed by the author and should be considered as representative and neither exhaustive nor necessarily endorsed by the entire user community:

1) Interface philosophy -- keep it standard and simple.

Table 6 International space utilization summary.

Mission Area	U.S.	ESA	JAPAN	CANADA
Material Science and Space Processing				
Manned R&D Lab	✓	✓	✓	✓
Attached Processing Facility	✓			
Co-orbiting Processing Facility	✓	✓	✓	✓
Life Science				
Manned R&D Lab.	✓	✓	✓	✓
Co-orbiting Res. Facility				
Space Sciences and Applications				
Earth Observation				
High Incl. F.F./P	✓	✓		✓
Attached Res.	✓			
Manned High Incl.			✓	
Astronomical Observation				
Attached Obs.	✓			
Low Incl. FF/P	✓	✓	✓	✓
High Incl. FF/P	✓	✓	✓	✓
Technology and Operations				
Large Structures	✓	✓	✓	✓
Energetics	✓	✓	✓	✓
Sensor Development	✓			✓
Maintenance/Service/Repair	✓	✓	✓	✓
OTV	✓	✓		✓

Table 7 Space station functional characteristics.

INITIALLY: 28.5° SPACE STATION

- Research Laboratory (Science, Applications, Technology, Industrial)
- Observatory
- Satellite/Platform Service Center (TMS)
- Communications and Data Processing Node
- Low and High Inclination Platforms

MIDTERM: ADD

- Transportation Node to Higher Energy Orbits (SBOTV)
- Storage Facility in Space
- Construction/Assembly Depot

FINALLY: ADD

- Enhanced Capability in All Areas
- Modest Replication in Polar Orbit

2) User charge policy -- establish it early and stick to it.

3) Establish a user counterbalance to the designers/buildings -- don't weaken the user accomodation impetus.

4) EVA is necessary -- establish the equipment, tools, and techniques early and get on with development, test, and training on the Shuttle.

5) High productivity is the key to cost effectiveness -- excessive management and ground control is the enemy of productivity.

6) The right mix of automation and crew is essential and must be established early.

7) To attract business the space station must be resource rich.

8) Satellite/platform servicing and OTV operations must be cost effective.

9) Onboard research labs should be equipped and operated like ground research labs.

10) Spare parts must be available when needed.

Conclusion

The time-phased mission set presented here represents the first step in a continuous activity to develop valid space station system mission requirements. The next phase of this activity will focus on continued refinement, analysis, and development of the commercial and technology development missions. Further development of the commercial user community is underway at NASA with two contracted studies and the establishment of a NASA inhouse team.

The estimate of the potential for commercial materials production is high, but is only our current best guess. Extensive shuttle sortie research is required prior to 1991 if this potential is to be realized.

There is no single mission or mission area that can justify development of a space station system. It is the totality of capabilities and auxiliary systems which will satisfy our space needs for the 1990s and provide the gateway to the future in the next century.

USE OF SPACE STATION FOR SCIENCE

T. M. Donahue*
National Academy of Sciences
Washington, D.C.

In this discussion, I would like to do several things. First, I plan to describe the kinds of missions that are needed to address the major questions that we believe need to be answered during the remainder of this century in the various disciplines that make up space science. Considering these mission sets for each of the disciplines, I shall try to identify the special requirements they place on the space program if their objectives are to be attained. Examples would be the kinds of orbital elements needed, the kind of accuracy and stability of pointing and control needed by the platforms on which instruments would be mounted, and the like.

Then, turning to a space station as a possible component in the space program, I shall try to identify the desirable qualities a space station might bring with it and that would enhance the attainment of the scientific objectives of that discipline. If there are any missions that actually would be enabled or any objectives that could not be attained with reasonable expenditure of resources other than with a space station, I shall try to identify them.

Then, I shall point out features of a space station that would tend to prevent or hinder the attainment of certain objectives, particularly if the manned space station alone were to be the only platform or the only transportation mode for space science missions. This kind of analysis will allow me to characterize the kind of space station system that would be most suitable for each particular discipline, to analyze how well some particular candidate system would serve our purposes, such as a manned station in a 28.5 deg inclination orbit, accompanied by one or more unmanned satellite platforms or free flyers, plus a collection of high inclination elements, manned or otherwise. I shall then point out what programs and what facilities, in addition to a space station, would have to be maintained in order to assure the health of each discipline.

From this analysis we should be able to determine what kind of space programs-- and what kind of space station, if one is in the program-- we need for a healthy space science program during the next couple of decades. The analysis will also tell us whether we need the space station in the program to assure this healthy program and allow us to make tradeoff studies between a program with a space station and without one for space science. This last exercise might, however, be gratuitous because I have not heard anyone in a responsible position state that its capability to do space science is a primary justification for a space station, any more than lunar science was invoked as the primary justification for the Apollo program.

*Chairman, Space Science Board.

For astronomy and astrophysics (Table 1) a manned space station in low Earth orbit (LEO) at low inclination (e.g., 28.5 deg) would be useful if it is equipped with a TMS as a base for servicing, change of instruments, and repair We fear that requirements for high pointing accuracy and freedom from contamination preclude clustering on the space station itself, and probably on an accompanying platform as well. Hence, these observatories probably should be independently orbiting free flyers. Certain payloads need to be in high-inclination orbits(sun synchronous in the case of solar observations, at 57 deg for OVLBI, and polar or outside the magnetosphere for low-energy cosmic ray studies), so the means of launching into those orbits must be preserved. The space station appears to be useful for servicing and repair, but is not mandated. Shuttle Orbiter servicing would suffice for the basic needs.

The needs of the earth sciences are similar to those of astrophysics, except that the need for high inclination orbits and GEO is emphatic (Table 2). OTV must be developed simultaneously with SS, or other means of getting to GEO and retaining high inclinations.

Launching from the space station involves plane change for planetary exploration. However, the stage can be fully fueled; there is no need for off loading of fuel. On balance, the space station has no clear advantages except for very ambitious missions such as sample return from Mars, asteroids, or comets. See Table 3.

Some space plasma physics experiments and observations could be carried out by instruments

Table 1 Astronomy and astrophysics

Missions
Observatories
Large: ST, AXAF, OVLBI, SIRTF
Small: IRAS, IUE, FUSE, EUVE
Special requirements
High inclination
Deep space
Constant thermal load
Sun synchronous
Environmental purity
Accurate, stable, independent pointing and control
Space station
Potentially desirable features
Servicing,refurbishment, and repair
Potentially undesirable features
Contamination
Motion
Thermal load
Slow manueverability
LEO

Table 2 Earth sciences

- Missions
  - Remote sensing observatories:
    - UARS, TOPEX, GRM, crustal dynamics
- Special needs
  - GEO
  - High inclination
  - Long term
  - Environmental purity
  - Accurate, stable, independent pointing and control
  - High-data rates (300 mb$\cdot$s^{-1})
- Space station
  - Potential desirable features
    - Servicing, refurbishment, recalibration
    - High power (SAR)
  - Potential undesirable features
    - Contamination
    - Competition for orbits
    - Inflexibility
    - LEO

Table 3 Planetary exploration

- Missions
  - Deep space probes: GLL, MGCO, HMP
  - Observatories
- Special requirements
  - High-energy stage
  - Accurate, stable,independent pointing and control
- Space station
  - Potential desirable features
    - For observatories, deep space probes, see Table 1
    - Sample return
    - Propellant
  - Potentially undesirable features
    - Plane changes

mounted on the space station and the long duration afforded would be very interesting. There is no conspicuous, identified need for servicing. There is a strong requirement for maintainance of a capability for missions to GEO, high inclination, and deep space. See Table 4.

Table 5 lists the needs of the life sciences.

What emerges from this discussion is the conclusion that a manned space station could be configured to serve the requirements of space science rather easily. For example, a manned space station in low Earth orbit at low inclination (around 30 deg) could be useful, especially for an orbit servicing, refurbishment, and recuperation of science payloads, provided no attempt is made to force science payloads on board the manned module itself. At the present time it is not clear how many elements of the various space science missions will lend themselves without serious compromise to incorporation on a single, large platform accompanying the manned module. We are a little edgy about the possibility of pressure to go into that mode and anxious that priorities not be distorted to accomodate missions that are adaptable, particularly by some kind of procrustean adaptation process, to being fitted onto a unified platform.

Table 4 Solar and space physics

- Missions
  - Observatories: SOT, SMM, solar seismology
  - Plasma probes: OPEN, DE
  - SpaceLab
- Special requirements
  - GEO
  - High inclination greater than 57 deg
  - Variable duration observing time
  - Subsatellites for plasma and CR
  - Large weight and volume
- Space station
  - Potentially desirable features
    - Large antenna arrays
    - Tethered subsatellite
  - Potentially undesirable features
    - Contamination and interference
    - Inflexibility
    - LEO

Table 5 Life sciences and space medicine

- No first-order biological need identified
- Space station studies in space medicine in preparation for long duration in space

In other words, we see a need for accomodating a squadron of free flyers in the fleet. There is also a clear requirement for a high-inclination platform plus free flyers, and the means to get adequate payloads to high altitude, to GEO, and into deep space must be continuously assured.

The decision to go with the IUS the last time around was once bitterly contested by scientists, particularly planetary scientists from the first days of STS planning. I am quite sure we were right then, and I don't want to see the analogous mistake repeated this time if we go forward with a space station.

It is also clear that we have not been able yet to identify missions that would be enabled by the space station, except possibly in the field of space medicine.

However, with a space station on orbit, the way we configure a spacecraft to carry out a particular mission could be drastically altered by the possibilities of on-orbit assembly and servicing. I must confess that our experience with the Space Shuttle, which led to the postponement of so many missions as ELVs were phased out and many capabilities touted for the Shuttle turned out to be virtual capabilities, tends to inhibit our consideration of these new, space station ways of doing business. There would be a lot of resistance to the suggestion that some of the missions in our plans wait until the space station has been developed so that they can be reconfigured. This statement is contained within the broader admonition that the highest priority be given to maintaining the health of space science while the space station is being developed, if it is to be developed. Otherwise, the scientists and the technologies required to do space science won't survive to the day when the space station becomes operational.

Whether the level of activity in our space science plans or in those of our colleagues in applications is high enough to justify a space station is very questionable, at least in the short term. However, as I said at the outset, I do not think any responsible person claims that the space station would be built to serve these missions primarily. It may well be that the discussion leading to a decision about a space station will identify some highly visible goal, such as manned exploration of Mars, as the national aspiration that a space station would serve. If the science advisor's views on this subject are being accurately represented in the press, this may be the case. If so, we in space science, have to ask whether the space station designed to support extraterrestrial human expedition contains the space station I have just been describing as a viable element.

Assurance of the health of the unmanned space programs in the presence of a large manned program is hardly a new theme. Space science has never been so healthy as it was during the heady days of Apollo. I trust that this precedent, rather than the Space Shuttle experience, is the valid one for the space station era.

USE OF A SPACE STATION FOR SPACE APPLICATIONS

G. A. Harter*
Vice President, TRW Inc.
Cleveland, Ohio

Abstract

The Space Applications Board conducted a one-week summer study on potential uses of a space station for civil or commerical applications. Six study panels identified future space application needs, and then assessed how well a space station might satisfy those needs. Some special requirements were identified to facilitate use of a space station for application purposes. Finally, the role of a manned presence was considered for each of the application areas.

I. Introduction

In the spring of last year, the Space Applications Board was asked to conduct a study about the potential uses of a space station for civil or commercial application purposes. The request came from NASA's Office of Space Science and Applications, and was to be only one of several agency and contractor studies aimed at developing and evaluating requirements for a manned, space station program. In requesting the study, NASA made it clear that the justification for development of a space station must be based on many considerations, of which space applications would be only a part. The Space Applications Board accepted the study assignment, with a goal of providing a completely objective assessment of a manned space station, or, alternately, a large unmanned space platform, for use in satisfying future application program requirements.

To conduct the study, our Board brought together about 30 widely-recognized experts from the various user communities. The group spent a week at Crestwood, in Snowmass, Colorado, engaged in an intensive effort to address the following study tasks:

- Review space applications requirements for the 1990s
- Identify desired technology and advances
- Identify requirements on a space station to facilitate Applications' programs
- Assess the utility of a manned presence to space applications

The group was not asked to evaluate a space station relative to other possible alternatives, so made no effort to develop an optimum approach for the 1990s. The assignment was to determine future application program needs, and then to assess how well a space station might satisfy those needs.

An important objective of the study was to identify requirements on the space station to make it most useful for applications purposes. In particular, NASA was interested in the potential utility of a manned presence to the operation of an applications instrument or system.

II. Panel Structure

The summer study participants were divided into panels addressing five user-oriented communities, and a sixth panel dealing with system design considerations. The system design panel assisted in examining requirements of the various applications missions from the standpoint of commonality, and with an objective of defining generic technical requirements that could be used for the conceptual design of a space station. This panel was tasked to insure that the user panels were considering practical approaches when identifying future mission requirements.

Each of the five user panels was structured to represent applications programs with similar characteristics. Earth Resources dealt with applications of remote sensing for resource assessment, such as agriculture, hydrology, and mineral exploration. The Earth Environment panel dealt with the earth's atmosphere and its impact on society, and gave consideration to upper atmosphere research, global chemical cycles, weather and climate. The Ocean Operations panel looked at both science and research oriented toward what might be termed maritime housekeeping, giving consideration to such things as coastal preservation, fisheries development, mineral extraction, ocean pollution, sea ice monitoring, oil and gas exploration, commercial shipping, etc. The remaining two panels, covering Satellite Communications and Material Science and Engineering, should, I think, be self-explanatory.

One early problem for the panels was lack of a clear definition of space station capability, or constraints. For purposes of the study, it was assumed that the station would be a large platform, capable of housing a wide array of diverse instruments, and could be either manned or unmanned.

With this background in mind, I will now discuss the findings of each panel and then summarize the overall conclusions of the study.

III. Earth Resources Panel

The Earth Resources Panel considered the application of remote sensing for assessing such resources as agriculture, forest and rangeland, minerals and petroleum, water, and land use patterns. They pointed out that satellite remote sensing is a relatively new information source

---

*Chairman, Space Applications Board of the National Research Council
Associate Fellow, AIAA

providing a means to survey and inventory a wide array of earth resources. It provides a capability to monitor changes and then, by predictive modeling techniques, to provide a forecast for the future. Utilization of this new capability is still in its infancy, and much further study and development will be required for its full exploitation.

The panel found a great diversity of requirements for the several application areas. However, some basic mission requirements seemed to be common for virtually all applications. The most important of these were:

- Having the space station, or platform, in a near-polar, sun-synchronous orbit
- Having a variety of sensors that can operate interactively full time
- Having a data format which integrates/merges the outputs from a disparate set of sensors
- Having timely delivery and archiving of data for the user communities
- Having a commitment to long-term continuity of data gathering

The panel concluded that on-board data processing is important for almost all remotely sensed data streams. They observed that a highly trained man or woman might be useful in the operation of the on-board processing system, but that this might also be accomplished by ground control. The panel observed that a manned presence could be most important during the research, or exploratory, phase of a new remote sensing program.

The panel also concluded that the current capability for data processing, handling, and delivery of remotely sensed data to the user is much too slow. This problem needs more management attention, independent of the space station considerations.

Finally, the panel recommended that the Space Applications Board establish a permanent committee for earth resource applications. The Board has accepted this recommendation and is in the process of selecting members for this standing committee.

IV. Earth Environment Panel

The Earth Environment Panel addressed four broad applications of remote sensing:

- Upper atmosphere research
- Global chemical cycles
- Weather
- Climate

Their approach was to identify status of current knowledge in each area, determine needs during the next two decades for continued development and expansion, and then derive the role of space observations in fulfilling those needs.

The panel foresaw several important roles for a space station. First, it would permit use of many different sensors to make simultaneous observations of a chemical species, with on-board processing and merging of the data. Second, a manned presence would enable a frequent recalibration of instrumentation. In this respect, the panel felt that the station could play a valuable role in comparing data from on-board sensors with that obtained by other free-flying satellites. The panel also saw an important role in the servicing of instruments, including the replenishment of cryogenic coolant and the periodic maintenance, repair, and changeout of sensor packages. Finally, it was thought that a space station would facilitate the use of active sensor payloads, such as LIDAR or microwave sounders, which because of their large size and power requirements, have been regarded as beyond the near-term capabilities of free-flying satellites.

The panel concluded that most of the applications considered would prefer a near-polar, and probably sun-synchronous, orbit. However, they did observe that gathering data on diurnal variations of geophysical parameters would require use of non-sun-synchronous orbits.

The panel was cautious, and undecided, about the benefits associated with a manned presence. On one hand, they foresaw a contribution in the calibration and servicing of instruments. On the other hand, they foresaw a number of potentially serious problems. There was concern about man's impact on sensor pointing accuracy and stability, and on the possibility of contamination in and around the vehicle. It was concluded that much more study would be required in this area.

V. Ocean Operations Panel

The Ocean Operations Panel looked at a wide variety of space applications, including coastal preservation, fisheries, oil and gas exploration, ocean pollution, sea ice monitoring, mineral extraction, and commercial shipping. They observed that the need is growing rapidly for more and better data about the vast ocean areas. The needs for data are very diverse, and are currently dispersed among many agencies, organizations, and individuals. A recent Congressional study identified more than 90 ocean programs, scattered across a host of government agencies.

The panel noted that only satellite-borne sensors can provide broad area, near synoptic data coverage of the world's oceans. Because of this, more emphasis should be placed on collection of data and making it available to ocean users in a timely manner. Such data is needed by many users on a daily basis, and the timely availability is most critical when used to support sea-going operations.

Data coverage is required for regions which extend to the earth's poles. Therefore, high inclination orbits are desirable for most applications. A sun-synchronous orbit would be quite satisfactory, although some existing satellites use orbits at 62 degrees and 108 degrees inclination.

With regard to the benefit of a manned presence, the Ocean Operations Panel came to the same general conclusion as did the Earth Observation Panel. They foresaw no meaningful role for man in the acquisition of data. However, they felt strongly that a manned space station could, and should, provide a capability for modification and repair of sensors in orbit. As an example, they cited the $125 million SEASAT oceanographic satellite, which could perhaps be restored to full operation if a means existed for retrieval and repair.

VI. Satellite Communications Panel

The Satellite Communications Panel surveyed all existing and projected new communications services for the period from 1990 to 2000. They concluded that a manned space station would offer little benefit during this timeframe from an operations standpoint. They pointed out that communication satellites need to be located in a synchronous, geostationary orbit, and that the launch capability provided by the Shuttle/Centaur combination is adequate for all envisioned requirements prior to the year 2000.

The Panel did envision some ancillary uses of a space station at low earth orbit, assuming that it would be placed there for purposes other than satellite communications. One thought was that the station might serve as a fuel depot and base of operations for launching satellites up to synchronous orbit, with the possible benefit of reducing launch costs or allowing launch of heavier payloads. The Panel also envisioned use of a low-orbit station for conducting some classes of communication experiments, such as the deployment and evaluation testing of a very large antenna. They also observed that techniques developed for assembling large structures in low earth orbit might some day find application for constructing very large communications platforms at synchronous orbit.

Finally, as a result of a panel recommendation, the Space Applications Board has decided to establish a permanent committee for Space Communications, and is in the process of setting up this important new capability.

VII. Materials Science and Engineering Panel

The Materials Science and Engineering Panel members were very supportive of a space station which would provide a laboratory capability staffed with technically trained operators. They felt that the materials community would make good use of a space station if in fact it were to be developed.

The Panel identified several key requirements for a manned space station laboratory. In addition to a low-gravity environment, materials processing would require a relatively large amount of electric power and would need facilities for cooling heated materials in a controlled manner. It was felt that the usefulness of a space station laboratory would be dependent, in large part, on the qualifications and experience of the technical staff manning the facility. The panel foresaw a need for one or more trained scientists to perform the materials experimentation, with at least one experienced technician available to maintain, repair, and modify the processing equipment.

The Panel noted that a space station could provide some important advantages over the current shuttle capability. Chief among these would be the much longer flight time, plus a potential for less interference by firings of the attitude control system thrusters. The ability to carry out weeks, or even months, of interactive materials experimentation in a space station would represent a great improvement over use of the shuttle spacelab.

VIII. System Design Panel Conclusions

The System Design Panel was unique in that it did not address a specific applications area. This panel focused on key system design and integration issues that spanned all aspects of the space station concept. Its initial charge was to work with user panels, and to assist them in developing a practical set of user recommendations. As the study progressed, this panel came forth with a number of their own recommendations about the future direction of the space station program.

As an example of their creative effort, two new concepts were identified for further study and analysis. One involved the use of real time communications coupled with remote control techniques, which would provide an operator on the earth's surface the capability to carry out complex operations within the space station. This real time, closed loop control was dubbed telepresence, and was put forth as a complementary role to manned presence for many space station operations. It featured two emerging new technologies -- wideband, high data rate telecommunications capability coupled with computer controlled robotics.

Another new concept was the development of a space service station, which would tend the in-orbit needs of unmanned satellites or space platforms. Functional capabilities would include repair, servicing, and replacement of expendables for various types of spacecraft and instruments. The panel observed that commercialization of space services could well be an important objective of the space station development program, right from the beginning.

The System Design Panel also made a few recommendations of a more general nature. First, it was suggested that furthering the study of man's role in space would be an important applications program in its own right. Determining the ultimate capability of man to perform useful functions in space, over an extended period, would contribute immeasurably to development of future space systems.

Second, the panel recommended that NASA develop and promulgate design rules for a number of important space station concepts, including system modularity, subsystem interfaces, and technology upgrading. Early definition of modular system architecture and subsystem interfaces would be very helpful in guiding the industry research and development effort. In addition, industry could use some guidance as to the most probable timing and direction of technology upgrading envisioned for future space systems.

As a final note, the System Design Panel suggested that development of a space station should encompass some significant advances in our current technological base. A program based in large part on existing, off-the-shelf technology would not be in our nation's best interest. To insure a balanced approach to incorporation of new technology, it was suggested that some form of top-level coordination be established among NASA, the Department of Defense, industry, and the university community.

IX. Overall Study Conclusions

This concludes my summarization of study results coming from the six individual panels. A more definitive set of conclusions is being reported back to NASA in two steps. The first step was a summary letter to the NASA administrator last fall, which I understand was used as an input to the several industry studies. A more comprehensive set of reports, one for each panel, is now being prepared for release this fall. As such, there will be a formal documentation of study conclusions for each panel, but there are no plans for a single, comprehensive report covering the total study.

Be that as it may, I would like to give you my personal assessment of the most important findings of the study. First, and not too surprising, there was the conclusion that most of the remote sensing applications require sensor payloads in a near-polar, and oftentimes sun-synchronous, orbit. This suggests that the space station program should include a multifunctional sensor platform in a high inclination orbit, with enough propulsion capability for accommodating mission-dependent orbit characteristics. The platform should include sensors in support of the many earth resource, meteorological, and ocean operations user communities. If cost or other considerations should preclude the early deployment of a multifunctional platform, either manned or unmanned, at near-polar orbit, then a fallback position would be to provide a servicing capability for more conventional satellites located in these orbits. It is my understanding that the most recent NASA plans do indeed provide for a multifunctional applications platform deployed in a high-inclination orbit.

As a second important point, members of the Space Applications Board feel that if a space station is developed, its effective utilization will need significant resources beyond those required for the space station alone. It is felt that there should be as much of a commitment to payload development as there is to the station itself. To sacrifice the development of one relative to the other would not be in the best interest of either the total mission or the total system concept. Our membership believes that if a decision is made to proceed with the development of a space station, it would be essential that NASA consider not only the anticipated costs of developing the station, but also the envelope of total system costs -- that is, costs related not only to the station itself but also costs for the application sensor systems for which the station will be designed to carry and support.

Finally, the study group recommended that NASA give serious consideration to two new system concepts -- Telepresence and a Space Service Station. Telepresence would appear to offer many of the same capabilities as a manned presence, but at much lesser cost than maintaining the same operator in space. At a minimum, telepresence would provide an important complementary capability to the functions made possible by a manned presence. The rapidly advancing technology in space communications and robotics seems to offer great potential for real time control of many space platform operations by an operator on the ground.

The concept of a space service station resulted from a self-generated study group objective to maximize the utility and financial payback of a manned space station program. A service station would seem to offer complementary benefits to the many unmanned satellite programs, and could lead to a large reduction in total operating costs for the United States space inventory. Although the general concept needs much more thought and development, the potential utility and benefits are enormous, and deserve further study.

That concludes my report of the SAB study on potential uses of a Manned Space Station for Application purposes. I appreciate your kind attention and hope that this summary serves some useful purpose in guiding the future direction of this potential, new national program.

MILITARY SPACE ACTIVITIES AND A SPACE STATION

Richard D. DeLauer*
U.S. Department of Defense
Washington, D.C.

In presenting the Department of Defense perspective on a manned space station, let me start with a few words on the U.S. military use of space. As you no doubt know, the DOD has made extensive use of space systems over the last two decades. Our military space assets have and will continue to support the overall objective of the DOD, which is to ensure the national security of the United States and its allies. While our strategy concentrates on deterrence as the best means of preventing war, we must constantly be prepared to wage war if we are unable to prevent it. U.S. military space assets provide support to critical and stabilizing functions such as communications, command and control, warning, navigation, weather, and space defense. In fact, the U.S. military use of space makes a major contribution to world peace. Space systems are, and will continue to be, designed to ensure secure command, control, and communications functions in order to prevent accidental war. Our launch-detection satellites and warning systems contribute immeasurably to strategic stability and deterrence. Likewise, arms control would be very difficult to monitor without the support provided from space systems. These advantages, coupled with the global access permitted by satellites and the efficiency and effectiveness of their use, both in terms of mission and cost, reflect the primary reasons we are in space.

Military space activities relate directly to DOD mission requirements. DOD mission requirements are formulated in an interactive and dynamic environment that considers U.S. national security policy and foreign policy, national security objectives, threats to U.S. national security, U.S. defense strategy and the prioritized military capabilities required to implement the U.S. national security strategy. As such, DOD space programs contribute to our overall military capabilities and are not simply a reaction to Soviet space activities nor an end onto themselves. DOD space programs are the result of a continuing DOD evaluation of alternatives for meeting national security objectives. As with all methods of satisfying national security requirements, military space systems receive funding to the extent that they demonstrate a competitive advantage over alternative means. The same is true among competing space solutions, be they manned or unmanned. From this, one can see that the use of a manned space station by the DOD would be based on the distinct contributions that a space station could make toward satisfying DOD mission requirements. A space station must also compete favorably on a technical and economic basis with alternative methods of satisfying these mission needs.

On the subject of man in space, the DOD does have a historical perspective. The Air Force was responsible for the launch of all of the Mercury and Gemini astronauts. Thus, it developed the capability of developing man-rated launch vehicles. In addition to the Mercury and Gemini missions, DOD astronauts have flown in NASA programs such as Apollo, Skylab, and now, the Shuttle. Additionally, DOD conducted two major programs in which man was integral to the system -- the Dyna-Soar and the manned orbiting laboratory (MOL). The Dyna-Soar and the MOL programs were both cancelled before flight. The current USAF studies on Spaceplane are a variation of the Dyna-Soar effort. The MOL was contemporary with Skylab. Being classified, MOL did not have the political appeal of the manned Skylab with its wide public participation. The cancellation of the MOL resulted from a combination of budgetary pressure on DOD combined with the fact that at that time all of the military missions of the MOL could be carried out more effectively in unmanned modes.

With the advent of NASA's current proposal that a manned space station be embraced by the United States Government as the next major civil space program, it was time for the DOD to take another long hard look at the role a manned space station could fulfill for national security. This study effort and internal DOD debate was led by the Air Force, but included all elements of the Department of Defense. The study objective was to determine if a body of DOD mission requirements existed that could be significantly enhanced or supplemented by a manned space station. This included an evaluation of whether a space station would best fulfill DOD requirements vs other means of satisfying them. To gain the benefit of the work of NASA's Space Station Task Force, the DOD supplied the NASA contractors addressing the possible requirements for a manned space station with additional funding to also look at the national security dimension.

Based on this DOD evaluation, we have concluded that there are no currently identifiable DOD mission requirements that could be uniquely satisfied by a manned space station. Further, no current DOD requirements were found where a manned space station would appear to provide a significant improvement to DOD over alternative methods of performing the given task. Over time, however, this situation may change. Therefore, we are devoting considerable attention to developing a better understanding of the potential future uses for the military role of man in space.

The principal future value of a manned space station to the DOD appears to be as a possible base for any RDT&E needing extended time on orbit beyond that possible with the Shuttle. Otherwise, military operations on a manned space station do not appear to offer the most efficient use of DOD resources. Studies to date have identified no unique, cost-effective contributions that man-in-the-loop can make to the execution of military missions such as surveillance, navigation, and communications. Fur-

*Under Secretary of Defense for Research and Engineering.

ther, considering the cost of developing and procuring one or more space stations and the difficulty in making a space station survivable, questions are raised concerning the reliance that could be placed on the availability of a space station in a conflict.

DOD does believe, however, additional space operational experience should be obtained to gain knowledge of man's capabilities in space, as well as developing technologies to take advantage of his capabilities. During the 1960s and early 1970s, NASA's manned space program gained considerable knowledge on man's utility in space. The Space Shuttle program was largely predicated on the fact that it would be manned and could use man in maintenance, repair, retrieval, etc., many of the same missions for which a space station might be useful. Thus far, the Space Shuttle has been used as little more than a booster to place satellites in orbit.

Rather than direct DOD participation in the development of a major new space station program, it appears more beneficial in the near term for the DOD to utilize Shuttle missions to continue to identify and evaluate the role of the military and to exploit the unique manned capabilities of the Shuttle. If full advantage is to be taken of manned opportunites for DOD in the Shuttle operations, near-term actions are required to ensure the availability of an appropriate technology base. Specifically, there should be a program to resolve the uncertainties associated with the utility of military man in space. Candidate efforts could include participating with NASA in enhancing the capabilities of the Shuttle. Examples of candidate enhancements are a power extension package (PEP) to increase the Orbiter mission duration; a teleoperator maneuvering system (TMS) that would enhance the low Earth orbit delivery, service, and retrieval of unmanned satellites and payloads; and other generic equipment such as hand tools, handling fixtures, etc., as required. These capabilities, in conjunction with the Shuttle, would enable DOD to pursue an evolutionary program capitalizing on the U.S. government investment in the Shuttle to gain first-hand experience with manned military space operations.

Another candidate effort consists of a dedicated DOD space test program "laboratory-type" Shuttle missions that could be pursued as a demonstration of the value of military crews in space. The payloads/experiments to be flown on these missions could emphasize RDT&E experiments that might benefit significantly from manned interaction (demonstrating man's role in space-based RDT&E), and also perform experiments that could help define the utility of military crews in performing operational military roles.

Such an evolutionary approach would better enable DOD to understand the potential role of military man in space and provide the necessary knowledge to make logical decisions pertaining to the proper future investments in manned space operations and capabilities.

In conclusion, the DOD is committed to the continued exploitation of space to improve the effectiveness of our defenses and to enhance the security of the United States and our allies. The significant resources the DOD applies to space activities reflect the high regard we have for space systems as a part of our force structure.

Even though our recent look at a national security role for a space station concluded that at this time a manned space station would not provide a significant improvement over alternative methods of performing a given military task, over time this situation may change. Therefore, we are devoting considerable attention to developing a better understanding of potential future requirements for the military role of man in space. To this end we believe that significant additional information can be obtained by using the Space Shuttle capabilities and by extending the Shuttle flights. Research efforts to extend the Orbiter's ability to remain in orbit, for instance, should be pursued in order to provide answers to these questions. Such a modified Orbiter could be used as an R&D platform leading to a better understanding of man's role in space. New technology, such as an orbital transfer vehicle or a teleoperator maneuvering system, should also be examined to perform on-orbit operations. With this knowledge, a space station may have a role for DOD at some future time.

Independent of any program to develop a manned space station, the United States must ensure that a viable Space Shuttle system is available to meet national launch needs, including those of national security. This nation must be certain that adequate resources have been dedicated toward developing the Space Shuttle to fully utilize its designed capabilities and to ensure the operation of a Shuttle fleet that is adequate to meet national needs.

Should a decision be made for NASA to develop a civil space station, the DOD would continue to consider the space station for use by military programs within our overall strategy for the use of space to enhance the accomplishment of our national security mission.

LUNCHEON SPEECH, JULY 18, 1983

Hans Mark*
National Aeronautics and Space Administration
Washington, D.C.

I would like to tell a story about the space station, which for me really focussed for the first time what a space station might be used for and why you might want to have one.

The idea, of course, of a permanent base in Earth orbit is not new. In fact, if you look at the literature, the notion was first broached seriously in the 1920s by Hermann Obert, who not only had a station but all kinds of things flying next to it and with it as well.

The story I will tell you was certainly not the first time I've heard of the idea of a space station, but it's really the first time it was brought home to me with an impact that stuck.

It happened about 10 years ago, when Wernher von Braun made one of his periodic visits to the West Coast. It happened that he was in San Francisco on the weekend of the Farallon Island race. The Farallon Islands are about 30 miles off the Golden Gate, and one of the better day races of the season is one around the islands.

I was entered in the race and as Wernher was a great sailor, he came along. Although the way out is usually rather rough, the way back is always pleasant, because the wind's behind you and you can just sit and let the boat go. Pretty much, that's what happened that year.

As we turned around the islands and the sailing became easier, some of the younger members of the crew became interested in the fact that von Braun was aboard. This being 1973, the announcement had just been made that Apollo 17 would be the last moon flight, so one of the crew asked von Braun when we were going to go back. His point was that we had gone through this business of going to the moon and now we were shutting it down and he wanted to know what's going to happen next. Why are we stopping, and are we ever going to go back?

The sail from the Farallons back to San Francisco took about 3 or 4 hours, and for those hours, Wernher held forth on what was going to happen and how we were going to go back to the moon.

What Wernher did was to tell the story of our eventual return to the moon in the context of the exploration of Antarctica. He started out by saying that Antarctica and the moon have several things in common. One, of course, is that they're both hard to get to. The second is that both require life support systems in order to survive. That is, you've got to have a place to live in both places that's more than the usual shelter. And the third is that both require large logistic tails in order to maintain any operations.

The similarities were to him so striking that he built his concept of how we would eventually go back to the moon by looking at the history of Antarctica.

He started out by saying that the existence of Antarctica was suspected for a long time but the recognition that there was a land mass there really invaded the consciousness of people only in the 19th century, when the first tentative landfalls were made.

In the early years of this century, Antarctica became an object of international competition, very much the way the moon became an object of international competition in the early 1960s. And there was indeed a race to the South Pole, just as there was a race to the moon.

The first point that Wernher made was that he would start out by making the analogy between the race to the pole in 1911, the race to the moon in 1969, and the things that motivated the people that started those races.

In 1911 there was international maritime competition. People were interested in reaching the poles first. Amundsen, the Norwegian explorer, decided to go to the South Pole only because Robert Peary beat him to the North Pole. He had tried to be the first there and failed, and he said, "Well, there's another pole."

The British already had a large expedition there, so the stage was set for the race to the South Pole in late 1911 between Capt. Robert Falcon Scott of the Royal Navy and Roald Amundsen of Norway.

It was an interesting thing, not only because of the circumstances, but also because of the technologies that each applied and because of the human element. Eventually, it was also exciting, but, in the end, tragic.

Amundsen used what we would probably today call the appropriate technology. What he did was to adapt the Eskimo way of doing business and used dogsleds. To use dogsleds long distances, you start out with many more dogs than will be needed. On the trip the weaker dogs die, which the stronger dogs then eat, thus eliminating the problem of carrying along dog food. If you think about it, that's a staged vehicle.

Amundsen used that method. Scott, on the other hand, used high technology. He brought the

From taped transcript.
*Deputy Administrator.

first gasoline-powered tractors to Antarctica. They didn't last very long, because people at that time didn't know how lubricating materials would work at low temperatures, or what happens to metals when they get very cold. And so the tractors failed very early. He also brought some Shetland ponies that didn't last very long either. Scott and his four companions actually had to drag their own sleds to the South Pole manually.

Of course, you all know the story: Amundsen got there first, and Scott got there about four weeks later.

Wernher told that story, and then made the analogy between it and the race to the moon between ourselves and the Russians in the 1960s. If you talk to the Russians today, they'd have you believe there was no race, but you know that there really was a very intense effort on the part of the Russians to put people on the moon and that they in fact tried and failed.

Wernher then asked us what happened after 1911 in Antarctica. And the answer was, not very much. Except for a few desultory expeditions, not much happened after the race to reach the South Pole in 1911--until 1929. And in 1929 something very unique happened: Richard Byrd first brought airplanes to Antarctica and demonstrated that they were usable under the conditions existing in Antarctica. It was a Fokker airplane, especially modified with skis and other equipment to be used in Antarctica.

Very quickly other people started using the airplane to explore the continent. In the next few years Lincoln Ellsworth went down every summer and mapped the entire continent in a matter of a few months flying. This really opened up the continent to serious exploration.

What Wernher said was that the airplane was in a certain sense the enabling technology that allowed the development of Antarctica beyond the first tentative steps of mapping the coast and then going to the South Pole as a sort of an international competitive enterprise. For example, the airplane was what allowed the establishment of the first permanent bases on the continent. There could be no permanent stations on the continent of Antarctica if you couldn't get there by air. Figure 1 shows the Fokker Byrd took to Antarctica and Fig. 2 portrays one of the airplanes in use today, a specially equipped Lockheed C-130. Figure 3 shows the U.S. base at McMurdo Sound. As you can see, it is a fairly extensive operation.

Wernher put all this into the following perspective. (This was 10 years ago, so the numbers are a little bit different, but the meaning is the same.) He said, "Look, we got to the South Pole in 1911, and 20 years later we developed the enabling technology, the airplane." That was circa 1930. "The interesting thing is that 1942 was the last year that no human being wintered in Antarctica. Since then the continent has been permanently occupied."

The permanent population of Antarctica today is about 2000-2500 and there are a dozen stations like McMurdo. About 15 or 20 countries now have permanent bases on the continent.

Summer populations approach 7000 or 8000 in terms of the number of people that go there to do various things. And as you probably know, they do all kinds of things. There's even an incipient tourist business.

Wernher went on, "Why don't we take seriously the analogy I made in the beginning, between the moon and Antarctica, and let's see where it leads."

Let's assume, just for the fun of it, that the time scales are comparable and see what conclu-

Fig. 2 Lockheed C-130 currently used in Antarctica.

Fig. 1 Fokker airplane, the first in Antarctica.

Fig. 3 U.S. base at McMurdo Sound, Antarctica.

sions we can draw. In that case, the year 1911 becomes the year 1969. So 60 years later, we had the race to the moon. Interestingly enough, after that race, not much happened on the moon either. Nobody went back. We haven't even sent unmanned probes or satellites to it. The reason, in Wernher's opinion was that we have to wait for the enabling technology in order to do something intelligent next. The enabling technology that we need to go back to the moon is a staging base. He then made the obvious point that every rocket and spaceship we've ever built, with the exception of the lunar landing module, was and is a vehicle designed to fly both in space and in the atmosphere. When you have to do that, your designs are rather severely limited.

What is needed is a base from which true spaceships can be staged, that is, the equivalent of the airplanes in Antarctica, to the moon and to fly around on the moon the way the C-130s fly around Antarctica.

What you would like to do, of course, is to have such a base where you take the Shuttle and leave it there. You don't fly to the moon with the Shuttle, because you've got a lot of weight on that bird that will eat up Delta v that you don't need. You've got tiles, you've got landing gear, you've got all kinds of stuff ---I don't have to make a list.

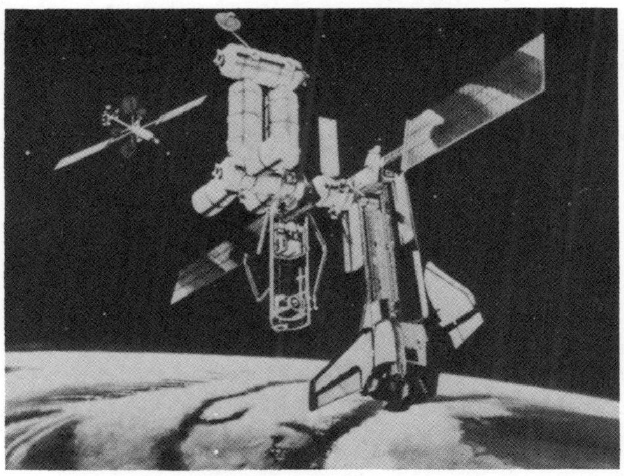

Fig. 4 Possible space station.

Fig. 5 Station on moon.

But what you'd like to have on the space station are some vehicles similar to the lunar module that are not built with the constraints of being able to fly in both the atmosphere and space, with which you will then go ahead and explore the moon.

Back in 1973, Wernher said, "Let's talk a little bit about time scales and enabling technologies. If you agree with me that the space station is indeed the enabling technology to go back to the moon, and if you accept the fact that the time scales that we're talking about are similar to the time scales that we had in the case of Antarctica, then it took about two decades to get from the initial race to the Pole, as it were, to the deployment of the enabling technology. My guess is that we will have the enabling technology to go back to the moon deployed by 1990."

"Let me draw the analogy a little bit further," he continued,"1942 was the last year when we had no permanent resident in Antarctica. That was 10 years after the deployment of the enabling technology. The year 2000 will be the last year that the moon is not permanently occupied by human beings." By the year 2020, he predicted that several thousand people would live on the moon. And he added that, as far as he was concerned, the predictions were perhaps slightly pessimistic.

What Wernher was talking about is obviously in a certain sense the subject of this Symposium. Figure 4 shows what the enabling technology might look like. This is a staging base. The Shuttle goes back and forth to the station on one side, and then something is deployed-- and I don't know whether it's a communications satellite or something else--but the point is that it will allow us to build and deploy true spaceships there, and thus to expand the operations in space enormously ...and, of course, ultimately to go back to the moon. That is clearly what will occupy our thinking as this Symposium proceeds.

I have a favorite place on the moon where I think we ought to build when we go back: the lunar poles, because it's the only place without the extremes of temperature and "climate" that characterize the rest of the moon. So what I think we need to do, once we have the space station that will enable us to go back to the moon, it to establish a permanent colony somewhere near the lunar pole, north or south, that looks something like Fig. 5.

A rather important statement regarding space exploration was made many years ago by Konstantin Eduardovich Tsiolkovski (1857-1935): "The Earth is the cradle of mankind but man cannot stay in the cradle forever." I think it illustrates the thought which governs why we're probably all here. The significant part of the statement is that it was made by a Russian. I, for one, believe that our race with the Russians is by no means over.

SPACE STATION UTILIZATION

Chairman: Burton I. Edelson
Associate Administrator for
Space Science and Applications
National Aeronautics and Space Administration
Washington, D.C.

The development of the space station concept, the historical evolution of the concept, and the configuration of the space station have been outlined in the session entitled "Overview." Also, several representatives of advisory groups (the Space Science Board and the Space Applications Board) and the military side of the house have reported.

In these sessions, the utilization of the space station is addressed. As glamorous, exciting, and challenging as the goal of achieving a "permanent manned presence in space" appears, it will not be really meaningful unless we can show that there is a purpose behind it. We need to insure that this manned presence will be helpful in conducting scientific missions, in providing space applications for beneficial use here on Earth, or in serving military, commercial, or other needed purposes.

These presentations cover some specific information on space station utilization, using the manned space station base to conduct scientific experiments, to service unmanned platforms, and to act as a transportation node and as a staging base for construction, checkout, and test of large and complex structures in space.

SOLAR MAXIMUM REPAIR MISSION

B. R. McCULLAR*
NASA - OFFICE SPACE SCIENCE & APPLICATIONS

1.0 Abstract

The Solar Maximum Repair Mission is scheduled for launch in April, 1984. The mission was made possible through a spacecraft designed for on-orbit repair, the availability of spacecraft repair hardware and the availability of the Shuttle System. Repair is expected to result in the availability of significant new science data and a demonstration of the capability for on-orbit repair of spacecraft. Considerations of the capability to maintain and repair spacecraft on-orbit for the future are also discussed.

2.0 Background

The Solar Maximum Mission (SMM) was launched in February 1980 to make scientific observations of solar flares and the energy output of the sun during a period of maximum sunspot activity. The spacecraft utilized three axis stabilization, solar pointing and no on-board propulsion. Seven scientific instruments were included on the spacecraft to observe the sun. Very importantly, the spacecraft was designed to be captured and serviced on-orbit or retrieved by the Shuttle.

While the SMM initial science objectives were met, two failures occurred in late 1980 which significantly reduced the continued scientific observations. The coronagraph polarimeter failed in October 1980 due to malfunctioning electronic parts and fuse failures in the Attitude Control System negated the fine pointing capability of the spacecraft in December 1980. The observatory is currently operating under coarse pointing control and gathering data from three of the seven scientific instruments.

3.0 Repair Feasibility

Availability of the Shuttle now makes it possible to repair spacecraft. Since SMM had been designed for repair on-orbit, the feasibility of repair was investigated.

The SMM anticipated orbit decay rates were reviewed to determine the expected altitude of the spacecraft at the time of repair. Considering the solar flux data and the spacecraft ballistic coefficient, a range of orbit decay curves were developed. The ability of the Shuttle to get from the nominal 160 n.mi. orbit to the expected SMM orbit was then investigated. Results of these investigations were that new Shuttle techniques could be employed to reach the anticipated SMM orbit of 265 n.mi.

Since the spacecraft is slowly rotating (1°/sec.), tests were then conducted to determine if the spacecraft could be stabilized so that it could be captured by the Shuttle's Remote Manipulator System (RMS). The concept was that a suited astronaut would be able to attach himself to the spacecraft and stabilize its motion using the Manned Maneuvering Unit (MMU). Tests using the MMU simulator indicated roll rates significantly greater than that of SMM could be stabilized.

4.0 Rationale

Based on the studies indicating that repair of the spacecraft was feasible, a plan was developed for the repair mission. The rationale for this repair mission was:

- o The Solar Maximum Spacecraft was this nation's only orbiting solar observatory
  - The spacecraft was partially disabled — 3 of 7 scientific instruments currently operating
  - Important new solar science could be done with a repaired spacecraft
  - Shuttle manifesting opportunities occurred in late 1983 to early 1984
  - Spacecraft and science repair kits could be made available by late 1983
- o The capability to conduct on-orbit servicing/retrieval was a planned and important capability unique to the Shuttle
- o Several important NASA programs include this capability: (e.g., Solar Maximum Mission, Space Telescope, Landsat, Long Duration Exposure Facility)
- o Timely demonstration was needed to encourage other users to incorporate future space repair/retrieval compatibility in their design specifications
- o A repair mission dramatically demonstrates the benefits of Shuttle over competing launch systems
- o The mission had high public and industry appeal
  - Necessary for effective space industrialization and future mission planning

Availability of spacecraft repair hardware, tools and the time necessary for mission integration and training resulted in a repair mission scheduled for early 1984. Congressional notification was begun in Dec. 1981.

5.0 The Plan

The mission objectives for repair of the Solar Maximum Mission are to:

* Program Manager, Solar Maximum Mission

- provide for an early demonstration of the Shuttle's capability to rendezvous with a spacecraft and repair it or return it to earth, and
- enhance and extend the science data return.

The plan for implementing these mission objectives follows five phases. The first phase begins with the Shuttle launch of the crew, tools and repair hardware. During this phase, new techniques will be used to deliver the Shuttle orbiter to the approximately 265 n.mi., 28.5° inclination orbit of the SMM spacecraft. The Shuttle will then rendezvous with the spacecraft to a parking position approximately 300' away. A suited Shuttle Astronaut will then maneuver from the Shuttle payload bay with the manned maneuvering unit (MMU) and attach himself to the spacecraft. Using the MMU thrusters he will stabilize the spacecraft so the Shuttle orbiter can be maneuvered to a position to capture the spacecraft using the Remote Manipulator System (RMS) located in the payload bay.

The second phase begins with the RMS placing the spacecraft on the Flight Support System (FSS) located in the payload bay. The FSS has been designed as a working platform for spacecraft repair and provides for locking the spacecraft into position and changing its position to allow access from various locations by the repair astronaut crew. Once the spacecraft is in position, the crew will begin the repair using the hardware and special tools which have been designed for this mission.

The third phase starts with repair of the Attitude Control System (ACS). This repair is accomplished by replacing the 4'x4'x18" standard module which houses the Attitude Control System. This module was designed to be replaced by removing two Acme threaded screws using a module service tool (MST). The MST can also be used to remove and replace the other two standard designed modules - the command and data handling (C&DH) and modular power system (MPS) modules. Disconnecting the single electrical connector for each module is accomplished automatically. The failed ACS module is then replaced with a new module which was brought along in the Shuttle payload bay.

The fourth phase begins with the spacecraft ground control crew checkout of the new ACS module with the rest of the spacecraft. This checkout will be conducted overnight while the repair crew is sleeping.

The fifth and final phase begins with the crew repair of the coronagraph polarimeter science instrument. Although the instrument was not designed for repair, ground tests indicate that repair can be completed on orbit. The repair consists of removal of the thermal insulation, retaining screws and electrical connectors. Replacement of the failed unit will follow the reverse of the removal. Once this repair is completed, the spacecraft is then checked out for deployment. Deployment of the spacecraft is accomplished using the RMS. The Shuttle and crew then return the failed spacecraft hardware for refurbishment and use on later spacecraft flights.

6.0 Status

Implementation of the plans to repair the SMM spacecraft are well underway. Ten of the twenty-one months necessary to prepare for this mission have elapsed. The spacecraft repair hardware, the tools and the training are on schedule for the April 1984 launch.

7.0 Future Outlook

Application of the lessons being learned from the SMM repair mission have led to activities directed at other future programs.

7.1 AXAF Panel

A panel was formed in the summer of 1982 to review the benefits to the planned Advanced X-Ray Astronomy Facility (AXAF) that could accrue from on-orbit maintenance and repair. The observations of this panel were:

- o From a Conceptual Standpoint,
  - AXAF would benefit from servicing
  - Redundancy, reliability, operations, spares and costs should be taken into account (should be balanced over the life-cycle)
  - Planned ground refurbishment is undesirable
  - EVA should be emphasized over robotics in the near term
  - Spacecraft should be "clustered" for servicing
  - "Don't fix it if it ain't broke"
- o From a Hardware Standpoint,
  - Either integral propulsion and/or Teleoperator Maneuvering System (TMS) for AXAF will be needed
  - A next generation space suit will be needed
  - Common support hardware should be developed across projects
  - Some servicing hardware and TMS should be baseline STS
  - There were no apparent standards for servicing
- o From an Operations Standpoint,
  - A "call-up Mission" capability should be developed
  - Transportation can dominate M&R costs
  - Contingency retrieval can impact manifest opportunities
  - Shared flight launch window can impact servicing missions

7.2 Astrophysics Programs

Based on these observations and lessons being learned out of the SMM repair mission the Astrophysics Program concluded that:

- Servicing could provide program flexibility and cost advantages
- The initial steps to establish M&R capability are underway
- An increase in STS baseline is needed

- Lessons learned from SMM repair should be focused for future programs
- Projects should not independently conduct M&R - there was a "Need for Maytag repairmen"
- "Maytag" approach should allow for Darwinian evolution not standard/low cost systems approach

7.3 Committee on Spacecraft Maintenance and Repair

In order to focus the spacecraft community activities in this area, NASA formed a committee on spacecraft maintenance and repair (M&R). This M&R committee was to review the current spacecraft and STS plans for on-orbit maintenance and repair for the purpose of making recommendations which would foster the development and acceptance of this new and emerging capability of the STS.

- Specific areas of focus were to:

  o Develop rationale for on-orbit servicing of future spacecraft
  o Develop criteria for determining what servicing functions are cost effective for various classes of spacecraft
  o Develop techniques, procedures and standards for on-orbit servicing (using ST, GRO, AXAF, and National Security Spacecraft as case studies)
  o Recommend requirements, modifications and designs for future spacecraft, the STS and future space stations

    - Catalog approved equipment which is used to conduct on-orbit spacecraft maintenance and repair
    - Recommend and prioritize additional spacecraft/STS equipment and capabilities for which the agency should budget
    - Develop manifesting requirements which would allow quick-reaction maintenance and repair missions to be conducted
    - Recommend initial standard spacecraft/STS interfaces

  o Identify areas which will require continued spacecraft/STS coordination to effectively conduct spacecraft maintenance and repair and recommend an organization structure to provide for such a resource

- Preliminary Ideas

  Preliminary ideas being considered by the committee are that spacecraft designed for on-orbit maintenance and repair (M&R) should consider:

  o Retrieval after spacecraft initial deployment and return to Earth by the STS
  o Spacecraft recapture/reboost
  o Changeout of spacecraft and scientific instrument modules
  o Replenishment of spacecraft consumables

  Additional considerations are that:

  o The extent of M&R should be determined by life cycle tradeoff studies
  o Spacecraft designs should be compatible with M&R based from STS, TMS and space station
  o The STS should develop and implement a quick response manifesting capability for spacecraft M&R

- Specific spacecraft considerations are that:

  o The extent of spacecraft M&R should be determined through life cycle/reliability tradeoff studies during phases A and B

  o Spacecraft M&R design implementation should consider:

    - Use of existing EVA design techniques and packaging document (NASA-STD-5000)
    - Use of common existing flight hardware [Multimission spacecraft modules, Space Telescope orbital replaceable units (ORU's)] when it provides an economic advantage
    - Use of common retrieval, docking interfaces
    - Include early considerations for sustaining engineering/operations
    - Plan for simulations to perfect M&R techniques

  o Spacecraft M&R interfaces design implementation should consider:

    - Utilizing satellite services catalog to determine the availability of STS equipment and STS documentation requirements
    - Utilize already developed spacecraft M&R hardware

- Specific STS considerations are that:

  o STS should develop and implement quick-response manifesting for spacecraft M&R which include:

    - Reserve STS capacity to conduct M&R missions
    - "Canned" ascent/reentry STS flight software and generic payload integration plans
    - Pre-integrated STS cargo elements
    - All STS orbiters equipped for M&R missions
    - Astronaut M&R cadre
    - EVA aids on-hand and qualified
    - The ability to implement back-to-back EVA capability
    - The availability of standard/flight qualified tools

  o STS should develop and maintain generic satellite services systems based on user requirements (tools, equipment, related logistics base)

- M&R Activities Relationship to Space Station

  The relationships between current M&R activities and Space Station M&R activities are considerable. The Space Station design/development considers that:

- The Space Station to be designed for on-orbit operational life of 10 years to indefinite

- It is presumed that over a 10 to 20 year period, virtually all on-board space station hardware will require servicing/repair/change-out

- Space station will be designed for on-orbit maintenance

- Near term agency activities to develop M&R technology/capability will feed directly into space station design/development

The spacecraft M&R activities assume that:

- The space station will provide significant augmentation to agency capability for accomplishing satellite servicing/maintenance and repair

- Approximately 25% of economic justification of space station derives from space station satellite servicing capability

- The agency activities to develop M&R technology/capability between now and space station initial operational capability will result in satellites that are design optimized for on-orbit servicing - thereby permitting early, efficient utilization of space station for this function

8.0 Summary

The benefits from the SMM repair mission are promising. A successful repair will restore this nation's only orbiting solar observatory for approximately 1/10 its replacement cost. The mission is already influencing many spacecraft designers to include the capability for on-orbit maintenance and repair using the Shuttle. The mission is also developing the hardware, tools and experienced people which will be required to take advantage of this new and emerging capability which is unique to the Shuttle.

SPACE STATION AND THE LIFE SCIENCES

Ronald J. White, D. B. Cramer*, Joel I. Leonard, and W. P. Bishop*

General Electric Company/MATSCO
Arlington, Virginia and Houston, Texas

Abstract

This paper presents a concise summary of past findings in the space life sciences and projects the future role that space stations might have in these fields. Although the wide scope of the space life sciences is discussed, this paper concentrates on matters related to basic space biology, the biomedical problems of man, and the elements involved in optimizing human productivity in space.

Introduction

Man possesses a spirit of adventure and an inner drive that forces him to continually question his world and to seek to extend it. It is clear, after 22 years of manned flight into space, that humans are readying themselves to leave Earth. The future deployment of a manned space station by the U.S. is but another step along this path, the next exodus. Continuation of this journey requires that humans develop an understanding of their tolerance to each of the elements of the environment of space, and a strategy to survive in that environment. Recognition of the unique characteristics of the environment of space leads, in addition, to the use of that environment as a potent probe into the workings of the Earth-bound world of biology. These considerations, centered about the environment of space, give rise to the two strongly connected foci of the space life sciences:

to promote the safety, well-being, comfort, and productivity of man for extended durations in space, and

to further man's ability to answer fundamental questions in gravitational biology through the use of the environment of space.

The life sciences thus play a unique role in manned space flight. They encompass a structure that spans the space between spacecraft design for human use and investigations into the importance of gravity to life on Earth. Only a part of that structure will be examined here. Issues related to crew and life support will be covered in other sessions of this symposium, while more detailed coverage of questions related to space station health maintenance are available elsewhere.[1-4] This paper will concentrate on fundamental and applied research in the space life sciences, and will attempt to present some of the major goals and tasks of the future in the context of our present level of ignorance.

Lessons of the Past

It is impossible to adequately summarize the results of previous space life sciences studies in this brief paper. Instead, only a few of the major lessons learned from such studies will be presented in an historical context. The interested reader is referred to more complete summaries which are available and to the original reports which are cited therein.[5-8] In interpreting the observations and measurements which have been made on man in space, it is important to recognize that the conditions under which they were obtained during different space flights were not the same in each case. Flights could, and many did, differ with respect to atmospheric compositions and pressures, levels of stress, personal diets, levels of preflight hydration, physical environments, and levels of personal exercise inflight. All of these factors play a part in the observed physiological responses of man. Even though the results of any one experiment may (and probably do) depend on many of these factors, a number of the observations which have been made are robust enough that they stand out across many flights, and, for the most part, these are the observations discussed here. We presume that they represent the basic responses to weightlessness.

Life sciences research in space dates back to the late 1940's when biological specimens, including primates, were borne aloft in balloons and rockets. Mainly, these early flights assessed the possibly adverse consequences of exposing organisms to brief periods of near weightlessness and to upper atmospheric radiation, matters which were largely unknown at that time. The data accumulated from these studies, which lasted through the early 1950's, firmly established the survivability of living organisms in space.

Beginning in 1957 with Sputnik I and continuing until man was launched into Earth orbit in 1961, a series of flight studies was conducted by both the United States and the Soviet Union. These studies, which involved chiefly primates in the U.S., and dogs in the U.S.S.R., had, as their major goal, either a refutation or a substantiation of the predictions which had been made of possible catastrophic failures in various vital functions of animals suddenly thrust into an environment with negligible gravitational force. The animals were instrumented to monitor a number of physiological variables related to cardiovascular function and metabolic balance. The observed increases in blood pressure, pulse frequency, and respiration rate were all well within physiological tolerance limits, thus demonstrating that weightlessness could be experienced without markedly adverse physiologic effects, at least for a short duration flight of a few hours.

Manned orbital flights into space began with the Vostok Program in the Soviet Union in 1961 and the Mercury Program in the United States in 1962. Table 1 summarizes the flight durations of each of the major programs of the U.S. and U.S.S.R. Data

*NASA Headquarters, Washington, D.C.
Members AIAA

Table 1 Manned flights into space (1961-1962)[a]

Program	No. of Flights	Average Flight Duration (Days)	Longest Flight Duration (Days)	Total No. of Man-Days in Space
VOSTOK (61-63)	6	2.6	5.0	15.9
MERCURY (62-63)	4	0.5	1.4	2.2
VOSKHOD (64-65)	2	1.0	1.1	5.2
GEMINI (65-66)	10	4.0	13.8	80.8
APOLLO (68-72)	11	9.4	12.6	311.4
SKYLAB (73-74)	3	57.0	84.0	513.1
APOLLO-SOYUZ TEST (75)	2	7.5	9.0	39.0
SOYUZ (67-82) (INCLUDES SOYUZ/SALYUT)	42	30.1	211.0	2581.2
SHUTTLE (81-82)	5	4.9	8.0	59.0

[a]Reference 4 summarizes all manned flights into space through 1982.

from the four Mercury flights made apparent some of the acute manifestations of exposure to microgravity. These included weight loss (part of which was due to dehydration), hemoconcentration, and orthostatic intolerance upon return to Earth. Interestingly, no postflight degradation in cardiovascular performance was reported by the U.S.S.R. until 1968, seven years after the first Russian manned orbital flight.

The ten flights of the U.S. Gemini Program during 1965 and 1966 confirmed and extended the biomedical findings of the earlier Mercury flights, in which pre- and postflight tests had been conducted. Some of the missions of the Gemini series included inflight measurements, as well. Particular emphasis was placed on the study of the cardiovascular system because of the known cardiovascular impairment occurring postflight in the Mercury astronauts. Further physiologic effects of space flight were observed in the Gemini astronauts, including moderate losses of red cell mass and exercise capacity, and small losses of bone density, bone calcium, and muscle nitrogen. The Gemini flights ranged in duration from almost 5 hours to almost 14 days.

Eleven more manned missions were completed in the U.S. in the five-year time span (1968-72) during the Apollo Program which brought man to the moon. Symptoms of space motion sickness were reported for the first time in a U.S. space flight during the early part of this program. Such symptoms reported first in 1961 by the Soviet cosmonaut Titov, in Vostok 2, seven years before the Apollo 8 mission clearly revealed this problem to the U.S. biomedical community. The biomedical studies of the Apollo Program were mostly limited to pre- and postflight testing. Biomedical findings from the Apollo missions were related to cardiovascular deconditioning, fluid-electrolyte metabolism, and space sickness, and generally confirmed observations made on previous flights.

Six months after the end of the highly successful Apollo series of flights, the U.S. launched a large space laboratory called Skylab. This orbital workshop had a volume of approximately 294 cubic meters, huge in comparison with the size of the earlier spacecraft. The three manned Skylab missions that followed provided a unique opportunity to conduct the first comprehensive program of biomedical investigations in the history of space-flight experimentation using a laboratory setting that was a true precursor to space stations of the future. Man's physiological adaptation to microgravity was studied in three separate missions of 28, 59, and 84 days duration. Each mission supported three crewmen. Unlike the previous missions in which few inflight tests were performed, Skylab missions used onboard equipment for extensive physiologic testing and continuous monitoring. A partial list of available equipment included a rotating chair for the assessment of vestibular changes, a bicycle ergometer for the study of cardiopulmonary and metabolic systems, a lower body negative pressure device for studying orthostatic intolerance inflight, collection systems for blood, urine, and feces, and devices for measuring body mass and body volume. The large amount of biomedical data gathered during these missions not only confirmed the previous findings of the Mercury, Gemini, and Apollo series, but also, more importantly, illuminated the possible adaptive processes that are set in motion upon entry into the nearly weightless environment of space. The single most important biomedical conclusion resulting from the studies of Skylab was that man can adapt and function effectively in space for periods of time, extending up to three months.

The American Skylab series ended in February 1974, but the Soviet Soyuz missions continued from 1967 to the present. This heroic set of Soviet space missions has accounted for nearly 2,600 man-days of space flight and has included missions of up to 211 days in duration. Many of the Soyuz missions included docking by the Soyuz module with one of the Soviet space laboratories from the Salyut series. The Salyut laboratories, which are smaller than the Skylab module, have been in existence since 1971. The longest space flight, lasting for a record 211 days, was accomplished in the latest laboratory, Salyut-7. Extensive biomedical investigations were conducted by the Soviets on these missions and the observed physiological findings were in general agreement with those obtained from Skylab.

Although the life sciences investigations of American and Soviet manned missions have been chiefly concerned with the physiological effects of microgravity in man, a number of manned and unmanned missions have included experiments on plants, bacteria, fungi and protozoa, tissue cultures, invertebrate species, and non-human vertebrates. Such non-human experiments have been extensively pursued on American Biosatellite and Soviet Cosmos flights. By and large, these experiments have shown that the microgravity of space *per se* had very little effect on simple cell functions, although some interesting changes in plants, related to plant orientation and reproduction, were discovered. In recent years, U.S. participation in the Russian Cosmos series of flights has produced other interesting and significant findings regarding the physiology of weightlessness in tissue and organ constituents of small animals, particularly rats. These findings have provided information concerning physiological function which would have been impossible to obtain in a terrestrial laboratory.

This brief historical sketch would be incomplete without some mention of the ground-based studies which have been used as analogs of the near weightlessness of space flight. Such studies have included supine bed rest, water immersion, head-down tilt (anti-orthostatic bed rest), and airplane flight using Keplerian maneuvers. Ground-based studies on animals have been conducted using immobilization, head-down tilt, dehydration, harness suspension, and lower body positive pressure. Ground-based studies on cells and plants have included both clinostat studies, which attempt to model microgravity, and centrifuge studies, which examine the effects of higher than normal gravity loads. All of these studies, which are conducted on Earth, have been extremely valuable by allowing the scientific tools used for space research to be sharpened, and by providing an opportunity for hypotheses to be formulated and tested in a preliminary way prior to their actual use during space flight.

One of the many lessons taught clearly by the past concerns man's demonstrated ability to quickly integrate a large amount of data, to derive understanding from evolving circumstances, to develop a strategy to cope with an emergent problem, and to improvise a successful solution for that problem within existing constraints. Time and time again, as problems have arisen from unforeseen circumstances, man has been able to solve them. It is this improvization of successful solutions in the face of unforeseen circumstances, more than anything else, that sets man apart from automated systems. The Apollo and Skylab astronauts provided a clear demonstration of many of these abilities. During the landing of Apollo 11, last minute human intervention avoided landing on an area of uneven terrain. In Apollo 13, the astronauts demonstrated the ability to fashion a make-shift arrangement to provide the necessary life support. Finally, in Apollo 17, astronauts demonstrated the ability to investigate and explore in such a way as to optimize the actual collection of samples. In Skylab, a micrometeorite shield was lost upon launch. In the process, a large solar array was also damaged in such a way that it would not deploy properly. Through the use of extravehicular activity, astronauts were able to fashion an effective solar shade for the Skylab and to free the damaged solar array so as to accomplish a successful deployment. In this case, it is probable that without such actions the entire Skylab might have been lost.[9] During the Skylab missions, astronauts were able, after equipment had been compromised, to modify experiments in such a way as to either collect data or to improve data gathering. Extravehicular activity was employed to open a telescope aperture door which had refused to open, and a jammed filter was repaired on an X-ray telescope. A more mundane repair involved a pedal on the bicycle odometer used for daily exercise. On yet another experiment, astronauts were able to optimize exposure times in order to conserve film and to significantly expand the amount of data collected. In all of these efforts, only the repair of the telescope aperture door was considered prior to the flight. Most of the experiments had been designed without considering the use of man in order to repair or enhance the experiments. When these experiments were compromised due to unforeseen circumstances, the astronauts were readily able to improvise solutions that generally preserved the integrity of the investigation.[10]

In the remainder of this section, the principal results of space-flight and ground-based studies will be examined in the context of the various life sciences disciplines. Then, those results which relate to man will be integrated into a composite picture which portrays much of our current knowledge concerning the physiological adaptation to space flight.

Major Space-Flight Findings

The major inflight changes in the cardiovascular system include shifts in blood distribution as reflected by a decrease in leg volume and by head congestion, decreases in circulating blood volume and hemoconcentration, a mean increase in resting heart rate, a diminished vital capacity, decreased orthostatic tolerance, and little impairment of the normal response to exercise (at least up to 75 percent of maximal aerobic capacity).[11,12] Measurements obtained immediately following flight indicate a diminished response to orthostasis as well as to exercise, a reduced cardiac size, and a reduction of blood volume. Neither flight nor supporting ground-based studies have revealed any apparent evidence of primary aberrations in cardiac function or any major circulatory control mechanism. All cardiopulmonary disturbances have been found to be completely reversible upon return to Earth. Cardiovascular dysfunction, to the extent noted above, can be attributed to an effective zero-g adaptation which is rendered inappropriate upon return to a one-g environment.

Research on animals during space flight has been more limited than on humans, particularly regarding circulatory responses. The Soviet Cosmos satellites have apparently not included inflight cardiovascular testing of inflight animals. An early U.S. effort to obtain detailed data from a fully instrumented primate (Biosatellite) was only partially successful, but results generally supported the hypothesis that fluid losses may be related to headward fluid shifts, increased central pressures, and altered renal excretion.[13] The one-g head-down tilt hypokinetic rat model and the lower body positive pressure primate model that have been recently developed appear to offer promising opportunities to perform detailed observations on circulatory adaptation resulting from headward fluid shifts. Studies with these models have already corroborated findings from similar head-down tilt and water immersion studies in humans.

Disturbances in most of the systems regulating fluids and electrolytes in the body are consistently found, during and following space flight.[14] This area of research, which will be called the renal-endocrine research area, includes a number of different body systems which regulate body fluid volumes, body electrolytes, renal and endocrine function, body biochemistry, and metabolism. The renal-endocrine response to space flight can be conveniently divided into an acute phase (lasting hours to days) and an adaptive phase (lasting from days to weeks). The acute phase is characterized by a significant reduction in body fluid volumes, while the adaptive phase is characterized by approaches to new homeostatic levels, especially for renal excretion, endocrine secretion, body fluid volumes and composition, and losses of intracellular electrolytes. During the acute response period, the major event appears to be a reduction in hydrostatic gradients in the body and a rapid headward shift of fluid. As a result of this headward fluid shift and central circulatory volume expansion, a complex set of reactions is believed to occur, all of which lead to a loss of extracellular fluid and electrolytes. Insufficient data are available from space flight to distinguish among the many redundant mechanisms which could be responsible for the frequently measured loss of plasma volume in astronauts returning from space, even after short-term flights. The acute phase of adjustment appears to be nearly complete after a few days in space, and includes the major part of the eventual reduction in extracellular fluids and salts.

The loss of other constituents takes place over a longer period of time during the adaptive phase, and the three Skylab missions have provided extensive data with which to characterize the renal-endocrine status during this phase. During the first several weeks of space flight, significant quantities of potassium and nitrogen appear to escape from intracellular compartments, reflecting the atrophy of lean body tissue. The role of hormonal regulators of renal function during the adaptive phase of flight is not yet clearly defined, but may involve long-term dietary responses secondary to actual space flight. Unraveling the role of the renal-endocrine regulators during this phase of space flight and achieving a consistent hypothesis entails a closer examination of each major feedback pathway.

The malady of space motion sickness has plagued man since his second flight into space.[15] The symptomatology of this affliction includes disorientation, pallor, nausea, anorexia, and vomiting. Usually appearing within three to five days of the start of a mission, it strikes nearly 50 percent of the crew. The symptoms of space motion sickness may compound other physiologic problems noted during the early phase of space flight, including dehydration and reduced food consumption, and may limit crew activity to some extent. It is generally accepted that motion sickness on Earth involves the balance or vestibular organs of the inner ear. Two primary theories have been advanced to account for space motion sickness: the sensory conflict theory, and the fluid shift theory. The sensory conflict theory, which is more generally accepted at the present time, postulates that space motion sickness manifests itself when the central nervous system receives a sensory signal which is unfamiliar in the context of previous experience. In weightlessness, the gravity-sensitive portions of the inner ear, the otolith organs, no longer provide familiar neural signals to the brain. More generally, orientation in space depends on at least four different sensory inputs. These include the otolith information mentioned above, information from the semi-circular canals regarding angular acceleration, visual information concerning body orientation, and pressure information from the body's proprioceptive receptors. In normal environments, previous experience dictates how these information components are related. When the environment is altered such that the different sensory system outputs appear not to be compatible, motion sickness may result. According to this theory, the etiology of space motion sickness is similar to that believed responsible for motion sickness on Earth. On the other hand, the fluid shift theory postulates that space sickness results from the body fluid shift associated generally with the acute phase of space flight and is either mediated by cerebral spinal fluid pressure effects on the central nervous system, or by exerting a direct effect of changing pressure on the labyrinthine fluids or circulation. Insufficient data exist at present to completely distinguish the causes of space motion sickness. In addition, effective countermeasures are not known at the present time.

The single most important hematological finding of significance in space flight is a reduction in the circulating red cell mass during the flight period.[16] Prior to the Skylab studies, the working hypothesis explaining the observed decrease in red cell mass was that the pure oxygen atmospheres in the space capsules represented a toxic stimulus and resulted in early death of significant numbers of red cells. For the Skylab crew, who lived in a supposedly normoxic environment, this hypothesis was no longer tenable to explain the average loss of 10 percent of the total red cell mass. In the absence of a consistent finding of increased red cell destruction, and in the light of an observed reduction in reticulocytosis upon return from space, it is now assumed that the loss of red cell mass was likely a result of suppressed erythropoietic activity caused indirectly by hyperoxia of a renal oxygen sensor, or by direct alteration of stem cell kinetics at the marrow controller. However, no conclusive proof of this is yet available based on measurements made inflight, and no one mechanism has been identified which is consistent with all the data. The concept, proposed shortly after the Skylab Program, that red cell mass will eventually recover following an initial decline, does not appear to be supported by data from the Russian manned flights of six months and longer duration.

Space-flight data regarding the human immune response have been ambiguous, since the effect of weightlessness itself is difficult to separate from the stress of landing and recovery.[16,17] Studies of the reactivity of the lymphocytes, the cells responsible for the immune response, have produced varying, conflicting results. During the Apollo missions, lymphocytes were studied both before and after flight. Though the data were not entirely consistent, no significant effect was observed in DNA and RNA synthesis measured in phytohemagglutinin (PHA) activated cells. In similar tests during Skylab, there was initially a decreased response to PHA when RNA synthesis was measured. Three days after incubation, the effects on DNA synthesis were less evident, and within one to two weeks after recovery, the lymphocytes had returned to preflight levels. Studies from the Apollo-Soyuz flight and the Russian-Soyuz flights also indicated a decreased postflight responsiveness of lymphocytes to mitogens. Other changes which have been observed postflight during various missions have included varied plasma immunoprotein concentrations and elevated white blood cell counts. Information from Cosmos revealed a decreased rat thymus and spleen size. While the data appear to be at least partially contradictory in nature, they do suggest some influence of space flight on lymphocyte reactivity, including a postflight decrease in responsiveness. A clinostat has been used to produce a simulated low gravity environment in which to study lymphocyte activation. Compared to controls, lymphocyte activation was reduced by 50 percent in cells incubated for three days with the mitogen concanavalin A. By contrast, lymphocytes exposed to 4 and 16 times normal gravity demonstrated accelerated activation and accelerated aging, indicated by ultrastructural alterations. Recent data gathered from Shuttle flights have demonstrated a clear, statistically significant decrease in the in vitro ability of lymphocytes to respond to a mitogenic challenge postflight. These data suggest a transient suppression of members of lymphocytes and their responsiveness immediately postflight. The overall effects on immunological function are unclear at present.

A significant amount of muscle atrophy has been noted to occur in man in response to space flight.[18] Specific changes which have been observed include a loss of lean body mass, decreased muscle mass in the calves, and decreased muscle strength. Despite an adequate protein intake, the effects of space flight appear analogous to those of the fasting state whereby muscle protein is broken down to its constituent amino acids. The major muscle loss is believed to be associated with anti-gravity (postural) muscle. Results of inflight investigations suggest an increased protein catabolism, which includes muscle breakdown. The data indicate negative nitrogen and potassium balances and increased total urinary amino acids. The most significant loss of nitrogen has been noted to occur during the first month of flight and appears partially independent of exercise. Cosmos data have revealed that the rat soleus muscle is susceptible to hypogravic atrophy, with 32 and 22 percent reductions in weight and cross-sectional area, respectively. Ground-based studies, which have included supine bed rest in humans and hind limb immobilization in rats, have obtained results which correlate with the findings from space.

Potentially, the most serious health hazard during long-term space flight could be a progressive and unabated loss of skeletal mass. During Skylab, a rapid increase in urinary calcium excretion was detected, which began soon after entry into the environment of space and which persisted throughout the flight.[19] On the longest Skylab mission of 84 days, the increase in urinary calcium excretion was accompanied by a rise in fecal calcium loss, due in part, perhaps, to a progressive loss in net intestinal calcium absorption. Efforts to avoid skeletal wastage through exercise regimens have not been successful, and it has not been possible to reverse inflight calcium or nitrogen loss. In bed-rest studies on Earth, calcium and bone loss has persisted in experiments of up to 36 weeks in duration. Rats flown on Russian Cosmos flights have shown a decrease or even a complete cessation of bone formation during flights of up to 20 days in length. Similar findings have been obtained in immobilized rats on Earth; however, measurements in space on man are not yet sufficient to allow an interpretation of the significance of the findings in rats. In the light of the available information, it is not clear whether the normal adjustments to gravitational and muscular stress which occur in space are primarily due to changes in the activity of osteoblasts that form bone, or osteoclasts that resorb it. It is likely that both cell types are involved. The consequences of bone loss during long-term space flight include the possibility that such loss is irreversible, that calcium and phosphate released from the bone may prove toxic to other parts of the body or may enhance the occurrence of such medical problems as kidney stones, and finally, that a diminished skeletal mass may pose a functional hazard upon return to Earth.

Studies in space which do not relate directly to man's adaptive changes have been relatively sparse.[6] All life on Earth has developed in the presence of a nearly constant gravitational field, and research techniques on Earth cannot remove that influence. Gravity is known or believed to affect life from fertilization through birth, maturation, and death, but how that influence is physiologically manifested remains highly speculative. Experiments in space have shown little or no effect of gravity on many simple cell functions, or on the growth of cells in cultures. Plants appear to develop and grow satisfactorily in space, although Russian studies have indicated that arrested development or early senescence may occur in some plants. But plants have never been grown in space for a full life cycle (seed to seed), nor have the roles of geotropic and phototropic mechanisms for orientation of plants been studied in space. The role that gravity plays in animal development has never been examined properly in space. Experiments which were conducted during the Gemini and Biosatellite Programs used frog embryos that had undergone several cell divisions prior to launch. Thus, the results, which demonstrated the lack of any notable space-flight effect, were not conclusive because the critical development period, insofar as gravity is concerned, may be during and shortly after fertilization. In fact, little is known concerning how gravity is perceived by plants or animals, how gravity affects and controls the physiology, morphology, and behavior of organisms, or what role gravity plays in development and evolution.

Integrated Effects on Man

The myriad results obtained from the various studies related to human adaptation may be organized in a variety of ways, but any approach to organization leads to the conclusion that all of the systems of the body are tightly coupled, and that physiological shifts in any one system produce complex, time-dependent, ripple effects in other systems. Interactions among some of the systems are portrayed in Figure 1, which is a simplified picture of the natural couplings between those systems which involve body fluids and those systems which do not. The tight coupling of cardiopulmonary and renal-endocrine systems, the bone and muscle areas, and overlapping effects of vestibular disturbances are all suggested in this figure. Such diagrams illustrate quite clearly the mandatory nature of interdisciplinary studies in developing a comprehensive understanding of the effects of space flight on man.

A different type of system integration is

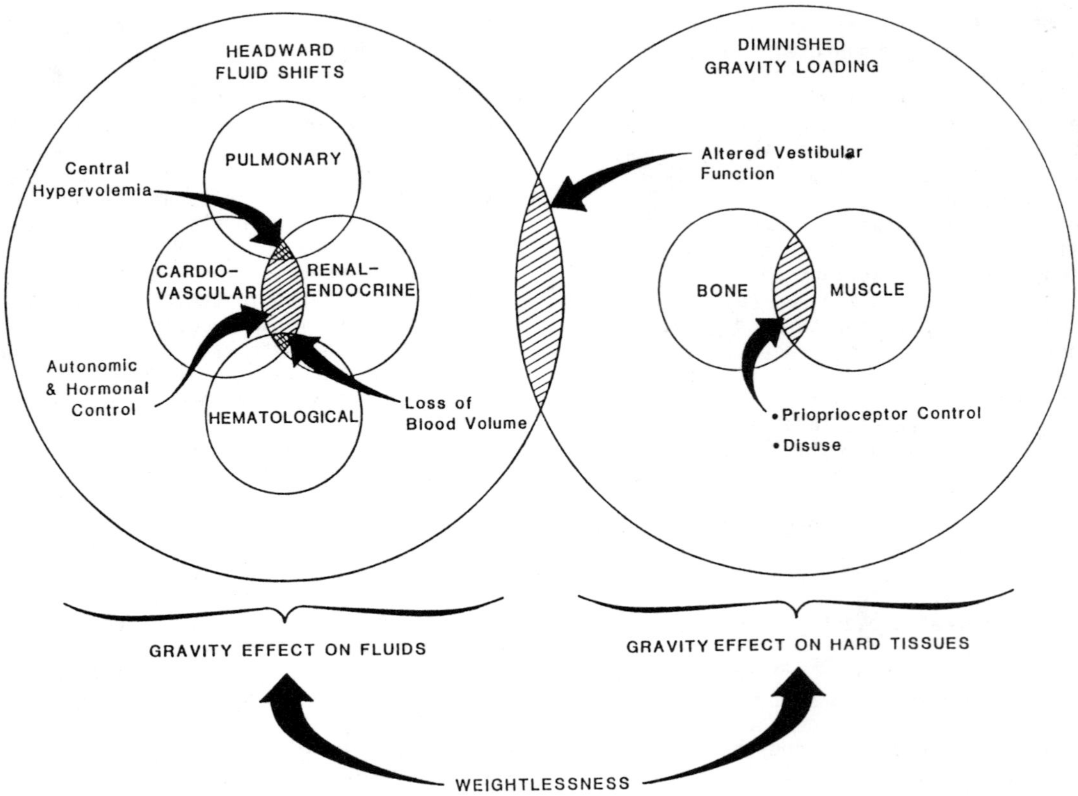

Figure 1 Interactions between major physiological systems during the adaptation to weightlessness.

depicted in Figure 2, which shows the overlap and interconnections among many of the competing and redundant pathways in the body. This particular unified view should be treated with caution, since some of the interconnections which are shown have not yet been confirmed by direct experimental evidence. Nevertheless, one can expect a similar set of pathways and events to actually function during space flight, and the following broad picture is likely to be generally accurate.

Just as in Figure 1, disturbances in the cardiovascular, fluid-electrolyte (renal-endocrine), erythropoietic, musculoskeletal, and metabolic systems during space flight appear to be attributed to at least two major effects of weightlessness. First, the absence of hydrostatic forces results in redistribution of the body fluids in a manner favoring upper body hypervolemia. Correction of this disturbance by the fluid-regulating systems leads to a reduction in extracellular fluids, most importantly, blood volume. Second, the absence of deformation forces results in degradation of normally load-bearing tissues with a major consequence being a reduction of intracellular constituents of bone and muscle mass. In addition, a third factor which enters the picture appears to be a long-term alteration of the metabolic state. This alteration is partly reflected by changes in dietary intake and is influenced by exercise levels and by weightlessness itself in some undefined manner. For example, a diminished dietary intake (both voluntary and involuntary) appears to be implicated, at least partially, in the loss of fat stores, muscle tissue, body water, and red cell mass, as well as in altered hormonal secretion and renal excretion. The stress of space flight also plays a role and this is reflected by significant increases in cortisol levels, which have widespread metabolic consequences. The occurrence of space motion sickness early in the flight and other vestibular disturbances manifested upon one-g recovery may be related to the headward displacement of body fluids, possible morphological changes in the vestibular apparatus, and perhaps most importantly, conflicting cues from multiple sensory stimuli. Whatever the cause of space motion sickness, it is associated with a significant voluntary reduction in food and fluid consumption with a possibility of wide ranging physiological consequences as noted above. All of these events have both acute and long-term effects which lead to the notable and consistent findings of loss in weight, change in body composition, altered blood biochemistry, decreased tolerance for orthostasis, and upon return to a one-g environment, a compromised response to physical activity. With only a few exceptions, no findings to date suggest that the responses of the body to weightlessness are pathological in nature. Rather, they appear to be appropriate and effective adaptations to the weightless environment. Many of the effects which have been observed can be tentatively explained in terms of normal, although complex, feedback regulatory processes.

The Experiments of Spacelab

Life sciences experiments have been selected or tentatively selected for the first four missions in which the U.S. Space Shuttle will carry the European-built Spacelab into orbit, for flights of about seven days each.[20] The first Spacelab mission, which is designed to verify the performance of Spacelab and is currently scheduled for launch in the latter part of 1983, will carry seven

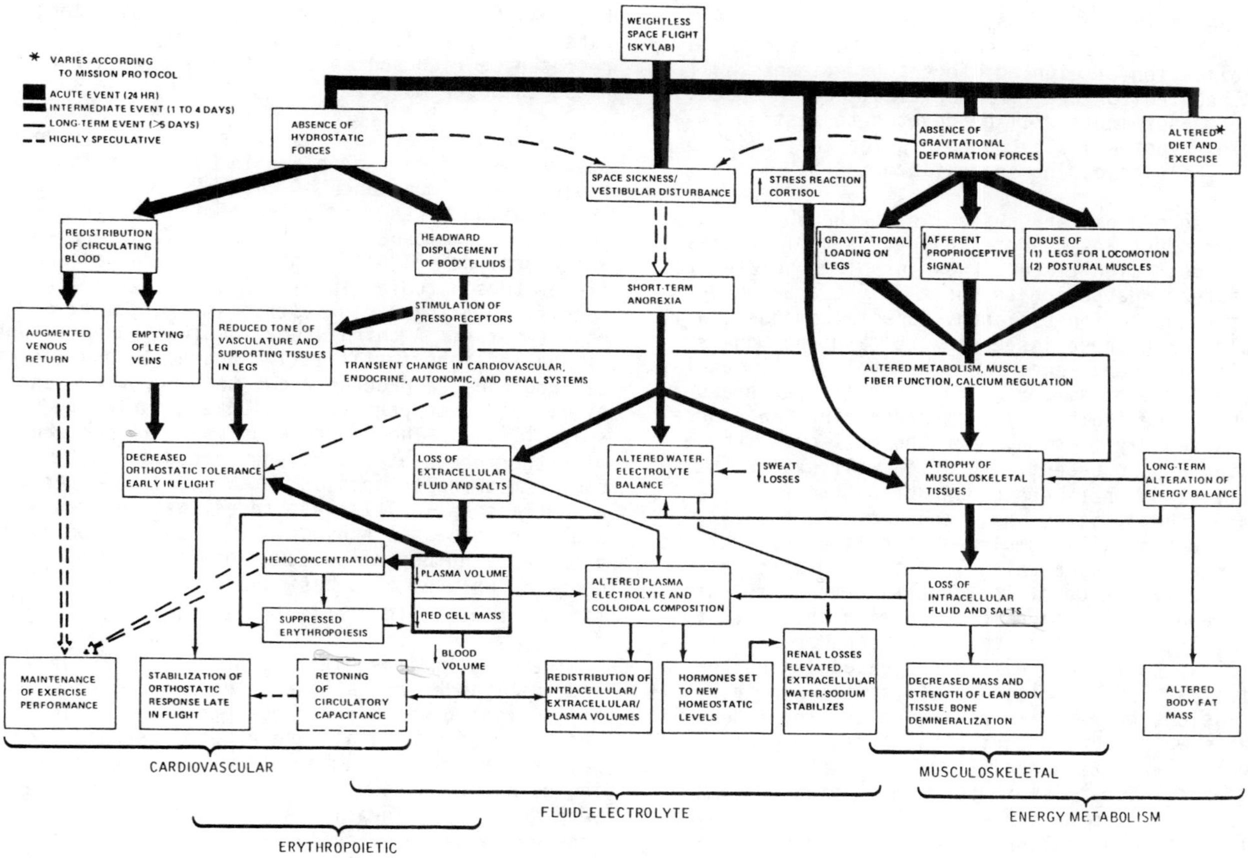

Figure 2 Integrated hypothesis of physiological adaptation to prolonged space flight.

life sciences experiments.[21] These experiments will examine plant nutation, human vestibular function, erythrokinetics, the humoral immune response, and the radiation environment within Spacelab. The second Spacelab mission is dedicated to astrophysics, but will have two carry-on life science experiments dealing with plant growth and calcium metabolism. Spacelab 3 will be a mixed discipline mission with a materials science emphasis. The life sciences portion of the payload for that mission is a five-rack mini-laboratory whose primary objective is the verification and evaluation of instruments and facilities that will be flown on the first dedicated life sciences mission of Spacelab. In addition, an experiment will be conducted on the use of autogenic therapy and operant conditioning (biofeedback) to reduce the symptomatology associated with space motion sickness.

The fourth Spacelab mission (SL-4) will be dedicated entirely to life sciences experiments and is, by far, the most complicated life sciences mission planned for the Shuttle era. The SL-4 payload will contain 24 separate investigations: 10 will utilize human subjects, 9 will share 48 laboratory rats, 2 will share 4 squirrel monkeys, 2 will involve plants, and 1 will concern frog eggs. These investigations were selected from almost 400 proposals submitted in response to a 1978 Announcement of Opportunity. The specific objectives listed in the Announcement of Opportunity were:

a. to investigate and understand physiological performance and biochemical changes which have been observed in humans who have flown in space;

b. to identify and investigate significant biological phenomena which may occur during and/or after exposure to the space environment;

c. to test and demonstrate, under operational conditions, equipment and procedures which are needed by the NASA Life Sciences Program to meet its overall goals.

The SL-4 payload was selected in such a way that maximal alignment was obtained with the Announcement of Opportunity objectives and with the individual experiment objectives. The elements of this mission are depicted in Figure 3, which shows the main areas of study and some of the interconnections among these areas.

The SL-4 mission has five major objectives or themes. The first objective of SL-4 is to conduct an interdisciplinary study of the human and animal responses which occur promptly upon achieving the near-weightless condition. The major emphasis of this portion of the mission will be on the cardiovascular and renal-endocrine responses to the documented acute fluid shifts, while a secondary emphasis will be placed on space motion sickness. For this objective, the most time-critical phase of the mission will be the first several days of flight, when the most dramatic physiological changes related to cephalad fluid shifts are believed to take place, space motion sickness becomes manifest, and other more slowly acting systems begin to exhibit the first signs of zero-g accommodation. While the majority of tests which will be conducted will be made on human subjects, a number of significant and crucial measurements will also be performed on small monkeys and rats. A systematic series of measurements, representing a consolidation of eight different experiments, will be scheduled for the first critical day in space. These measurements will be devoted to testing a complex hypothesis that accounts for the rapid loss of body fluids during space flight.

The second primary objective of the SL-4 mission is to conduct an interdisciplinary set of studies of the most significant, known problems related to short-term adaptation to space flight and subsequent readaptation to Earth. Special emphasis will be placed on those issues in the cardiopulmonary, vestibular, renal-endocrine, hematological, muscle, calcium metabolism/bone, and immunological areas which can be fruitfully examined during the course of a seven-day mission. For the most part, these studies of short-term adaptation will extend the observations that were begun during the first few days of the mission, and, together with those measurements, will complement the findings from the Skylab Program concerning long-term adaptation. With the exception of the central venous pressure measurement to be made on man during the first day of the mission, all of the tests conducted during the first three days will be repeated during the latter part of the mission. This will permit conclusions to be drawn regarding the longitudinal effects which obtain during the first week of space flight. Taken as a collection, the inflight studies related to short-term adaptation in humans and animals form an impressive group. Areas which will be studied include circulatory and cardiopulmonary adaptation during rest and exercise; the approach to new homeostatic levels for renal function, renal regulating hormone secretion, body fluid volumes, and body fluid biochemistry; the adaptation of the vestibulo-neurological system and changes in otolith morphology; the search for a suppressive mechanism of erythropoiesis; the dynamic behavior of calcium metabolism; the pathways for muscle degradation and changes in muscle tissue; and, in vitro tests for alteration of immunological function.

The third SL-4 objective is to develop valid experimental zero-g animal models and one-g analogs of weightlessness which can be used to study biomedical problems in space, or in Earth-based laboratories. It is to be expected that animals will have an increasing role on future missions, especially when ethical or safety considerations preclude direct study of human subjects. Therefore, it is important that animals be qualified for flight studies as soon as possible. The degree to which accurate inferences regarding human physiology may be drawn from animal models is a function of the degree to which the animal model is validated for the use in question. If successful comparisons between humans and animals can be made, then both the rat and the monkey may qualify as suitable animal models for studies of human circulatory, renal, endocrine, hematological, muscle, and calcium responses to weightlessness. The data from humans during the SL-4 mission will represent information that is necessary, not only for the validation of animal models, but also for the assessment of the validity of several important ground-based analogs of weightlessness. These include head-down tilt studies in man, lower body positive pressure studies in primates, and animal suspension models. Finally, it should be mentioned that mathematical models of physiological function have been, and will continue to be, used to aid in the analysis and interpretation of space-flight data. The interdisciplinary studies of SL-4 represent a rich source of validation data for these computer models.

The next objective of SL-4 is to investigate several significant problems in basic gravitational biology, the importance of which is recognized on this mission by the inclusion of four experiments which address issues of a fundamental nature and which provide scientific balance to the payload. Two of these studies involve plant development with specific emphasis on gravitropism, nutation, and phototropism. A third study involves observations of the inflight fertilization and early development of frog eggs, a process known to be gravity-sensitive, but never previously studied completely in space flight. The fourth experiment is concerned with thermoregulation and the circadian rhythms of body temperature, both of which could be altered by weightlessness. This last investigation uses a small primate as a subject. It should be pointed out that a number of parts of other experiments contain features of a basic nature which have a strong relation to gravitational biology.

The final major objective of SL-4 is to develop and test, under operational conditions, procedures and equipment which are needed to conduct biomedical research in a Spacelab environment. The payload contains a number of unique aspects, including new methodologies that have only recently been developed, invasive studies which have never been performed previously in space flight, and specialized equipment for studying animal and human physiology. Examples of new physiological techniques that have been adapted for SL-4 include a non-invasive rebreathing method for measuring cardiac output, a baroreceptor sensitivity test using a neck suction device, an automated analyzer for administering a spectrum of cardiopulmonary tests, a rotating dome for studying visual-vestibular-tactile interaction, an in vivo study of microcirculatory function in rats using a transparent skin flap chamber, and microsampling techniques for the assay of body fluids. New invasive studies include the use of radioactive and stable tracers (for measuring a variety of fluid compartment volumes, renal clearances, calcium and protein fluxes, and red cell production), indwelling circulatory catheters and probes for measuring circulatory pressures and cardiac output in humans and animals, and animal sacrifice and dissection techniques. Some of the specialized equipment that will be onboard includes a mass spectrometer, a two-dimensional echocardiograph, automated urine collection devices for humans and primates, complete blood processing equipment, a research animal holding facility (RAHF), and the hardware for studying plant development. It is safe to say that many of these tests and a majority of this equipment, or their direct successors, will become part of the general purpose research tools for other life science missions.

This brief summary illustrates that by the time Spacelab 4 flies (late 1985, or early 1986), research conducted in the space environment will be unlike that ever performed before in the U.S. space program, and quite different from that conducted in any traditional terrestrial laboratory. Concurrent operation of 24 experiments on humans, animals, and plants within a single roomsize remote laboratory; the use of surrogate investigators to perform the experiments and to also act as subjects; and the detailed advanced planning, engineering, and preflight simulation studies required to undertake such a mission are all unique components of a research program in space. It is likely that the Spacelab 4 mission will become the strongest biomedical mission ever flown in space.

Space Station and the Future

We stand at the edge of space, peer into the darkness, and ask questions. The questions concern man, life, and the universe; they are important questions and the simplicity with which they are expressed belies their intrinsic difficulty. In this paper, such questions will be assumed to hover in the background, and we will turn our direct attention to other, more practical, matters which lie on the surface of our sphere of ignorance.

Prediction of the future is fraught with difficulty, except in very special cases where the boundaries of the system in question and the evolutionary model for that system are known. Nevertheless, certain broad features of tomorrow's terrain have the appearance of clarity, at least in today's light, and these will be sketched here in broad strokes. Architecture and issues related to technology will not be addressed, in spite of their importance, as these are the subjects of other sessions of this symposium. We will discuss the future of the space life sciences on a basic level in terms of those questions which are likely to be unanswered prior to the existence of a space station and which are particularly suitable for study using such a station. On an applied level we will focus on issues related to human productivity.

At the most basic level, it is clear that gravity plays a key role in much of biology. Life itself has been shaped by the environment of Earth, and a nearly constant downward force of gravity has always existed on Earth. In space, much of the environment of Earth can be reproduced relatively easily; artificial gravity similar to that on the surface of the Earth is more difficult to create. The nearly complete absence of gravity has been studied for such a short time that we do not yet even understand the interactions between life and gravity, much less whether we need to create an artificial gravity if man is to stay in space for long periods. Gravity is believed to influence life from fertilization through birth, maturation, and death, but how that influence is physiologically manifested remains highly speculative. Plants grow upward partly in response to gravity. Animals and plants develop circulatory systems to distribute nutrients, and systems of physical support in Earth's gravitational field. The genetic expression of the bone tissue of each developing vertebrate may be adapted, in form and in quantity, to that animal's needs in a one-g environment; in space, the microgravitational environment may alter the normal pattern of bone development. The mechanisms by which higher organisms discern direction and orient themselves on Earth develop soon after conception. It is not known if gravity is necessary for the embryonic development of those mechanisms. Phenomena such as the parental rotation of bird eggs, the self-generated rotation of developing frog eggs soon after fertilization, and the orientation of embryos in the rat uterus, all support the notion that gravity is a significant factor in the production of viable offspring. This notion deserves adequate testing in space, since the issue involved may be fundamental to the generation of life in space. The microgravity of space provides a means for examining the extent to which gravitational loading influences the functional and morphological scale relationships which have developed on Earth. All such studies require exposure to the environment of space for much longer periods than is possible using the Shuttle alone, and many of the fundamental studies which should be done in space also require more adequate laboratory facilities than can be presently provided, even using the Spacelab module.

Earlier portions of this paper outlined some of the physiological changes which have been observed in man in previous space flights. Much remains to be done to characterize, throughout the time span required for adaptation of the various physiological systems, the extent and ultimate effect of such changes in man. Furthermore, we do not know whether changes in physiological systems of potential significance to man's health and safety during long-term space habitation stabilize, recover, or acquire medical significance following several months in space, and especially, whether or not an adaptive state appropriate for return to Earth even exists. Related to this issue, investigations should be undertaken to determine the physiological limits, if any, of man's stay in space. The cumulative effects of multiple exposure should be assessed, and limits of safety regarding multiple and total exposure to microgravity should be developed. Special consideration should be given to the influence which various individual factors such as age, gender, and physical condition have during space flight. The processes of readaptation to Earth's gravity after space flights of various duration should be carefully studied, and ways should be sought to minimize any adverse aspects of this readaptive process. In fact, an important result of the life sciences studies both on a space station and, subsequently, on Earth, will be the development of the most effective countermeasures to use to preserve and protect the health and safety of the crew. Even the words "healthy and normal" need to be defined for the space traveler, and defining the normal state must be done with care and by design.

An active program of clinical studies will be required on any manned space station in order to develop methodologies for onboard preventive health care, diagnosis, and treatment of space station personnel. Such medical treatment should be prepared to support a crewmember for up to three weeks before a rescue from Earth could be mounted. Various types of inflight trauma need to be examined, characterized, and analyzed to determine the most appropriate therapy in space. The efficacy of drugs must be assessed in space, since in microgravity, certain aspects of pharmacokinetics, such as gut absorption, distribution in body fluids, and excretion, may be altered. Finally, the development of techniques and instrumentation necessary to a major surgical facility (e.g., surgical tools, anaesthesia, containment of bleeding, and fluid handling) should proceed, even though such a facility may not be cost-effective during the early stages of a space station.

Much of the work to be done in the space life sciences will involve man, as a subject, but a significant and crucial component of the research will be conducted using non-human species. Animals have been used for biomedical research throughout the ages because they offer the means to accomplish special and critical experimentation which is not possible on human subjects. Just as animal studies lead the way down the path of life-saving medical advances on Earth, so must every step of man's journey into space be preceded by carefully conceived, thoughtful use of animals in order to acquire the data needed to protect the astronauts. In many cases, this means conducting sufficient parallel studies on both humans and animals to establish the validity of the animal in question as a human surrogate. Then, using properly validated models, the long-term effects of space flight may be examined in animals, both with and without countermeasures of various types, and conclusions drawn relative to the level of danger facing man. Similarly, data gathered from appropriately conducted studies on man and animals in space may be used to determine the quality and value of various ground-based analogs of space flight in use on Earth. Such Earth studies are valuable because they allow extensive research to be conducted in a more cost-effective manner than for comparable studies in space.

Finally, it should be remembered that most physiological responses observed in man (and other life forms) are actually the result of an integration of many responses from several different systems, each reacting simultaneously to the multiple challenges of space flight. The individual systems involved are themselves complex, containing both competing and redundant pathways and multiple interconnections. The problem to be solved involves unraveling the integrated response to reveal the roles played by each of the major components of the systems following the transition to and from space. One tool which is eminently suitable for this task involves the use of mathematical models to test hypotheses directly related to physiological function in space. Results from such studies have been particularly encouraging, and have led to integrated analyses of the type embodied in both Figures 2 and 3.[22,23] Comprehensive computer modeling has not been widely used in past space life sciences research, but this is likely to change as computers become more routine laboratory elements and as sophisticated programs are developed for use by non-mathematically oriented researchers.

On an applied level, research during the space station era will acquire a new focus: the enhancement of human productivity in space. Human productivity, in this sense, is the employment of automated systems and human factors engineering in order to allow man to make the greatest use of his intellectual and sensorimotor skills to obtain utilitarian objectives on a space station. It involves the use of man and automated systems in a complementary fashion; using the automated systems to obtain the best performance from man and using man to enhance the performance, flexibility, and reliability of the automated systems. During the Apollo and Skylab space programs, man functioned primarily as an explorer, repairman, and a scientist. It is clear that in the space station era, man will make a transition to new roles, including that of serviceman, worker, and builder of large structures.[24] In that era, there will be a new emphasis on extravehicular activity (EVA). EVA itself must change from being a contingency activity to a routine activity that can be accomplished on a daily basis, or whenever necessary. The kinds of activities to be accomplished by EVA will involve satellite servicing and repair, checking and calibrating satellites prior to their being launched into geosynchronous orbit, and upgrading or improving the systems operating on existing satellites. Another prominent activity employing extensive EVA will involve the assembly of various structures in space. These structures, for the most part, will be considerably larger than those which fit in the STS cargo bay. Presently, it is necessary to configure scientific instruments in such a way as to survive the stress of launch. Being able to assemble instruments in space means that it will no longer be necessary to design them so as to withstand launch stress. Thus, assembly in space not only promises the availability of less expensive instruments, but also presents new opportunities for larger and more sensitive devices. In order to accomplish these EVA activities, it will be necessary to develop a new high-pressure space suit. This suit must be more durable than the existing suit; it must be serviceable on orbit, capable of automated check-out, and easily donned with no pre-breathing required. In addition to the suit, other devices which constitute an EVA workstation will be developed. This workstation will include the use of standarized tools and procedures, ample power, illumination, communication, and the appropriate fixation aids to allow the astronauts to work in an efficient and effective way. Lastly, it will be necessary to develop a new maneuvering unit with greater range capabilities and which is, at the same time, serviceable on orbit.

A great deal can also be done to promote human productivity within the pressurized volume of a space station. The intravehicular activities (IVA) are more varied than those involved in EVA and include a number of different elements. Commercial development will be an important activity on a space station, and will involve such disciplines as materials processing, earth and ocean observations, and possibly, the support of EVA to service, repair, and launch satellites into higher orbits. In addition, science and applications will be a major activity in a space station. Astrophysics, solar system exploration, and earth sciences missions will all be accommodated through external ports of the space station, while materials science and the life sciences will be actively explored within the pressurized laboratory of the space station. Also, the data flowing from these many investigations may be collected and partially reduced prior to transmission to Earth. Other IVA activities include teleoperator support, parts and repair support for all EVA activities, and the supervision of launch operations into higher orbit. In order that these varied tasks be accomplished both well and efficiently, considerable human factors engineering must be included in the design of the various components that will make up any future space station.

In the long run, it is likely that enhanced human productivity will be dependent on improved habitability of future space stations. These improvements in habitability should not only include the traditional individual concerns, but should also consider the crew members as a group. From an architectural viewpoint, it is important that individuals be provided both private space and communal areas for recreation and dining. A varied diet of customary foods is also an important capability that requires considerable development. As the stay-time on orbit increases, personal

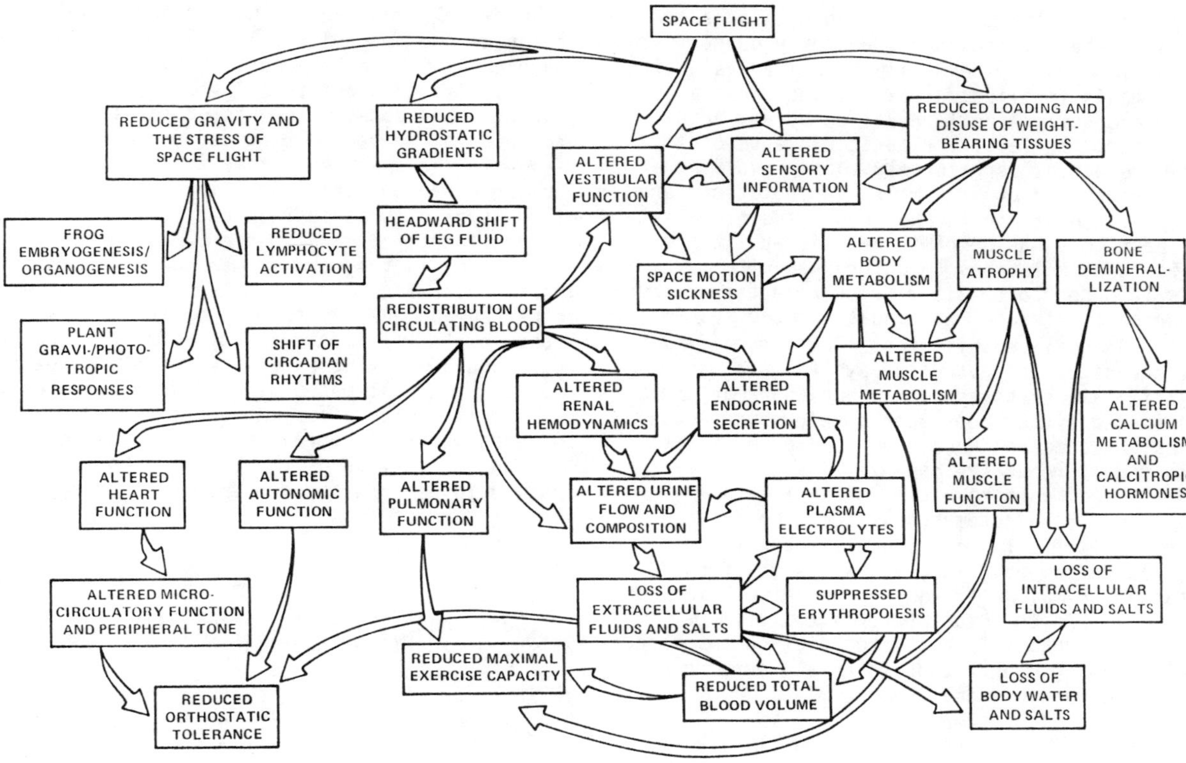

Figure 3 The SL-4 mission science overview.

hygiene will become of increasing importance. The ready availability of clean, customary clothes along with the ability to frequently shower are both desirable objectives. Housekeeping must be arranged in order to provide the crew a clean and quiet environment in which to work. Lastly, there should be a reasonable level of medical capability on the space station to diagnose and treat acute disease and occupational injury.

From a group viewpoint, it is essential that an organizational structure be developed which psychologically supports long-term human productivity. Hopefully, more efficient physiological countermeasures will be developed during the next few years which will further free the crew for productive work. Workloads will need to be carefully planned, to optimize long-term considerations of crew productivity, not necessarily just the short-term output of the crew. Lastly, data will need to be collected on crew behavior so as to incrementally improve the human engineering and space station system.

A formal process, whereby it would be possible to determine a near-optimal role for man, would be very desirable. The conclusions of such a process might then be integrated into the engineering trade studies in order to determine those tasks and procedures that man might assume in order to optimize the performance of a space station. A conceptual model of such a scheme is displayed in the Figure 4, which is based on work currently underway at the Marshall Space Flight Center.[25] This process views either the man or the automated system as a series of generic control functions. Spacecraft systems are then divided into a series of related controlled functions which are examined to determine the degree to which they might possibly benefit from some combination with man's capabilities. The performance benefits of man's participation are then weighed against the cost of man's involvement. This is necessarily an iterative process and is sensitive to the prioritization of the string of control functions representing the various spacecraft systems. It is important to recognize that the cost of man's presence represents a step increase in the overall cost of the spacecraft. When this cost is distributed only over a few tasks, the benefit/cost ratio of man's involvement may not be favorable. However, as the basic cost of man's presence is distributed over the larger volume of tasks associated with the space station, then man's role becomes cost effective.

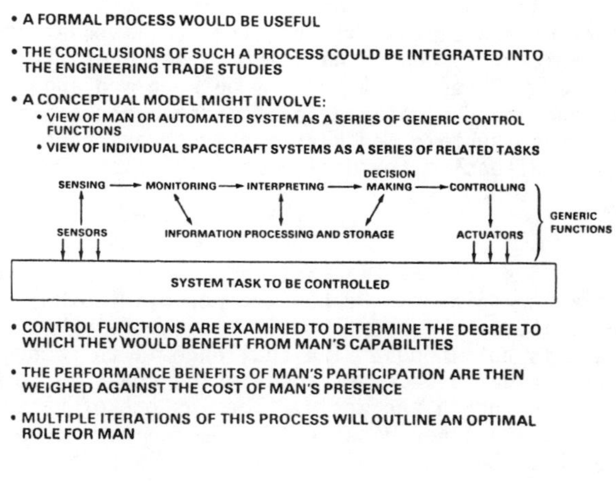

Figure 4 Determination of the optimal role for man.

What are the alternatives to man? In a narrow sense, automated systems can frequently outperform man. However, when a more comprehensive evaluation is made, man's intellectual skills and sensorimotor capacity are not presently equalled by any automated system. In fact, the basic elements of life sciences research, particularly those involving animal surgery and other complex manipulations are impossible without man in the loop. Additionally, in the past, man has demonstrated his ability to improvise good solutions to unforeseen circumstances in order to obtain mission objectives. For the foreseeable future, man and automated systems will have unequal capabilities. The central issue is to recognize that automated systems are not necessarily alternatives to man but that man and the automated system represent complementary capabilities. Automation should be used primarily to enhance man's productive capabilities and man should be used to improve the reliability and flexibility of the automated system. On a space station, it is likely that man will steadily put himself out of a job. When a new task is first undertaken, it may well require extensive human involvement. As this task becomes better understood, it can gradually be turned over to automated systems, freeing man to address novel, more complex tasks that demand his involvement.

The best way to promote human productivity in space, is not presently clear. However, there are number of things that might be done in order to provide better insight into this task. Initially, we need a more detailed understanding of man's intellectual and sensorimotor skills and, in particular, how these might best be used in the space station environment. Such information would make it possible to do better engineering trade studies involving the design of automated systems both aloft and on the ground. The second major issue is to gain practical understanding of the marginal costs in habitability and human engineering necessary to achieve a particular level of long-term performance from the crew on a space station. Through the use of the Shuttle, a great deal might be learned about human factors engineering and weightlessness, and these results could be readily incorporated into the space station design.

In summary, human productivity in space is best served by combining man's unique capabilities with automated systems in a complementary fashion, to optimize the accomplishment of utilitarian objectives on a space station. Sustained high performance on man's part will require new improvements in habitability which have yet to be carefully defined. These trade studies between man and machine and the related level of required habitability to sustain man's productivity are central issues affecting the design of a space station.

Conclusion

In this century, man has sprouted wings and left his mother planet. He has widened his foothold on the Universe ever so slightly. But merely being in space is not enough. The challenge is to recognize those unique aspects of space which are able to be mined to enrich the world. The lessons of history teach us that new discoveries lead to new riches, and that the progression of mankind is along a trail emblazoned by the highlights of exploration. It is no accident that man has journeyed into space.

The life sciences are intricately woven into the fabric of the technological world enfolding the space program, and they contribute strongly to the overall list of benefits to be derived from man's journey into space. Elements of the life sciences are involved in determining the most productive use of man in the space environment, in the design of habitable manned facilities, in the determination of the type of systems required to support the various facets of man, in the maintenance of good health, and in the care of man during sickness and injury. Other components of the life sciences seek to use the unique environment of space as a tool to probe some of the mysteries of life, particularly regarding how living things perceive and react to gravity. The ultimate value of such research to man is difficult to estimate, but there is no way to learn about microgravity environments except by leaving the surface of the Earth. The questions that could be answered are important enough to warrant the effort.

Acknowledgments

The authors gratefully acknowledge the support of the National Aeronautics and Space Administration through contract NAS W-3685.
In addition, the authors wish to thank Ms. Dolly Bingham and Ms. Linda Gust for their assistance in the preparation of the manuscript for this paper. Without their devotion to duty, this paper would not have been possible.

References

1. P.C. Johnson, and J.A. Mason (Eds.). Medical Operations and Life Sciences Activities on Space Station. NASA TM-58248, 1982.

2. S. Furukawa, A.E. Nicogossian, P. Buchanan, and S.L. Pool. Medical Support and Technology for Long-Duration Space Missions. 33rd Congress of the Int. Astronautical Federation (IAF-82-174) pp.1-8, 1982.

3. C.L. Fischer. Medical Aspects of Space Flight. Proc. of the 34th ACEMB p.233, 1981.

4. Life Sciences Considerations For Space Station. NASA Life Sciences Division, Washington, D.C., 1982.

5. A.E. Nicogossian and J.F. Parker. Space Physiology and Medicine (NASA SP-447). Washington, D.C.: U.S. Government Printing Office, 1982.

6. H.P. Klein. U.S. Biological Experiments in Space. Acta Astronautica 8:927-938, 1981.

7. J.A. Rummel and S. Deutsch (Eds.) BIOSPEX, A Summary of Life Sciences Experiments Carried on U.S. Spacecraft. NASA TM-58217, 1979.

8. M.D. Buderer. Russian BIOSPEX: Biological Space Experiments, A Space Life Sciences Bibliography. NASA JSC-17072, 1981.

9. J. Von Puttkamer. Skylab: It's Anguish and Triumph - A Memoir. Journal of the British Interplanetary Society 35:541-549, 1982.

10. O. K. Garriott. Skylab Report: Man's Role in Space Research. Science 186:219-225, 1974.

11. G.W. Hoffler. Cardiovascular Studies of U.S. Space Crews: An Overview and Perspective. In Cardiovascular Flow Dynamics and Measurements (Ed. by N.H.C. Hwang and N.A. Normann). Baltimore, University Park Press, 335-363, 1977.

12. H. Sandler. Cardiovascular Effects of Weightlessness. Prog. in Cardiology 5:227-270, 1976.

13. J.P. Meehan and R.D. Rader. Cardiovascular Observations of the Macaca Nemestrina Monkey in Biosatellite III. Aerospace Med. 42:322-336, 1971.

14. C.S. Leach. A Review of the Consequences of Fluid and Electrolyte Shifts in Weightlessness. Acta Astronautica 6:1123-1135, 1979.

15. J.L. Homick. Space Motion Sickness. Acta Astronautica 6:1259-1272, 1979.

16. S.L. Kimzey. The Effect of Extended Space Flight on Hematologic and Immunologic Systems. J. Am. Med. Women's Assoc. 30:218-232, 1975.

17. A. Cogoli, M. Valluchi-Morf, M. Muellar, and W. Briegleb. Effect of Hypogravity on Human Lymphocyte Function. Aviat. Space Environ. Med. 51:29-34, 1980.

18. C.S. Leach, P.C. Rambaut, and N. DiFerrante. Amino Aciduria in Weightlessness. Acta Astronautica 6:1323-1333, 1979.

19. P.C. Rambaut and R.S. Johnston. Prolonged Weightlessness and Calcium Loss in Man. Acta Astronautica 6:1113-1122, 1979.

20. R.M. Farrell, D.B. Cramer, and D.H. Reid. Life Science Research in Space: The Spacelab Era. Proceedings of SPIE 284:158-168, 1981.

21. C.A. Larson. NASA-Selected Life Science Experiments For the First NASA/ESA Spacelab Flight 1980, Proceedings of the European Symposium on Life Science Research in Space (ESA SP-130), pp.29-34, 1977.

22. J.I. Leonard, R.J. White, and J.A. Rummel. An Integrative Approach to Space-Flight Physiology Using Systems Analysis and Mathematical Simulation. Proc. 11th Space Simulation Conference, NASA CP-2150, 149-162, 1980.

23. R.J. White, J.I. Leonard, J.A. Rummel, and C.S. Leach. A Systems Approach to the Physiology of Weightlessness. J. Med. Systems 6:343-358, 1982.

24. J. Von Puttkamar. Roles and Needs of Man in Space. Working paper, NASA Headquarters, Office of Space Flight, Washington, DC, 1983.

25. G. Von Tiesenhausen. An Approach Toward Function Allocation Between Humans and Machines in Space Station Activities. NASA TM-82510, George C. Marshall Space Flight Center, November 1982.

THE ROLE OF SPACE STATION IN EARTH SCIENCES

Lawrence R. Greenwood*
Fairchild Space Company
Germantown, Maryland

Executive Summary

In August 1982, the Space Applications Board of the National Research Council sponsored a summer study to examine the technical requirements that should be considered in the design of a space station/platform. This paper will present some of the more important findings of the Earth's Environment Panel, the membership of which is shown in Table 1. The entire Panel report is scheduled for publication shortly.

During recent years significant progress -- some of which has been the result of practical applications of space systems -- has been made in our ability to observe and forecast changes in the Earth's environment. In preparing this report, the Panel examined the present state of scientific understanding in the four major subject areas of upper-atmospheric research, global atmospheric chemistry, weather, and climate for the purpose of defining requirements for future observational data. The final Panel report includes a section on each of four areas and indicates the characteristics that would be needed in a space station or space platform to meet the data requirements of each. This paper will focus on the recommendations that concern future measurements and the role that orbiting platforms may serve in satisfying this need. Following is a summary of the conclusions derived by the Panel.

The capability would be needed to:

1) Combine a large number of sensors for simultaneous observations of many species in the atmosphere.

2) Recalibrate instruments frequently in order to use these instruments to intercalibrate satellites.

3) Refurbish and repair instruments and maintain cooled detectors for an extended period.

4) Introduce new facilities in space, such as lidar or microwave observatories (larger than currently used).

Advances in technology that should be explored are identified at the end of the sections on upper-atmospheric research, global atmospheric chemistry, weather, and climate. The advances recommended by the Panel are as follows:

1) Development of remote-measurement techniques for OH and NO_2.

2) Higher precision or higher resolution for new kinds of observations.

3) Data-system-handling techniques for multiple-data-stream operations, including onboard processing.

4) Larger spaceborne power systems.

5) Development of tropospheric remote-sensing technology.

6) Development of a global wind-measuring system.

7) Improved microwave radiometry on low-Earth-orbit satellites.

8) Development of a method for measuring rainfall from satellites.

9) Development of a high-spatial-resolution microwave sounding system.

In addition to the recommended advances listed above, the final section of the paper presents a detailed discussion of major technological developments such as a space lidar facility or a microwave facility required for observations of the Earth's environment in the 1990s.

While low-inclination, low Earth orbits are useful, especially for observing tropical features of weather and climate, by far the most important orbits for environmental observations are low-altitude, near-polar orbits (for rapid-response, synoptic weather forecasting requirements) and geostationary orbits (for general environmental observations and for atmospheric sounding needs). If a space platform having a number of sensors is contemplated, it should be placed in a low-altitude, near-polar orbit.

The issue of human presence is addressed where appropriate throughout the report. The general conclusion is that human presence would be useful on a periodic basis rather than on a continuous basis. It should be noted that this conclusion, by necessity, is based on a limited examination. The possibilities of a more creative and innovative role for man in space should be examined in more depth.

Upper Atmospheric Research

A major use of space systems now and for the foreseeable future is to observe and measure the characteristics of the upper atmosphere (the stratosphere). The general state of understanding of this area has been recently reviewed in a report by the World Meteorological Organization (1981).

*Formerly, Assistant Director, Aerospace Systems Division, Ball Corporation.

During the past decade, the stratospheric ozone layer has become a focus of concern and a matter of regulatory and legislative activity. One concern is that humanly caused pollutants in the stratosphere could decrease the average amount of ozone and thereby admit increased amounts of biologically harmful ultraviolet light to the surface of the Earth. After atmospheric scientists predicted that chlorofluorocarbons diffuse to the stratosphere and contribute to ozone depletion, the Environmental Protection Agency (EPA) regulated some uses of these substances.

A second area of concern arises from the fact that, in addition to screening out harmful radiation, the ozone layer plays a part in controlling the thermal structure of the stratosphere and mesosphere. Both dynamic and radiative mechanisms have been identified through which long-term changes in ozone could possibly affect weather patterns of the lower atmosphere -- and hence climate.

Thus, the study of the stratosphere has taken on a sense of urgency. Clearly, there is need for the ability to foresee reliably the impact of various substances generated by an industrial society on stratospheric chemistry.

Characteristics Required of a Space Station and/or a Space Platform

There are several ways in which a space station or space platform could prove useful in achieving long-term upper-atmospheric measurements. One involves cooling systems. Should cooled sensors prove essential for long-term observations, on-orbit replenishment of cryogens may be imperative. On the other hand, active cooling systems may be more effective, even though they are heavy and high in energy consumption. Another problem of long-term observations with any quantitative system would be the need to validate instrument calibration periodically. Direct access to the instruments would greatly improve the reliability and credibility of long-term observations. Repairing failed instruments on orbit would perhaps prove to be cost effective, but further study would be required to establish this. All of these activities would require novel features in instrument design and an on-orbit infrastructure that would make them both possible and affordable.

Human presence on-orbit could well prove cost effective for servicing, replenishing, and calibrating. However, caution would be required in designing the human interface, since observations of the type described in this section would be quite sensitive to the chemical or particulate contamination and the vibration that would be associated with human presence.

Acquisition of the needed long-term global observations of the upper atmosphere may require a unique low orbit to provide complete limb-scanning coverage at various local times. There is also the possibility that the needed measurements would require a number of special-purpose remote sensors making simultaneous, collocated observations. Requirements of some or all sensors for a clear field of view could affect the size or configuration of a platform or a station used for upper-atmospheric observations.

Technologies Needed

Beyond those to be used on UARS, technologies needed in upper-atmospheric observations, can be divided into three categories: remote sensing, data systems, and operational support.

Examples of the remote-sensing technology needed include the development of remote-measurement techniques for OH and NO_2 and the measurement of HCl with better coverage (not using occultation). Various techniques, including lidar, need to be considered to make possible higher precision or higher resolution or new kinds of observations.

Table 1 Panel on the Earth's environment

Lawrence R. Greenwood, Chairman Assistant Director Aerospace Systems Division Ball Corporation	Frank R. Malinowski Head, System Analysis Electro-Optical Instrumentation Santa Barbara Research Center
William L. Chameides School of Geophysics Georgia Institute of Technology	Abraham Schnapf (former manager, Satellite Programs, RCA Astro-Electronics)
John Firor Executive Director National Center for Atmospheric Research	Thomas H. VonderHaar Department of Atmospheric Sciences Colorado State University
Gordon Little Director Wave Propagation Laboratory Environmental Research Laboratories NOAA, Department of Commerce	

The Panel gratefully acknowledges the assistance of many persons, especially Dixon Butler and Dudley McConnell of the National Aeronautics and Space Administration, James Purdom and Harold Yates of the National Oceanic and Atmospheric Administration, and Norman Phillips of the National Weather Service.

The capacity of data systems must be increased to handle the multiple-data-stream mode of operation that will be increasingly required if simultaneous measurements of a large number of species and parameters are to be made. The need for high-spectral-resolution measurements and the need to survey broad spectral regions to determine upper-atmospheric composition have the potential for creating large data streams consisting of new kinds of data (for example, complete interferograms). Data systems that included onboard processing might be essential in order for such devices to be used with maximum effectiveness.

Operational support technologies would be needed for the new generation of observations discussed here to become a reality. Larger spaceborne power systems may be required by lidar techniques, and stable, large structures may have to be developed to permit flying a large collection of sensors on one spacecraft.

Global Chemical Cycles

This section focuses on the application of space systems to the study of the lower-atmospheric (i.e., tropospheric) portion of global chemical cycles. The preceding discussion of upper-atmospheric chemistry is separate from this section on the troposphere only for convenience -- these systems are in fact coupled and research in the two areas needs to proceed in parallel.

Cycling of chemical species in the troposphere is frequently characterized by emission from the Earth's surface of chemicals in partly oxidized states (e.g., CH_4, NH_3, H_2S), transformation of these compounds by the atmospheric free radical OH to highly oxidized states (e.g., CO_2, HNO_3, H_2SO_4), and subsequent removal of the oxidized compounds from the atmosphere by rainout and washout. Global chemical cycles are closely coupled with the global hydrological cycle. Thus, a complete understanding of hydrology as well as of the chemistry of the air and the oceans is needed.

Global chemical cycles affect many areas of human and environmental concern, including human health, agriculture, weather, and climate. Concerns range from global questions (e.g., will increases in carbon dioxide in the upper atmosphere cause climate change?) to more localized problems (e.g., how does acid rain occur and what are the most effective steps to reduce its occurrence and resulting damage?). Developing a quantitative understanding of the biogeochemical cycles and their temporal variability will provide an essential tool in determining the impact of human activities on the biosphere and in designing ways to prevent or ameliorate the unintended consequences of human activity.

Possible Uses of a Space Platform and/or a Space Station

Once a remote-sensing technology for tropospheric chemical species has been successfully developed and tested, the instrumentation could be placed very usefully on a space platform. The advantages of a space platform would include: 1) the ability to have a large number of instruments on a single platform so that simultaneous measurements could be made of several atmospheric species and related meteorological parameters such as temperature, winds, cloud cover, and precipitation; 2) the ability to have a crew periodically visit, service, and calibrate the instrumentation, making it possible to obtain continuous data over a long time period and thus to study atmospheric temporal variability; and 3) the ability to use large, powerful, and heavy active sensors (such as lidar) to measure species that cannot be sensed by passive techniques.

For remote sensing of the lower atmosphere, a space platform should be placed preferably in a near-polar orbit that would afford complete global coverage. However, a high-inclination orbit (e.g., 57 deg) would probably yield a large body of valuable information. Spacecraft in low-inclination low Earth orbits would be useful for tropical measurements; however, the usefulness of a low-inclination orbit (e.g., 28.5 deg) alone would be limited.

Beyond the 1990s, another very helpful configuration would be a series of platforms in various geosynchronous orbits that would permit study of the effects of episodic events (such as forest fires and rain systems) on chemical composition, and that would also allow for monitoring the dispersion of anthropogenic emissions in the global atmosphere.

At this time, the Panel has identified only periodic maintenance and calibration tasks for which human presence in space would be helpful in the operation of remote sensors for environmental observations.

Technologies Needed

The greatest technological need in the study of tropospheric chemical cycles is the development of remote-sensing technology. The current state-of-the-art and likely near-future advances in this technology have been summarized in several recent reports (including Keafer, 1982; NASA, 1979; and Levine and Allario, 1982).

Based on these reports, Table 2 indicates the remote-sensing techniques that presently look the most promising for measuring data on global chemical cycles.

Weather

For more than 100 years, the United States has operated an observing system for weather forecasting. Today this system includes data from passive radiation sensors on geostationary and polar-orbiting satellites. These satellite data include information on clouds (from visual and infrared

Table 2 Remote-sensing technology needed

Technique	To Measure
Optical filter radiometer	Temperature, H_2O, O_3, aerosols
Gas filter radiometer	CO, CH_4, and possibly HNO_3, NO, H_2CO
Laser hetrodyne	H_2O, CH_4, and possibly O_3, HNO_3
Orbiting spectrometer	
Interferometer	Spectral survey
Lidar	O_3, aerosols, temperature, winds, possibly NO_2 and SO_2

images), on vertical profiles of temperatures and moisture (infrared supplemented by some microwave on the polar-orbiter spacecraft), and on sea surface temperature (infrared). Half-hourly cloud images from the geostationary satellites enable wind estimates to be made in those parts of the atmosphere where clearly identifiable cloud elements exist and are trackable from one image to the next.

Under the auspices of the World Meteorological Organization and International Council of Scientific Unions, the Global Atmospheric Research Program, culminating in the Global Weather Experiment in 1979, provided the framework for development and testing of global remote-sensing techniques, for improvement of models, and for acquisition of the most advanced computer systems by meteorological research groups and weather services. The improved understanding and capability that resulted from these activities are increasing the ability to forecast large-scale features of the weather.

Forecasting for 6-8 days is done routinely. A recent study by Lorenz (1983) concludes that further improvement of initial analyses (i.e., data) could extend this predictability limit to about 2 weeks.

On the nonsynoptic scale (the mesoscale and storm scales), there is neither the long history of continuous observations existing on the synoptic scale, nor the ability to model realistically the complex and nonlinear processes that become more and more important as one moves to progressively smaller scales in the atmosphere.

This section discusses the possible roles for space stations and/or space platforms in producing further improvements in synoptic-scale forecasting and in undertaking the range of activities required to develop a national capability in mesoscale observing and forecasting.

Spaceborne System Advances

Spaceborne system advances needed in the field of weather can be divided into three aspects:

1) Increased accuracy and vertical resolution of satellite-measured profiles of temperature and moisture are needed. Present operational systems are not competitive with the conventional radiosonde, especially in the measurement of convective instability (the latent instability of the atmosphere in the presence of moist air convection).

2) There should be a greater exploitation of microwave sensors, not only to provide temperature and water vapor measurements in cloudy areas, but also to increase the accuracy of measurements of precipitation and ground wetness from satellites.

3) Direct measurement of wind by satellite-borne Doppler lidar would greatly enhance the accuracy of large-scale meteorological analyses. In typical midlatitude situations, a temperature measurement with a 1 deg error is easily matched in information content by a wind measurement with an error of 3.5 m/s (Lorenz, 1983). At low latitudes, the temperature measurement error must decrease significantly to match the wind information (e.g., see Phillips, 1983).

New and different sensing systems and spacecraft will be required to make the needed measurements. For example, a possible observing system might include the following:

1) Spacecraft in high-inclination (polar) low Earth orbit (to provide data advanced over that provided by TIROS N, one of the satellites in the National Weather Service operational weather observation system).

2) Spacecraft in low-inclination (tropical) low Earth orbit (especially for the measurement of winds and precipitation).

3) Spacecraft in geostationary orbit (carrying imaging and profiling systems, if possible modified to use microwave channels as well as infrared and thus providing profiling capabilities through clouds).

Space platforms and/or space stations may be useful for some of these sensor systems, because they may require large amounts of power (more than 1 kW for a satellite-borne lidar designed to measure winds), their weight may be high, and they may need large antennas (for active or passive microwave sensors of adequate spatial resolution).

The operational payloads at low Earth orbit and geosynchronous orbit that will be required to monitor the atmosphere adequately will be considerably larger and more expensive than existing systems. It seems probable, therefore, that if manned space stations were placed in low Earth orbit for other purposes, it would be cost effective to use the stations to update sensors or other system components, to repair and/or retrieve the payloads, and to construct larger structures for transfer to geostationary orbit.

Technologies Needed

Currently, in the lowest levels of the atmosphere over land and in regions of clouds, satellite sounding systems are not able to define the three-dimensional fields of wind, temperature, and humidity with the accuracy desired for numerical weather prediction. Therefore, for the present, the satellite observing systems should be complemented over land by continuous, automatic, ground-based profiler systems, which are able to measure the desired fields in these regions with the necessary accuracy (Hogg et al., 1983).

In the future, an improved global wind-measuring system is needed to measure the tropospheric wind field. This is particularly necessary in the tropics because of the time required there for geostrophic adjustment processes. The infrared Doppler lidar concept is one such system.

Improved microwave radiometry on low-Earth-orbit satellites is needed to enhance satellite temperature and humidity profiling capabilities under cloudy conditions.

A method of measuring the rainfall rate from satellites needs to be developed and tested. No generally accepted method of achieving this has yet been identified.

A high-spatial-resolution microwave sounding and precipitation monitoring system needs to be developed for use on geostationary-orbit satellite sounders. This will require a large microwave antenna system, for which some concepts already exist.

Climate

Climate is usually defined as the weather that occurs at a particular locale over an extended period of time. For the purposes of this report, a 30-day period is used as a lower limit to distinguish climate from weather.

The Sahelian drought and the oil embargoes of the 1970s focused worldwide attention on the effects of climate on our utilization of scarce resources. In 1979, recognizing the critical role of climate in societal and economic affairs, Congress enacted the National Climate Program Act. More recently, a World Climate Program has been established jointly by the World Meteorological Organization (WMO), the International Council of Scientific Unions (ICSU), and the United Nations Environment Programs (UNEP).

Improved forecasts of climate would permit more efficient use of scarce resources. For example, they would contribute to more effective planning of irrigation, more cost-effective distribution of fuels, and better management of commodity inventories.

Forecasting large-scale climate depends on progress in understanding a range of phenomena such as stratospheric ozone, cloud/radiation processes, ocean temperature, ocean/air interface conditions, and ocean ice cover. For instance, knowledge of stratospheric ozone chemistry depends on laboratory studies of chemical kinetics and spectroscopy. Certain cloud/radiation processes in climate should be studied in the laboratory. Also, more work is needed to develop stable radiation detectors and calibration standards for trace-gas analyses. Although direct laboratory studies of the overall climate are at present very limited, laboratory research in related areas is vital.

Many studies of climate and of the advances required to predict it have been made by the National Research Council (U.S. Committee for GARP, 1975a, 1975b) and others (Rasmusson and Carpenter, 1982; National Center for Atmospheric Research, 1980; and NASA Goddard Space Flight Center, 1977). These studies led to the Panel's suggestions for field measurements.

Possible Role of a Space Platform and/or a Space Station in Climate Research and Applications

As noted earlier, a climate measurement program requires long-term, stable, reliable measurements. A space platform implies a long-term effort and thus is consonant with climate objectives. Several possible uses of a platform for climate research and applications are specified below.

Intersatellite Calibration Facility

Because climate measurements require very careful attention to time and space sampling, systems of satellites are now being planned for climate purposes. The International Satellite Cloud Climatology Project is proposed to gather data in 1983-1987 using six satellites (five geostationary, one sun synchronous) to obtain the required information about clouds. During 1984-1986, two or three satellites (one with a 57 orbit and the others sun synchronous) are planned for use in the U.S. Earth's Radiation Budget Experiment to measure the energy exchange between Earth and space and to monitor direct solar energy output. Such systems of satellites are essential to meet climate (and some weather) requirements. Often such satellite systems may include some non-U.S. satellites. In such cases, there is a special need for each satellite system to be normalized so that calibration shifts of individual sensors, should they occur, do not invalidate the overall system. A sensor package on board a space station would serve as an excellent calibration standard because it could be routinely tested in space or even returned to Earth if necessary.

The sensors' footprints on the Earth would have to be matched and a space platform (for, example, on an orbit inclined at 28.5 deg) would have to underfly free-flying satellites in the system to permit intercalibration at each encounter.

Low Signal-Above-Background Measurement

Some climate-related measurements, such as those for very detailed winds, certain volcanic gases and aerosols, and low-level moisture have a very low signal-above-background level. To detect this fine-scale, yet important, climate features, special instruments (e.g., lidars) may be needed. Power and other requirements for these sensors may preclude their use on a free-flyer that would perhaps be too small to support such sensors; thus, a space platform may be the optimum satellite for this purpose. Normally, global coverage would be required for climate modeling and monitoring. However, certain important parameters such as tropical winds could be measured from a low-inclination platform.

Assembly of Large Antennas

Climate measurement requirements could benefit greatly from higher-resolution microwave measurements. This will call for the deployment in space of antennas with diameters up to 10m in low Earth orbit, and much larger in geostationary Earth orbit.

Role of Man in Space for Climate Research and Applications

In general, the Panel on the Earth's Environment finds no requirements for human observers in space, for either climate research or applications of climate-related remote sensing. If a manned observatory is planned for solar physics studies, certain climate-related measurements of total and spectral energy from the sun might be included. Of course, the accuracy and precision of the measurements would have to match or exceed those from unmanned sites.

Facilities and Capabilities Needed in Space

The technology needed to support the requirements of major users of observations of the Earth's environment in the areas of upper-atmospheric research, global chemistry, weather, and climate have been highlighted in the previous sections.

During the SAB study, cross-disciplinary technology needs were identified by the Panel on

Earth's Resources and the Panel on Ocean Operations and they are further discussed in the report of the Panel on System Design. The experts in these disciplinary fields agree that the principal developments needed are: space-based lidar facilities, space-based microwave facilities, on-orbit calibration equipment to support the various payloads, precision stabilization of payloads, erection and deployment of large antennas, and capability for on-orbit servicing and for modification of payloads on a periodic basis.

Space-Based Lidar Facility

Development of a space-based lidar facility is essential for providing users with a wide variety of unique and important observations about the processes governing the composition, transformation, and dynamics of the Earth's atmosphere.

As has been indicated, the layers of the atmosphere are strongly coupled, and an understanding of any one layer requires knowledge of the others. A space-based lidar facility could perform many types of measurements simultaneously over an extended range of altitudes with global coverage. Table 3 lists items that a space-based lidar facility might measure experimentally and that are key to user needs in the areas of Earth's environment, Earth's resources, and ocean operations.

Many of the elements needed for the lidar facility discussed here have already undergone laboratory testing and preliminary conceptual studies of them have been conducted by NASA field centers, the Jet Propulsion Laboratory, universities, and industry.

Microwave Facility

Microwave equipment in space can be used to substantially enhance observation of the Earth's surface in a number of disciplines.

The two general classes of microwave measurement systems (active and passive), take advantage of the capabilities uniquely provided by a large platform in space -- including large antenna structures, availability of large amounts of power, the possibility of being serviced by people, and the possibility of having experimental equipment changes.

Because microwave techniques could bring about advances in remote sensing and because both active and passive systems with modular features are required, it is recommended that NASA initiate development of a general-purpose microwave facility.

Table 4 shows some of the demonstrated capabilities and candidate experiments for active and passive systems.

Replacement and Servicing of Measurement Instrument Subsystems

A method of replacing and/or servicing measurement instruments on a space platform or space station would be needed. Replacement of expendables (e.g., cryogens for cooled detectors and telescopes) and short-lived components (e.g., closed-cycle cooler and lidar flashlamps and dyes) would provide a means of achieving a long-term measurement program using generically short-lived instruments. In addition, better utilization of large telescope structures could be achieved by periodic replacement of the detection subassemblies used for different measurement experiments in the focal plane of large telescopes. Methods of telescope alignment, focusing, and instrument calibration would also be needed when instrument subsystems were serviced or replaced.

Table 3 Measurements by a space-based lidar

Aerosol winds
Cloud-top heights
Cloud-top winds
Tropospheric temperature profile
Tropospheric pressure profile
Surface and cloud-top pressure
Water vapor profiles
Tropospheric cloud and aerosol profiles
Stratospheric aerosol profile
Stratospheric aerosol composition
Trace-species (tropospheric/stratospheric)
Stratospheric ozone
Stratospheric ozone profiles
Upper-atmospheric trace-species profile
Surface albedo
Cirrus ice-water distribution

Table 4 Microwave measurement capabilities and candidate experiments

Active systems
- Demonstrated capabilities
  - Surface wave field imagery
  - Surface wind speed and direction
  - Significant wave height
  - Surface current field dynamics (warm core shedding)
  - Sea ice mapping
- Candidate experiments
  - Rain rate
  - Geoid
  - Geostrophic currents
  - Surface pressure
  - Surface wind stress

Passive systems
- Demonstrated capabilities
  - Sea surface temperature
  - Surface wind speed over ocean
  - Atmospheric temperature and humidity sounding
  - Polar ice boundaries
  - Sea ice extent (percentage of open water, first-year ice, multiyear ice)
  - Precipitation mapping over oceans
- Candidate experiments
  - Air/sea temperature difference
  - Rain rate over land and oceans
  - Soil moisture
  - Snow and ice extent and characteristics
  - Trace-gas profiles (O_3, NO_2, C_{10}, etc.)
  - Higher-spatial-resolution revisions of prior experiments

Large Microwave Antenna Assembly, Test, and Calibration

To achieve temperature sounding of the atmosphere in the presence of clouds and precipitation microwave instruments are required. These instruments, in turn, require large antennas to obtain, from both low Earth orbit and geostationary Earth orbit, the spatial resolution needed. Therefore, a method for assembling, testing, and calibrating such large antenna systems in space, before placement into free-flyer orbits, is required.

Space Platform Stability

Some of the measurement programs described above require simultaneous operation of many scanning instruments, each requiring high pointing accuracy. To avoid disturbing the accurate pointing of one instrument by correcting or changing in the pointing of another, a means for maintaining platform stability with simultaneous operation of many scanning sensors is required.

References

Atlas, David, and Otto W. Thiele, eds. (1981) Precipitation measurements from space, NASA Workshop Report NASA TM-85329, NASA Goddard Space Flight Center, Greenbelt, Md.

General Electric Company (1980) Atmospheric lidar multi-operational instrument-system definition study, NASA CR 3303, Valley Forge, Pa.

Hogg, D. C., M. T. Decker, F. O. Guiraud, K. B. Earnshaw, D. A. Merritt, K. P. Moran, W. B. Sweezy, R. G. Strauch, E. R. Westwater, and C. G. Little (1983) An automatic profiler of the temperature, wind and humidity in the troposphere, Journal of Climate and Applied Meteorology, Vol. 22, No. 5, pp. 807-831.

Keafer, L. S., Jr., ed. (1982) Tropospheric passive remote sensing, Proceedings of a workshop held July 20-23, 1981, NASA CP 2237, Accession No. 82N-26637, National Technical Information Service, Springfield, Va.

Levine, Joel S., and Frank Allario (1982) The global troposphere: biogeochemical cycles, chemistry, and remote sensing, Environmental Monitoring and Assessment, Vol. 1, D. Reidel, ed., Dordrecht, Holland/Boston, pp. 263-306.

Lorenz, E. (1983) Atmospheric prediction experiments with a large numerical model, Tellus, Vol. 34, pp. 505-517.

National Aeronautics and Space Administration (1979) Shuttle atmospheric lidar research program, NASA SP 433, Washington, D.C.

National Aeronautics and Space Administration, Goddard Space Flight Center (1977) Proposed NASA contribution to the climate program, Greenbelt, Md.

National Aeronautics and Space Administration, Marshall Space Flight Center (1982) Meteorological Satellites: Past, Present, and Future, NASA CP 2227, Accession No. 82N-24781, National Technical Information Service, Springfield, Va.

National Center for Atmospheric Research (1980) Space-based observations in the 1980s and 1990s for climate research: a planning strategy, Committee on Space Research, International Council of Scientific Unions, Boulder, Colo.

Phillips, N. (1983) An accuracy goal for a comprehensive satellite wind measuring system, Monthly Weather Review, Vol. 111, pp. 237-239.

Rasmusson, E. M. and R. A. Arkin (1981) Precipitation data for climate diagnostics, Precipitation Measurements from Space, D. Atlas and O. W. Thiele, eds. NASA Goddard Space Flight Center, Greenbelt, Md., pp. D10-D18.

Rasmusson, E. M. and T. H. Carpenter (1982) Variations in tropical sea surface temperature and surface wind fields associated with the southern oscillation/El Nino, Monthly Weather Review, Vol. 110, No. 5, pp. 354-384.

Seinfeld, J. H., F. Allario, W. R. Bandeen, W. L. Chameides, D. D. Davis, E. D. Hinkley, and R. W. Stewart (1981) Report of the NASA Working Group on Tropospheric Program Planning, NASA RP 1602, NASA Langley Research Center, Hampton, Va.

Staelin, D. H. and P. W. Rosenkranz, eds. (1978) High Resolution Passive Microwave Satellites Accession No. 81N-71117, National Technical Information Service, Springfield, Va.

United States Committee for the Global Atmospheric Research Program (1975a) Elements of the research strategy for the United States climate program, National Academy of Sciences, Washington, D.C.

United States Committee for the Global Atmospheric Research Program (1975b) Understanding climatic change: a program for action, National Academy of Sciences, Washington, D.C.

University Corporation for Atmospheric Research (1983) The National STORM Program: Stormscale operational and research meteorology, scientific and technological bases and major objectives, Boulder, Colo.

VonderHaar, T., S. Cox, T. McKee, and E. Raschke (1982) The spaceborne global climate observing system, Final report on Contract NAS 5-26343, NASA Goddard Space Flight Center, Greenbelt, Md.

World Meteorological Organization, Global Ozone Research and Monitoring Project (1981) The stratosphere 1981: theory and measurements. Rept. 11, WMO Case Postale No. 5, Geneva, Switzerland.

MANNED SPACE STATION RELEVANCE TO COMMERCIAL TELECOMMUNICATION SATELLITES: A PROSPECTUS TO YEAR 2000

Melvyn Savage*1, Clarence Catoe*2, and Philip Caughran*3
COMSAT General Corporation
Washington, D.C.

Abstract

This paper discusses the relevance of the Space Station to future commercial telecommunication satellite development. In the "Prospectus", projections are made of future commercial communication satellite development up to the year 2000 in three different categories: international, U.S. domestic, and other free world domestic and regional satellites. Assessments are made of future technology trends that will impact commercial communication satellite development. The factors affecting Space Station utilization are presented, along with industry's comments on the potential usefulness of the Space Station to their business.

Introduction

It is a pleasure to have the opportunity to share with you some of our thoughts on the potential usefulness of the Space Station to the commercial communication satellite industry.

The question of usefulness of a future new capability to a mature industry that has grown and prospered without it must be addressed with care. It would be all too easy to conclude:

1. Existing and planned launching systems will adequately cover the needs of the commercial telecommunications industry.

2. Currently projected commercial satellites do not need any of the orbital services that a low earth orbit space station might provide.

We suspect that this is in fact the commonly held view throughout the commercial communications satellite industry.

These conclusions are reinforced by the continuing increase in geostationary orbit launch vehicle performance and in satellite communications capacity per unit of orbital mass.

It therefore appears that space stations must offer financial benefits if they are to be used by the commercial communications sector. These benefits can come from either reduction in the actual costs of space circuits or from enabling new communications services and therefore new business opportunities.

It is with these thoughts in mind we have addressed the issue of space station usefulness to our industry. The results discussed herein are from a contract we had with the Grumman Aerospace Corporation in support of Grumman's study for NASA on Space Station Needs, Attributes, and Architectural Options. In the course of this study, we made a forecast of how we saw the industry developing up to the turn of the century and the key technologies that will influence its growth, and the conditions under which the Space Station could be beneficial. We then circulated this "Prospectus" within the communications industry in order to get feedback from carriers, manufacturers, insurers, and users that would be helpful to NASA in its planning. The comments we received both complemented and modified our thinking about the future usefulness of the Space Station.

As a result of this activity, we have developed a healthly respect for the complexity of the issues involved in defining and implementing a National Space Station Program, serving both the public and private interests.

Space Station Relevance to Commercial Telecommunication Satellites

The text of the subject "Prospectus to Year 2000" that we prepared and sent out for review and comments is presented herein. The prospectus contained:

1) A forecast of telecommunications satellite quantities by mass class in three categories:

   A. International satellites (INTELSAT and INMARISAT)

   B. U.S. Domestic satellites

   C. Other free world domestic and regional satellites

2) A summary of future technology trends that impact future commercial satellite size, performance, and cost

3) A discussion of Space Station capabilities of potential interest

4) An indication of the factors affecting utilization of Space Station capabilities and resulting potential benefits

---

*1 Division Director, Engineering and Systems Integration
*2 Senior Staff Engineer, Spacecraft Systems
*3 Future Systems Definition Manager

Future Satellites Forecast

In developing a forecast of commercial telecommunication satellites for the free world to the year 2000, a variety of different data sources was used. The Battelle report entitled, "Outside User Payload Model," dated July 1982, served as the strawman. This report contained future launch projections up to the year 1997 for International, U.S. Domestic, and Regional Commercial Communication Satellites. These projections were then modified by information obtained from various spacecraft manufacturers, space system operators, foreign organizations and NASA Shuttle mission planners. The resultant information was further refined by the projected launch schedules for Shuttle, Delta, and Ariane. In formulating the model it was assumed that satellites would be replaced at 80 to 90% of expected life.

(A) International Communications

International communications services are provided by the INTELSAT and INMARSAT satellites.

For INTELSAT, it has been estimated that the average annual growth, based upon user demand, will be approximately 15 percent per year up to the year 2000. During this period it is estimated that between 44 and 48 new and replacement INTELSAT's will be launched to meet the demands of the international market.

In the mid-1990s, INTELSAT VI's capacity will be saturated in the Atlantic Ocean Region and more than 100,000 additional half voice circuits will be required. By that time, it is estimated that approximately 30% of the INTELSAT satellite capacity will be leased for domestic and regional services and the total capacity of all INTELSAT satellites in orbit will increase by a factor of five. It is anticipated that the capacity of a single INTELSAT VII satellite may be as high as a quarter of a million one-way voice channels.

The increased capacities of the INTELSATs during the 1990s will be achieved through the use of frequencies no higher than K-band and their reuse with multiple spot beams, together with increasingly sophisticated signal processing techniques such as source coding, digital speech interpolation (DSI), forward error correction (FEC), and satellite-switched time division multiple access (SS-TDMA).

The INMARSAT organization was established to improve maritime communications utilizing satellites. The first generation space segment is planned for operation over the period from 1982 to 1990. It will consist of two MARECS satellites and 5 specially equipped INTELSAT V satellites, which will service maritime (mobile) communications requirements until 1990. Beginning in 1990, nine new second generation INMARSAT satellites will be launched to meet the demands of the international market in each ocean region up to the turn of the century.

The International Communications Satellite Launch Forecast is given in Table 1.

TABLE 1. INTERNATIONAL COMMUNICATIONS SATELLITE LAUNCH FORECAST

LAUNCH SCHEDULE

Approximate S/C Mass (kg)	82	83	84	85	86	87	88	89	90	91	Total No. of S/C
- 700	-	-	-	-	-	-	-	3	3	1	53
- 1200	2	5	3	2	-	-	-	-	-	-	
- 2300	-	-	-	-	2	3	2	3	5	4	
- 2800	-	-	-	-	-	-	-	-	-	-	
TOTAL	2	5	3	2	2	3	2	6	8	5	
Approximate S/C Mass (kg)	92	93	94	95	96	97	98	99	00		
- 700	2	-	-	-	-	-	-	-	-		
- 1200	-	-	-	-	-	-	-	-	-		
- 2300	-	-	-	-	-	-	-	-	-		
- 2800	-	2	2	2	2	2	1	2	-		
TOTAL	2	2	2	2	2	2	1	2	-		

(B) U.S. Domestic Communications

Within the U.S., the demand for telecommunications will exhibit a relatively high growth rate from now up to the year 2000. The projected annual growth rates, during this period for voice, video, and data services should be 10%, 11%, and 17%, respectively. In order to meet the demands of the U.S., market approximately 148 new and/or replacement DOMSATs will be launched.

The trend toward heavier, higher capacity satellites promises to provide an abundance of high speed data and video channels at increasingly lower cost. Between 1986 and the turn of the century, a total of three or more new ventures in direct television broadcasting may be established, each with several satellites for time zone and regional coverages.

In the 1990s, enhanced services, including electronic banking, electronic mail, teleconferencing, and land mobile vehicle communications, may have sufficient demand to justify their own satellites. However, there is not likely to be routine use of satellites for personal communications services such as radio paging and radio.

In the 1990-95 time period, C and Ku-band satellite capacity for fixed-satellite service will be saturated and the next higher band will have to be utilized. In addition, orbital spacing of satellites in the early 1990s down to 2.5° or 2° is anticipated.

Large U.S. domestic systems are likely to be interconnected with INTELSAT satellites by intersatellite relay links in order to avoid double-hop connections and to increase the efficiency of earth-to-space spectrum use.

One or two percursor large geostationary platforms servicing diverse communications payloads may be constructed by the year 2000 to achieve efficient connectivity and conservation of the frequency spectrum and geostationary arc. This approach could provide operational flexibility and, eventually, cost savings over the traditional single mission approach.

In Table 2, U.S. Domestic Satellite Launch Forecast is given.

TABLE 2. U.S. DOMESTIC SATELLITE LAUNCH FORECAST

LAUNCH SCHEDULE

Approximate S/C Mass (kg)	82	83	84	85	86	87	88	89	90	91	Total No. of S/C
- 700	3	3	7	1	2	1	1	-	2	4	148
- 1000	4	1	3	3	-	-	1	-	1	-	
- 1500	-	-	2	1	6	7	5	4	4	4	
- 2300	-	-	2	2	-	-	-	2	2	-	
- 3100	-	-	-	-	-	-	-	-	-	-	
TOTAL	6	5	14	5	8	8	7	6	9	8	
Approximate S/C Mass (kg)	92	93	94	95	96	97	98	99	00		
- 700	4	2	3	2	2	3	2	1	-		
- 1000	-	-	1	1	1	1	-	-	-		
- 1500	4	4	6	7	3	0	3	2	1		
- 2300	1	2	4	-	2	2	1	2	2		
- 3100	-	-	-	-	1	-	-	-	-		
TOTAL	9	8	14	10	9	6	6	5	3		

(C) Other Countries Domestic and Regional Communications

This category includes satellites launched by single countries (private or governmental sponsored) and satellites sponsored by a regional group of countries. The types of services include voice communications, TV distribution, maritime communications, data transfer, and direct broadcast. It is estimated that 176 foreign and domestic communications satellites will be launched between now and the year 2000. Approximately 60% of these satellites will provide conventional fixed services, e.g., voice communications, TV distribution, and data transfer and the remaining 40% of the satellites will be used for direct broadcast services.

In Table 3, Other Countries Domestic and Regional Communications Satellite Launch Forecast is given.

TABLE 3. OTHER COUNTRIES DOMESTIC AND REGIONAL COMMUNICATIONS SATELLITE LAUNCH FORECAST

LAUNCH SCHEDULE

Approximate S/C Mass (kg)	82	83	84	85	86	87	88	89	90	91	Total No. of S/C
- 700	3	8	4	5	5	4	2	1	4	2	176
- 1000	-	2	1	-	4	1	5	5	2	1	
- 1500	-	1	1	5	2	4	5	4	3	6	
- 2300	-	-	-	2	1	1	3	3	0	1	
TOTAL	3	11	7	12	12	10	16	13	9	10	
Approximate S/C Mass (kg)	92	93	94	95	96	97	98	99	00		
- 700	3	4	4	1	4	3	2	2	1		
- 1000	1	1	1	3	3	4	3	1	2		
- 1500	5	4	3	2	4	3	1	3	-		
- 2300	-	-	1	2	0	1	-	1	1		
TOTAL	9	9	10	8	11	14	6	7	4		

(D) COMSAT General's Conclusions on Mission Model

1. All of the 377 spacecraft identified in the traffic projections up to the year 2000 can be launched by the Space Shuttle, or Ariane III, or IV.

2. Various signal processing techniques and frequency reuse schemes, as well as the use of higher frequencies, will be utilized to reduce the impact of saturation at C- and Ku-band in the mid-1990s.

3. Intersatellite link development is important for improving satellite connectivity during the 1990s.

Future Space Technology Trends

Future technology trends that will impact commercial communications satellite development from now to the turn of the century are identified and listed below:

1. Efficiency and lifetime of microwave amplifiers between now and the year 2000 are as follows:

Table 4. EFFICIENCY AND LIFETIME OF MICROWAVE AMPLIFIERS

Frequency Band	Efficiency, %		Life, Years	
	1983	2000	1983	2000
C	40	60	7-10	15
Ku	40	55	7-10	15
Ka	25	40	3-5	10

2. Development of multibeam antenna technology for 30/20 GHz fixed services in the mid-1980s.

3. Development of intersatellite relay technology in the optical and millimeter band in the mid- to late 1980s.

4. Development of technology for 50/40 GHz in the late 1990s for fixed-satellite service.

5. Development of feed arrays using distributed solid-state monolithic module amplifiers at C-band and TWTA modules at Ku-band for phase and/or amplitude control of antennas in the mid to late 1980s.

6. Development of solid-state field emission cathodes capable of current densities of 50 $Amps/cm^2$ in the 1990s.

7. Development of Ni/N_2 batteries that have energy densities of 45 W-H/kg in the 1990s and sodium sulfur batteries that have energy densities of 80 W-H/kg by the year 2000.

8. Increase in heat pipe thermal transport capacity between now and the year 2000 from $\leq$ 10000 W-m to $\leq$ 30,000 W-m.

9. Development of lightweight gallium arsenide solar cells which have higher conversion efficiency (36% as compared to silicon cells - 22%) and less susceptibility to radiation damage.

Effects of Future Space Technology Trends

1. Key technology improvements, including development of GaAs solar cells; more efficient, longer life microwave power amplifiers, and high energy-density batteries, will increase in-orbit reliability and average spacecraft lifetimes, thus reducing satellite replacement rates.

2. Other technology developments, including use of multiple spot beams, frequency reuse, intersatellite links, advanced materials, monolithic antenna feeds, and VLSI circuits, allowing full exploitation of the C- and K-bands, will yield very substantial increases in the traffic capacity of communications satellites with only modest increases in satellite masses.

3. All of the 377 spacecraft identified above can be launched by the Space Shuttle and complementary upper stage rockets, either mated to the communications payload (e.g., PAM, Centaur) or integrated with it, as planned for INTELSAT VI or Ariane III or IV.

4. NASA is expected to enhance its multiple spacecraft launch services once the Centaur is integrated into its Shuttle-based transportation system in the late 1980s. Since the Centaur places satellites into geostationary orbit without requiring apogee motors, this will provide more flexibility in the design of commercial satellites in the 1990s.

Space Station Utilization

(A) Potential Capabilities

The low earth orbit manned Space Station, as envisioned by NASA, may well be part of the overall space transportation system which includes the Shuttle, various expendable upper stages, a Teleoperated Maneuvering System (TMS), expendable and reusable Orbital Transfer Vehicles, and ultimately a Manned Orbital Transfer Vehicle (MOTV). The facility could have the following capabilities:

1. An orbital transportation base or harbor for assembly of upper stages, including orbital fueling and checkout of satellites before their deployment to higher orbit.

2. New commercial satellite services to provide narrowband broadcast or land mobile communications, because of their possible requirements for large antennas (50-150 meters) and large power systems (10-20 kW), could benefit from the use of a space station for in-orbit assembly and test before transfer to geosynchronous orbit. This requirement is not likely to occur before the late 1990s.

3. A facility to permit large satellites to be assembled, tested, repaired, and even modified, if necessary, before launch to geostationary orbit.

4. A space-based R&D laboratory for test and evaluation of advanced communications components and subsystems to establish proof of performance prior to their use in an operational satellite.

(B) Factors Affecting Utilization of Space Station Capabilities

Three key factors are likely to influence planned utilization of Space Station capabilities by the commercial satellite community. There are uncertainties regarding:

1. Reality of the capabilities that will be provided.

2. Facility user charges (vis-a-vis known costs of existing means to assemble and launch satellites).

3. The practical value of anticipated benefits.

The following are observations pertinent to these factors:

1. Communications satellites for fixed and broadcast services over the next 10-15 years can be directly launched to geosynchronous orbit by STS with appropriate apogee stages or by Ariane III or IV or other U.S. expendable vehicles.

2. New commercial satellite services to provide narrowband broadcast or land mobile communications, because of their possible requirements for large antennas (50-150 meters) and large power systems (10-20 kW), could benefit from the use of a space station for in-orbit assembly and test before transfer to geosynchronous orbit. This requirement is not likely to occur before the late 1990s.

3. Advanced technology demonstrations by NASA of large antennas may be required around 1990 to secure commercial support for new services.

(C) Costs for Use of Space Station Capabilities

The commercial use of Space Station capabilities will require that the costs for such services be carefully controlled. The following observations should be recognized:

1. The currently envisioned STS, Ariane and U.S. expendable systems can adequately provide the launch capability required for projected future fixed and broadcast satellites, at least to the mid-1990s. Thus, use of the Space Station as a low-orbit transportation node should not result in increased costs for launch services to geosynchronous orbit.

2. Without a significant cost or risk advantage, the only operational missions likely to stop at the Space Station are those of the late 1990s that have physically large and complex communications or power subsystems. In such cases, the space station could enhance the possibility of providing new commercial services.

3. The total pricing for use of the Space Station as an orbital R&D laboratory by the commercial sector must be kept low to encourage its use.

Summary of Potential Benefits

1. A low earth orbit U.S. space station program may have a major effect on future satellite orbital operations, orbit-to-orbit transportation, and eventually the configurations of future specialized satellites. While the space station capabilities described earlier are not required to meet future commercial communications needs currently anticipated through the late 1990s, the ultimate benefits could be significant. It is therefore believed that the wide range of operational services that will be developed as part of the U.S. space station program will ultimately have a favorable impact on communications satellite configurations, capabilities and costs, to the benefit of the commercial sector. The commercial sector should maintain an awareness of, and an involvement in, the determination of orbital services that will be developed.

2. Assessment of benefits and future uses are influenced to a large degree by the costs of space station services, such as in-space assembly, repair, checkout, and launch, to the commercial sector. NASA must define such economic philosophies and pricing structures.

3. There must be demonstrations of orbital services and cost benefits before the characteristically conservative commercial communications industry will commit to their operational use.

4. Until transportation and service cost benefits are demonstrated, the low earth orbit space station is likely to be used mainly as an R&D test facility.

5. By the late 1990s, the benefits available from satellite servicing, and satellite fueling could allow much greater flexibility to the industry in developing new communications satellite configurations, architectures and services.

Industry Comments On Prospectus

The prospectus was sent to 42 organizations (including 15 representing the insurance industry). Replies were received from 25 organizations. By and large, the responses constituted an endorsement of COMSAT General's assessment of the future satellite projections, and the potential uses and benefits that could be derived from the existence of a Space Station.

Some of the significant reactions and comments received from the communications and insurance companies were as follows:

1. General agreement with forecasts, observations, and conclusions (70% of respondents).

2. Space Station benefits and requirements:

   A. Skepticism that the Space Station will benefit the commercial satellite community before the turn of the century.

   B. The view that the Space Station is not useful or applicable in any respect to the commercial communications satellite business as it is now known.

   C. The view that a requirement for communications satellite missions "stopping" at the Space Station for checkout may exist in the early 1990s.

   D. Confirmation of the view that the primary value of a station will be as an orbital R&D laboratory until such time as very large and complex systems, especially antennas, need to be assembled in space.

   E. The view that orbital demonstrations of large antenna deployment may be required before 1990 in order to develop technology for frequency reuse at C-band via multibeam antennas.

   F. The view that, in the 1990s large, complex, multiantenna K- and C-band satellites with extensive frequency reuse and very precise tolerance requirements for beamwidth, etc., may require orbital assembly to achieve acceptable risk.

   G. The view that a large aperture land mobile satellite may not be required, if ground cellular systems are implemented to reach all but mobile users located in the most are implemented to reach all but mobile users located in the most remote regions, in which case a conventional satellite with a 4.5-m antenna could suffice for national coverage.

H. Doubt that precursor large geostationary platforms will be required and/or constructed by the year 2000.

3. Importance of Space Station use costs:

A. Reinforcement of the idea that economic access to geostationary orbit is of overriding importance to commercial communications satellite owners and that the impact of low-orbit Space Stations will depend on whether they contribute to lower total system transportation costs and reduced mission risk.

B. Doubt that there is any cost or risk advantage of using a Space Station as a way station for operational satellites.

4. Dependance on NASA/Government

A. Confirmation of an industry "show-me" attitude, strongly suggesting that NASA should take the initiative with exploratory development of Space Station concepts and demonstrate their value to prospective business users.

B. Acknowledgement of the apparent ambiguity between the eagerness with which the industry talks about pushing new technology and the cautiousness with which it actually takes risks.

5. Insurance Cost Implications

A. The view that the capacity of the world insurance industry to underwrite the commercial satellite launch forecast at affordable rates depends on continued development of technology to reduce mission risk and that the Space Station capabilities cited could be beneficial in this respect.

B. The view that the insurance industry will be less inclined to acknowledge the Space Station mission reliability enhancement benefits until it has been fully tested supporting non-commercial satellites (i.e., government or military).

C. The view that effective use of the manned Space Station for communications satellites will in time have favorable effects on insurance costs related to launch, deployment and in-orbit failure insurance.

Conclusion

The consensus within the telecommunications industry is that the Space Station could provide future benefits if:

1. There is a national commitment to the program.

2. The program contains provisions for satellite and orbital transfer vehicle servicing in low earth orbit.

3. Demonstrations of these capabilities are planned.

4. The costs of the capabilities provided are competitive with other alternatives.

Based on the above, we would like therefore to propose a two-part challenge:

The first part of the challenge is directed at NASA to formulate a program that will demonstrate, at the earliest possible date, that Space Station capabilities, such as payload assembly, pre-launch checkout and other services for satellites bound for geosynchronous orbit can in fact reduce mission costs and risks.

The second part of the challenge is directed at the commercial satellite industry to cultivate an open-minded position toward the proposition that successful employment of the Space Station can enhance our capabilities and competiveness, both in the U.S. and internationally.

Only through the cooperative and creative marriage of NASA and industry talent, can the space station effectively support the nation's development and exploitation of space technologies.

ROLE OF A SPACE STATION IN PHARMACEUTICAL MANUFACTURING

James T. Rose*
McDonnell Douglas Astronautics Company
St. Louis, Missouri

This presentation is a quick look at the EOS program (electrophoresis operations in space), which is a commercial thrust to manufacture a product in space for later use on the ground. We've been working on this program since early 1977, and we believe we're making great strides in arriving at a state of commercialization. Let me first start by stating our intent:

1) McDonnell Douglas Astronautics (MDAC) intends to develop a major new commercial business for processing high-value pharmaceutical/biological products in space using electrophoresis, based on MDAC-STL proprietary technology and exclusive agreements with NASA.

2) The business is currently in a product development phase that will extend into 1987.

McDonnell Douglas Corporation intends to develop a major new commercial business using the space environment to produce natural biological materials that can be used as pharmaceuticals and depending heavily on the technology that we have developed in continuous flow electrophoresis over the last five years. The program, obviously, is in what we call a development phase. The extent to which we believe this industry can grow and become is ultimately dependent upon the availability of a space station. To explain that, I should give you a little background into some of the initial strategies that we employed to get to the state we are today--well into the middle of a flight test program.

In order to bring this program about, we had five major objectives to accomplish. First, we needed the support of an active pharmaceutical company. The Ortho Pharmaceutical Corporation, a subsidiary of Johnson & Johnson, joined us in 1978 to do those parts of the job unique to the pharmaceutical business. In other words, McDonnell Douglas has no intention of getting into the pharmaceutical business per se. By actually taking the expertise of two corporations such as ours and applying those expertise in the right areas of work, we feel we bring the necessary skills to this program to make it viable. In other words, McDonnell Douglas would develop the process; McDonnell Douglas would integrate the product with the process; McDonnell Douglas would design and build the flight hardware, conduct the flight testing, and be the principal interface with NASA. Ortho Pharmaceutical will do the marketing research for products and all product testing through clinical and animal trials. Plus, they will be the interface with the FDA. They understand the FDA. We believe we understand NASA. We think the combination works. And if we are successful commercially, we will do the processing of the intermediate materials in space and operate the plants and facilities; they will do the marketing, packaging, and distribution of the products.

In setting out on such a venture as this, well before the feasibility of the technology has been established, we needed some form of incentives or support from the government. NASA came forth with their joint endeavor policy, which we believe is an excellent way for cooperation between industry and government. It probably has applications well beyond the scope of materials processing. In essence, McDonnell and Ortho are responsible to develop all of the technology, to develop all hardware, and to develop the products. For that NASA will provide flights and services while we are verifying the technology and proving commercial feasibility. Once we start a commercial enterprise, we are a regular paying customer like any other user of the Space Transportation System.

We believe the incentives are in the right place in this type of joint endeavor arrangement. NASA is interested in seeing technology developed and brought to the public at the earliest possible time. Since we're putting up front-end money to develop and verify the technology, the only way we can recoup one penny of that investment, which has now been going on since 1977, is to sell something. And it's the only way we will ever return that investment. We are therefore highly motivated to get this technology to the marketplace as rapidly as we possibly can.

And, of course, in order to go to space, we had to develop the technology on the ground to totally understand what we were talking about, and also to ascertain if there really were advantages in space for this type of technology. As a result, we believe we have advanced the state-of-the-art in preparative biological processing with this process we have now developed. In fact, we are using it extensively in supporting research at a number of medical universities and other biological clinics.

We announced officially in July 1983, that the Washington University School of Medicine in St. Louis and McDonnell Douglas have entered into a long-term agreement to work jointly in developing the pancreatic beta cell as a potential treatment or ultimately a cure for diabetes. We will be flying beta cells on the Shuttle flight in August.

After developing a process, it's obvious we need to have a product. Ortho has worked very diligently and a main product is now in development. We are very encouraged with the progress we

*Director - Electrophoresis Operations in Space.

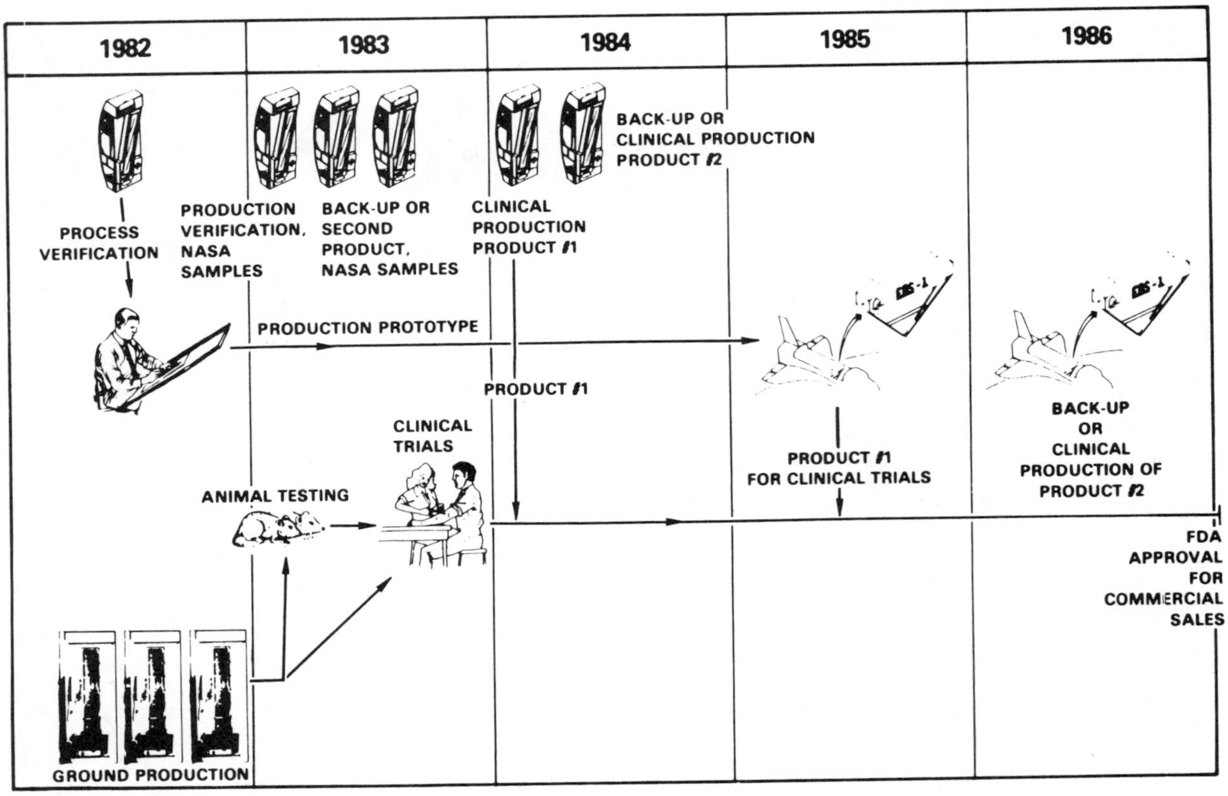

Fig. 1 Steps to commercial operations.

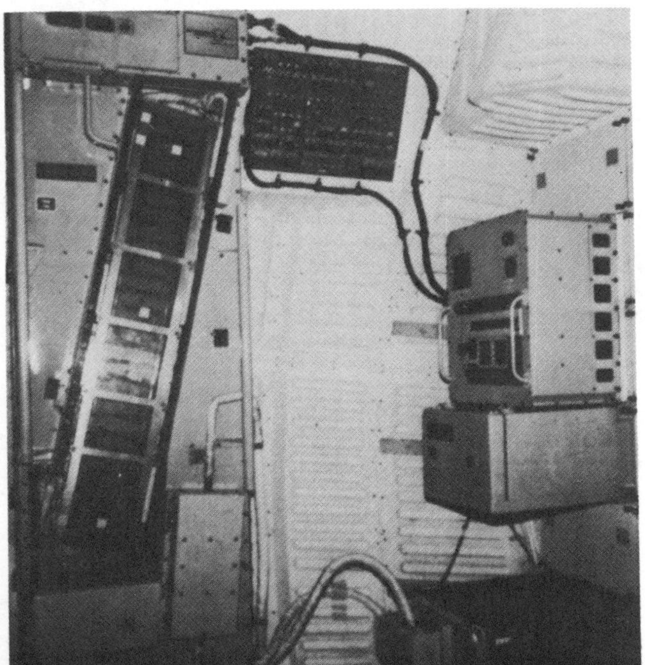

Fig. 2 Mid-deck hardware for first commercial production unit prototype.

are are making to that end, which I will say more about later.

With all of this in mind, we were then able to look at the economics. In 1981, the corporation officially reviewed our business plan. Although we were a high-risk program, they thought the rewards certainly offset the risks, and we should press on officially. EOS became an officially recognized program by the Corporation in June 1981. We're now well on our way to achieving our commercial goals. We have an integrated program, which I will go over briefly, to show how we intend to arrive at initial commercial operations. (See Fig. 1.)

McDonnell Douglas has planned two major flight test programs. The first we call the mid-deck flight program, which is ongoing at the present moment. This will phase into a flight program using what we call the production prototype unit. It is a unit with 24 times the capacity of the mid-deck unit, weighs some 5000 lbs, is totally automatic, and will fly in the payload bay of the Orbiter scheduled to be flown on STS-24 in midyear 1985. In the meantime, Johnson & Johnson is beginning to carry out its product development program. The testing will be with animals and then move to humans. We are looking forward to having a program that consummates in a FDA approval sometime in the late 1980's.

But let me show you the interaction of this program as we now envision it. We don't call these flights experiments. We call them flight tests because we had certain objectives that we wanted to accomplish with each of the flights we've flown to date. Our first two flights, STS-4 and STS-6, were primarily set up to verify the process and understand how the process works in space relative to the ground. Achieving success on those two flights, the later two flights on STS-7 and STS-8 have been geared to characterize certain products of interest that we are pursuing at the present time. We now have completed STS-7; we are moving on to STS-8, which, as I said before, will be dedicated to processing living cells. We will also be processing cells for NASA for their research program. This will be a flight dedicated to three different types of cellular material.

As a result of the success on STS-4, one of our major milestones, we immediately moved forward to start the detailed design of the production prototype unit. The preliminary design review was made in March 1983. We are now physically cutting hardware on that particular unit, and we expect to have it completed and into testing by the latter part of 1984.

In the meantime, we are using our ground laboratory electrophoresis unit to provide material to Ortho to start their animal testing. We also felt that, with the success of STS-4, it made good sense to advance the program. We felt that if we could modify the unit to make it fully capable to produce five days worth of material over the length of the flight--turning it on when the Shuttle hits orbit and turning it off before returning--then we could bring back sizable quantities of the product to feed into the Ortho test program. This would mean we will be using space-derived material nearly a year and a half earlier than we would have if we had to wait for the production prototype to come on line in mid-1985.

By converting the mid-deck unit to physically producing clinical quantities of material, it becomes an integral part of the product development program and makes its flight very significant to the program's overall success. As a result, we requested NASA to consider the possibility of letting us fly our own payload specialist to run this equipment, to make sure it stays running for the full five days and make repairs as necessary. NASA has now concurred. It is jointly agreed that it makes good logic to fly a payload specialist rather than a mission specialist. We will be flying a McDonnell Douglas test engineer on the STS-12 mission.

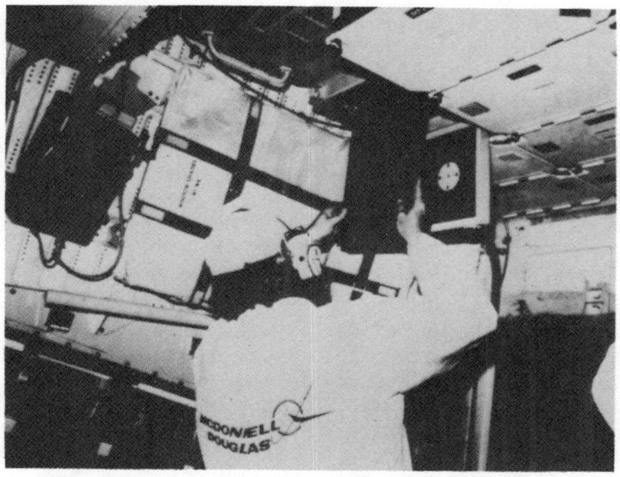

Fig. 3 McDonnell Douglas engineer installing collection trays on the Shuttle.

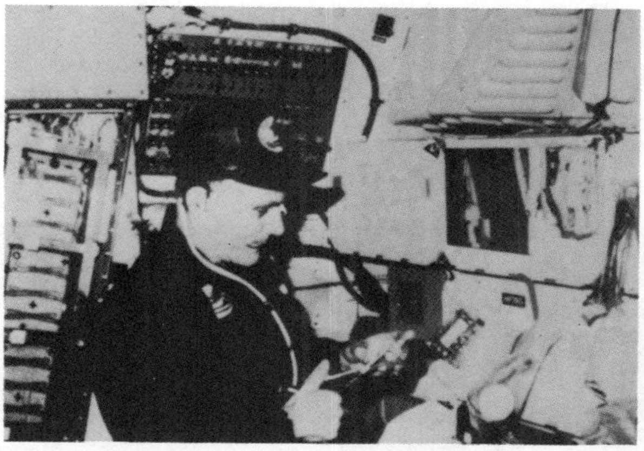
Fig. 4 Commercial prototype equipment on STS-4.

The first production prototype flight in 1985 is a very important one as well, in that we will verify the total automation of the first production plant. It will produce all of the necessary material to finish a clinical program as now defined. And if the FDA says "go", we're ready to go.

The mid-deck hardware now flying includes a separation chamber that is 6 ft tall and weighs about 500 lb. Also aboard is a 75 lb computer, and a 70 lb refrigerator used to store samples and collection trays, and a 50 lb module allowing water cooling of the system. Figure 2 shows the unit as it was installed in Challenger. The computer is a pretty smart one in the sense that it basically controls all functions of this unit. The astronauts use the computer to start sequencing, to look for parameters that go out of tolerance, and to do monitoring functions. They've done a very superior job in working with this particular piece of equipment for us. Our engineers installed the collection trays in the refrigerator on the launch pad about a day before lauch of STS-4 (Fig. 3). Figure 4 shows Astronaut Mattingly working in the mid-deck area on STS-4. You can see where our equipment is mounted behind him. In Fig. 5 Astronaut Musgrave loads one of the sample trays in the unit during the STS-6 mission.

Figure 6 shows a stream of protein material flowing through the seperation chamber during the STS-4 flight--and it is a space phenomenon. That little stream of red, which is dye-marked (black in this reproduction) represents a material that has 463 times more protein in it than we can possibly flow through the machine here on the ground. Figure 7 is a typical example of what a separation looked like during the STS-4 mission. One interesting thing is the bubbles along the cathode formed by electrolysis. They are neutrally buoyant. A series of such pictures show that the bubbles don't move--they are always in the same position.

Moving on to the state of commercialization, at this point we are remaining extremely flexible in how we commercialize this operation, see Fig. 8. From a good business standpoint, we do have the fall-back position of being able to operate effectively in the payload bay if we can achieve, as a payload of opportunity, some 30 days of operation during the course of a calendar year. We can separate a large enough supply of material to establish a positive cash flow position in that mode while waiting for a transition into some form of long-duration operation. And obviously that's where we want to go. No plant is efficient working 30 days out of the year.

But do we want to move into an unmanned dedicated satellite mode or some form of a platform. I'm very encouraged by NASA when I'm hearing that the date for their manned space station might be moving forward. Obviously, the ultimate solution to the growth of such an industry is the manned space station or support by a manned space station. If our projections are anywhere near correct, this continuous flow unit in a dedicated free-flying mode will hit only about 10% of the U.S. market. Obviously for a first product, we are not going to build one plant; we're going to add a number of plants as the market grows. For what we are working on now, there is as yet no market; but the market projections for the United States are extremely large and, because Johnson & Johnson is international, we will sell internationally very quickly. So the market is even larger and we believe it will be our job to try to keep up with it. Table 1 shows the dedicated unmanned spacecraft scenario. By the mid-1990s, we hope to have nearly 190,000 lb of upweight going into space to support a 10-12 plant operation. That's over three dedicated Shuttles annually just to do the logistic support work.

The scenario that we want to develop is a second or third or fourth product for an expanding ongoing business as shown in Table 2. How do we develop the second or third product? If we use the current mode of operation, we would start to characterize new materials by using either the mid-deck or the space lab going up and down, getting manifested, going through the integration procedures, the planning procedures, etc. We obviously have to develop clinical quantities of the materials that have to go into the payload bay with a device similar to the production prototype unit. If we're successful through the clinicals and FDA approval, then we move --as an interim measure again--into either the sortie mode of operation or onto a dedicated free flyer.

If we had a manned module operating as a factory for the first product chosen for space manufacture, all we would need is a couple of our electrophoresis chambers operating separately to routinely characterize new materials and produce new materials for clinical testing. Then for production quantities with a manned station, we could actually vary the plant from producing one product to a second product on a weekly or a daily basis by simply changing the buffer and tuning the system differently. What we believe this means is the ultimate development of a major new industry. The possibilities are compared in Table 3. It is our best estimate that in a manned environment perhaps 15 new materials could be brought to the marketplace over a 10-year time period. In the unmanned mode, we believe we could bring only three new materials to commercial viability. That is an improvement of 5:1 in terms of new biological materials that will be extremely beneficial to the public.

Since I'm in a commercial business, the patient is going to pay my development costs. I will pass those on as part of the selling price of the material. The time saved in developing new materials in a manned space station is dramatic in reducing the front-end costs for that development. Using the mid-deck or payload of the Shuttle or some unmanned sequence would require several years. A space station could reduce this to months. The point is that, against my baseline for the unmanned mode, there is nearly a 70% savings in development and testing costs of a new drug to get it to the marketplace.

The savings in operation costs is not as great in the actual production of materials even if we are on a manned station. But there are some advantages, even though a manned module is more expensive from a leasing or operation point of view.

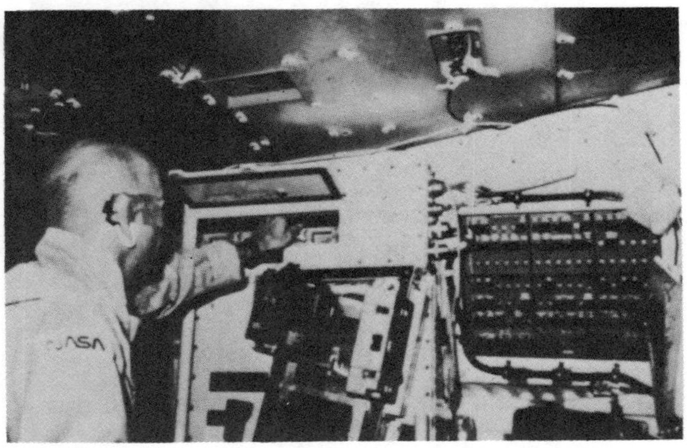

Fig. 5 Sample trays of commercial prototype equipment being monitored on STS-6.

SAMPLE No. 1
25% PROPRIETARY
AT INLET

Fig. 6 Operation of separation chamber on STS-6.

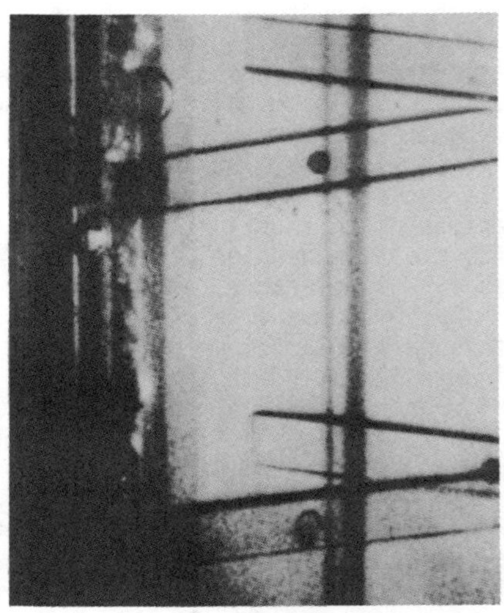

SAMPLE #4
1% ALBUMIN
SEPARATION
AT 30 cm

Fig. 7 Separation of protein material during STS-4 (note bubbles at left).

Table 1 EOS Shuttle utilization

	86	87	88	89	90	91	92	93	94	95
JEA ⟶										
SPACECRAFT LAUNCHES AT 11K LB			2	2	2	2	2	4	3	2
FACTORY MODULE LAUNCHES AT 5K LB			2	2	2	2	2	4	3	2
RESUPPLY MODULE LAUNCHES AT 5K LB		2	5	9	13	15	19	23	26	26
FLIGHT SUPPORT SYSTEM LAUNCHES AT 3K LB		2	2	2	2	3	4	4		4
TOTAL MASS UP — K LB		16	63	83	103	116	139	191	190	174
SPACECRAFT RETRIEVALS AT 6K LB						1		2	2	2
FACTORY MODULE RETRIEVALS AT 5K LB						1		2	2	2
RESUPPLY MODULE RETRIEVALS AT 5K LB		2	3	7	11	14	17	21	25	26
FLIGHT SUPPORT SYSTEM RETURN AT 3K LB		2	2	2	2	3	4	4	4	4
TOTAL MASS DOWN — K LB		16	21	41	61	90	97	119	159	164

NOTE: ASSUMES 5 YEAR LIFE FOR SPACECRAFT AND FACTORY MODULE

Table 2 Steps for new biological product development

	UNMANNED FREE FLYER MODE			MANNED SPACE STATION MODE
	MIDDECK OR SPACELAB	PAYLOAD BAY	UNMANNED FREE FLYER	
CHARACTERIZATION	✓			✓
CLINICAL TRIALS MATERIALS		✓		✓
INITIAL COMMERCIAL PRODUCTION		✓ (INTERIM)	✓	✓
EXPANDED PRODUCTION			✓	✓

Table 3 Improvement of manned space station over unmanned platform on product development price to patient

	UNMANNED OPERATIONS MODE	MANNED OPERATIONS MODE
RELATIVE COST OF FRONT END EXPENSE	1	0.38
RELATIVE COST OF OPERATING EXPENSE	1	0.77
RELATIVE NUMBER OF NEW PRODUCTS DEVELOPED	1	5.0

What we save in logistics costs, i.e., being able to truck raw materials as opposed to big heavy modules into space (Shuttle costs as a logistics vehicle), more than offsets the additional cost of using a habitable station as a method of production. Also, we can produce a lot more material. In an unmanned mode, we assume that over a 5-year production period, we could lose six months production of the product because, if the plant breaks down, we've got to go get it and then we may have to bring it back to fix it--or it's going to take us some time to get a manifest to get up there. In the manned mode, we can maintain the system routinely and thus do not anticipate those types of losses.

Overall, we believe we can produce the materials, operationally and slightly less expensively than is the unmanned mode. We believe it will be quite practical to produce and develop new products using the space environment. We are doing it. And we're putting money in the project because we believe in it. However, to go into space and expect to have a money-making capability using the Shuttle in a sortie mode is not going to be practical except on an interim basis. But it is a method with which to at least begin some positive operations until the transition can be made to something better.

Obviously, to expand the industry to any size, long-duration capability is needed. There are two choices: unmanned or manned. In the unmanned mode, we believe we can bring three products into being over the next 10 years, and then we're ultimately going to be limited by the supporting services of the logistics systems--the Orbiters. Will we have

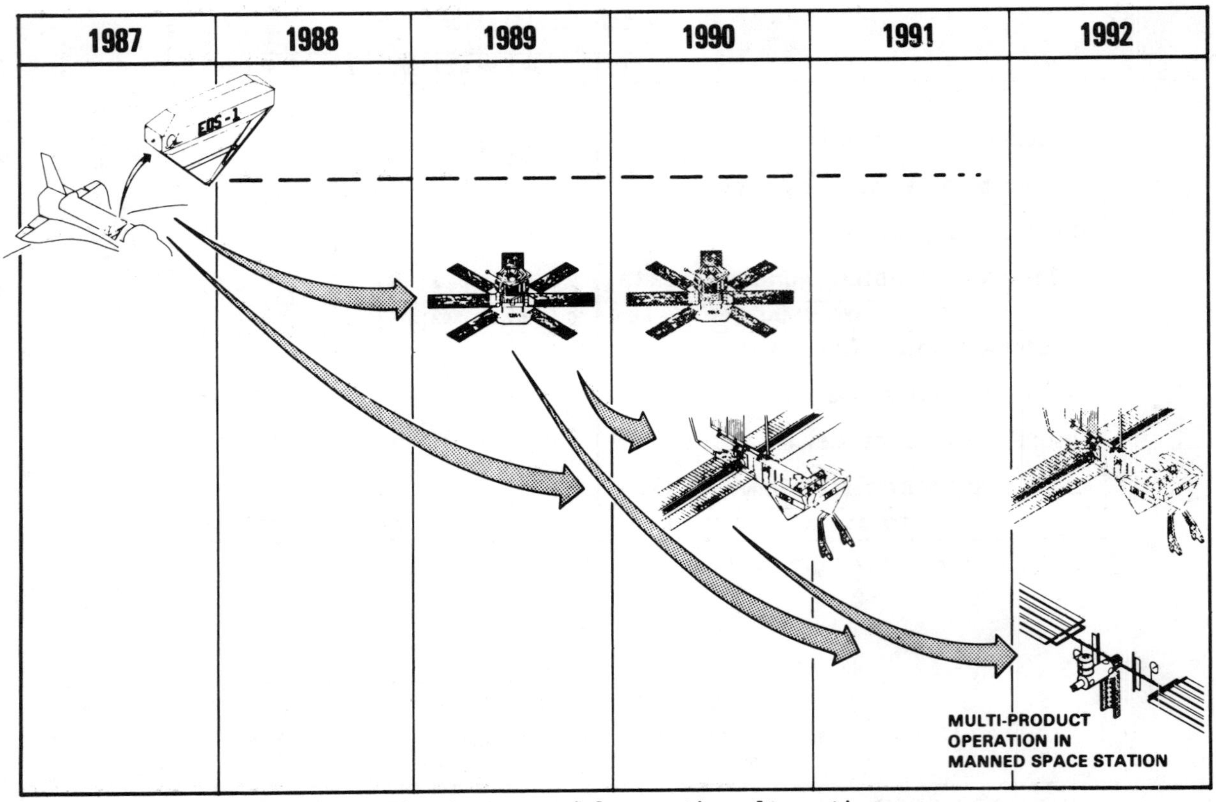

Fig. 8 Commercial operation alternatives.

enough Orbiters in the fleet to support it? For much more than a two- or three-product operation, I think not if the plants are all on individually supported spacecraft.

To summarize:

1) Potential for manufacturing new and improved products in space is real.

2) Without long-duration capability, market penetration for any one product is limited.

3) Unmanned free-flight support will allow market development for one or more products within the limitations of the Space Transportation System.

4) Manned long-duration facility can provide the basis for industry growth with improved economics.

So we have concluded that if this industry is to grow and become the type of industry that we believe it can be, it will not be only McDonnell Douglas playing in the field. If we're successful, we expect to have competition and a lot of it. We are going to need a large manned presence in space to make this type of industry the size we believe it could ultimately become.

There is no question of our strong support as NASA moves forward as rapidly as possible with the facilities that can support us and the many other industries we believe are going to materialize.

ROLE OF A SPACE STATION IN MATERIALS PROCESSING

Charles F. Yost
National Aeronautics and Space Administration
Washington, DC

Introduction

The unexploited and primarily unused property of space is weightlessness. The high vacuum that is possible in space is another primary property but little use has been made of it to date. Continuous weightlessness is a phenomena that occurs only in space-there is no earthly equivalent. As earthlings we have become accustomed to a one-g environment and everything we do is conditioned by this. With the advent of the space shuttle and eventually, of a space station we are offered the possibility of investigating the role of micro-g on materials processing and on the properties of materials that can be derived in this new environment. The purpose of this paper is to provide a brief history of the materials processing in space (MPS) program and to outline the role that the space station is expected to play in its future.

U.S. Space Policy Goals

On July 4, 1982, President Reagan announced the following space policy goals for the United States:

- o Strengthen security of U.S.
- o Maintain U.S. space leadership
- o Obtain economic and scientific benefits through the exploration of space
- o Expand U.S. private sector investment and involvement in civil space and space-related activities
- o Promote international cooperative activities in the national interest
- o Cooperate with other nations in maintaining the freedom of space for activities which enhance the security and welfare of mankind

The role and goals of MPS throughout its brief history continue to be consistent with these announced policy objectives.

Early History of MPS

While MPS is a small but unique part of the total field of materials R&D, it is a key element of NASA's space program. MPS is concerned principally with examining the role of gravity on materials processing and subsequently on materials properties. In contrast to communications systems or remote sensing devices, MPS will result in materials being made in space that cannot be made on the ground. In addition, knowledge obtained from MPS R&D in low-g will result in improvement in materials processes on earth. Space processing of materials will probably be restricted to high-value-low- weight materials such as pharmaceuticals and electronic materials because of the inherent weight limitations of the shuttle and the high cost of getting to space.

Goals of MPS

In order to understand the missions and payloads listed later in this paper and how they have been derived, it is necessary to understand that Materials Processing in Space is currently an infant science and that there is a non-existent industry based on space processing. Prior to the fourth and sixth flights (STS-4 and STS-6) of the orbiter the MPS science base was developed from about eighty-five experiments conducted in a drop tube, on a KC-135 aircraft flying a parabolic trajectory, SPAR rocket flights and limited experiments on Apollo-Soyuz and Skylab. The total accumulated low-g flight time from all of these flights is numbered in hours. Table 1 and Figure 1 provide pertinent information on these capabilities. Table 2 contains a listing of major MPS milestones.

NASA MICROGRAVITY CAPABILITIES

	SIMULATION			SPACE		
	DROP TUBE/ TOWER	AIRCRAFT	SPAR ROCKET	SKY LAB	SHUTTLE	FUTURE SPACE STATION
	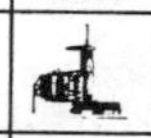					
APPROX. TIME AT "0"g	5 SEC	1 MIN	5 MIN	~ 3 MTHS	1-4 DAYS	3+MO.
APPROX. SAMPLE SIZE	~.1-1gm	10 gm	300 gm	.1-1kg	~.1-1KG	~10-100 KG
POWER AVAILABLE	BATTERIES-- POWER LIMITED			3-5kg	3 KW	25 KW+

TABLE 1

Although the total low-g time is small at this time, evidence has none-the-less been collected to indicate that materials in the fluid state have altered characteristics upon being resolidified than they do in the one-g environment of earth. The elimination of gravity- driven convection in molten materials can reduce the sometimes undesirable stirring and mixing encountered during the growth of crystals, the casting or solidification of alloys and composites, chemical reactions or the separation of biological materials. For example, the elimination of gravity-induced sedimentation and buoyancy can broaden the spectrum of alloys and composites that may be formed by permitting particles of vastly different density to remain in suspension until solidification occurs.

From this early research has emerged techniques for levitating liquids and molten materials by acoustic, electrostatic and electromagnetic methods. This permits the study of materials without the use of containers and the resultant container contamination. Moreover, containerless processing opens the possibility of

LOW-GRAVITY CAPABILITY

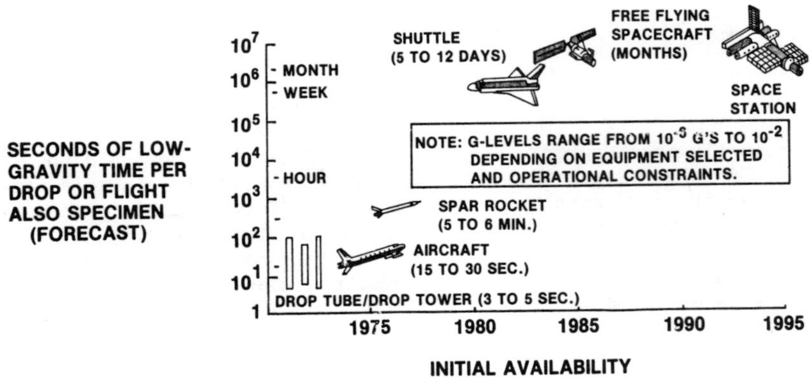

FIGURE 1

NASA HQ EN83-2700 (1) 6-30-83

CHRONOLOGY OF MAJOR MPS MILESTONES

1969— NASA INITIATES "MATERIALS SCIENCE AND MANUFACTURING IN SPACE" PROGRAM

1971— MPS EXPERIMENTS ON HEAT FLOW AND CONVECTION, ELECTROPHORETIC SEPARATION AND COMPOSITE CASTING PERFORMED ON APOLLO 14

1972— ELECTROPHORESIS EXPERIMENTS PERFORMED ON APOLLO 16 AND HEAT FLOW AND CONVECTION ON APOLLO 17

1973— 11 MPS EXPERIMENTS PERFORMED ON SKYLAB

1974— 5 MPS EXPERIMENTS ON SKYLAB (160 HOURS)

1975— 12 MPS EXPERIMENTS ON APOLLO-SOYUZ (125 HOURS)

1982— MPS EXPERIMENTS ON SHUTTLE: MONODISPERSED LATEX SPHERES, ELECTROPHORESIS SYSTEM TEST

1983— MPS EXPERIMENT ON SHUTTLE: ELECTROPHORESIS

TABLE 2

investigating a host of high temperature materials.

In handling and studying fluids at low gravity, many unique features arise that are primarily related to liquid surface behavior and role as a communication interface to the surrounding environment. Wetting phenomena became important in containered liquids because they assist in determining the internal configuration adapted by the liquid. Surface tension gradients induce flows along free surfaces. In cases where control over such convective behavior is desired, a more complete understanding of such behavior is necessary. Substantial interest has been stimulated in convection caused by surface-tension -gradient flow and has resulted in increased appreciation for the importance of such flaws in terrestrially grown crystals. MPS Processes are shown in Table 3.

Thus, the early research of MPS under the restrictive condition of short duration flight time on ground-based facilities has lead to some very interesting insights into the role of gravity.

Current Status of MPS

Current areas of research which have emerged are:

- Crystal growth and solidification
  - Solid solution IR detectors (HgCdTe, PbSnTe)
  - Vapor growth (HgI_2, alloy type)
  - Solution growth (growth environment vs. morphology)
  - Float zone (Marangoni convection, radial segregation, interfacial stability)
- Metallurgical materials and processes
  - Immiscible alloys
  - Magnetic composites
  - Metal foams
  - Solidifcation at extreme undercooling
- Glasses
  - Glass fining
  - Optical glasses with unique properties
  - Metal glasses
- Chemical processes
  - Monodispersed latexes (polystyrene microspheres)
  - Colloidal interactions
  - High temperature properties of reactive materials
- Separation sciences
  - High-volume, high-resolution electrophoresis cell separation
  - Protein purification by continuous flow isoelectric focussing
- Fluid studies
  - Non-buoyancy driven convections
  - Wetting and spreading studies

MPS PROCESSES

	FURNACE PROCESSING	CONTAINERLESS PROCESSING	FLUIDS/CHEMICAL PROCESSING	BIOLOGICAL PROCESSING
RESEARCH PROCESSES	• ISOTHERMAL GRADIENT FREEZE • FLOAT ZONE • DIRECTIONAL SOLIDIFICATION • VAPOR CRYSTAL GROWTH • ELECTROEPITAXY ETC.	• ACOUSTIC • ELECTROMAGNETIC • ELECTROSTATIC		• ELECTROPHORESIS • ISOELECTRIC FOCUSING • ISOTACHOPHORESIS • ETC
TYPICAL EXPERIMENTS	• CRYSTAL GROWTH FROM MELT • CRYSTAL GROWTH FROM SOLUTION • CRYSTAL GROWTH FROM VAPOR • SEMICONDUCTOR CRYSTALS • IMMISCIBLE ALLOYS • IR DETECTORS • EUTECTICS–MONOTECTICS • COMPOSITE MATERIALS	• ULTRA HIGH TEMP. MATERIALS • GLASS PROCESSING • IMMISCIBLE GLASSES • FUSION TARGETS	• LARGE SCALE CHEMICAL PROCESSING–POLYMERIC PROCESSING • LOW TEMPERATURE CRYSTAL GROWTH • MEASUREMENT OF THERMODYNAMIC PHENOMENA • HIGH VOLUME BIOLOGICAL PURIFICATION & SEPARATION • MEASUREMENT OF TRANSPORT PHENOMENA • CHEM. DEPOSITION • CATALYSIS • FLUID DYNAMIC STUDIES	• SEPARATION OF PHARMACEUTICAL MATERIALS • BLOOD PROPERTIES

NASA HQ EN83-2701 (1)
6-30-83

TABLE 3

- Role of convection in processes (electrokinetic separation, electroplating, corrosion, etc.)

In addition to the above research on materials it has been discovered that low gravity affects combustion and other fluid flow processes. NASA studies as well as work by recognized experts in the private sector have provided strong advocacy for combustion research in space. The importance of this work is clear when one considers its potential application to coal combustion in fluidized beds, droplet combustion in internal combustion engines, gas and dust explosions, oil and gas burner pollution, space safety and fire spreading on solid and liquid surfaces. The underlying theme behind this research is that for particular processes there is a set of important theoretical questions that are currently unanswered, can be wholly or partially attributable to confusing gravitational influences and can be potentially answered by microgravity experimentation.

The opportunity now available for longer duration flights on the orbiter to further develop and expand the fundamental science of materials processing in the space environment is the first goal of MPS. In addition, expanded use of the ground based facilities for short-duration low-g experimentation is anticipated. A space station will further enhance the ability for long-duration experimentation particularly in materials growth processes where currently existing capabilities do not provide a long enough time for the growth process to proceed. Opportunities that may present themselves for research as well as for commercial exploitation are shown in Table 4.

Commercial Goal

The second goal of MPS is a commercial one and is directly and intimately connected to the first goal. The commercial goal of NASA's MPS program is the encouragement of private sector establishment and operation of self-supporting space processing industries. The strategies for achieving this are to:

- o Create awareness of possible commercial opportunities to industry through publications, seminars, visits to NASA Centers

- o Encourage private sector participation in space processing by:
  - Utilizing Joint Endeavor Agreements,(1) Technical Exchange Agreements and Industrial Guest Investigator Agreements
  - Identifying barriers and disincentives to industrial exploitation of space and taking actions necessary to overcome them

(1) See Appendix Number 1.

OPPORTUNITIES UNDER STUDY

OPPORTUNITIES	TECHNICAL BARRIERS	PAYOFFS IF SUCCESSFUL	U.S. MAJOR SPACE GOALS ADDRESSED
• METALS	HETEROGENEOUS NUCLEATION/ GROWTH/ DECOMPOSITION	NEW ALLOYS & CONTROLLED MICROSTRUCTURES & PROPERTIES	• LEADERSHIP • ECONOMICS • SECURITY
• NON-METALS – SEMI CONDUCTORS	DENSITY/GRADIENTS PRODUCE IMPERFECTIONS & REDUCE SENSITIVITY & PERFORMANCE	FASTER/HIGHER POWER SWITCHES/CIRCUITS	• ECONOMICS • LEADERSHIP • SECURITY
– FIBER OPTIC GLASS	"	"	
– DETECTORS	"	INCREASED SENSITIVITY	
• BIOLOGICALS	THROUGH PUT & PRECISION OF SEPARATION	USEFUL QUANTITIES OF RARE/UNAVAILABLE PROTEINS, CELLS, ENZYMES, ETC	• LEADERSHIP • ECONOMICS • INDUSTRY
• FLUID PROCESS & COMBUSTION	MULTIPLE VARIABLES HARD TO DIFFERENTIATE	ACCURATE MODELS & GUIDANCE FOR EARTH-BASED PROCESSES	• LEADERSHIP • INTERNATIONAL COOPERATION

NASA HQ EN83-2702(1)
6-30-83

TABLE 4

The rationale for materials processing in space commercial benefits is:

- o Advances in the science and technology of materials processing would benefit terrestrial processing methods
- o The demonstration of products with unique and valuable properties produced in space experiments would provide a powerful spur to the develpoment of Earth-based, alternative production methods
- o The production of unique materials products in space ultimately could lead to a space-based materials processing industry

Figure 2 approximately illustrates how NASA plans to achieve the two MPS goals. By involving researchers from universities, other government agencies and from industry in experimental activities on the orbiter as well as on the ground based facilities listed earlier the MPS research base should expand and opportunities for commercialization should develop. The 1980's will be a time where the institutional relationship as initiated by the Joint Endeavor Arrangements of Appendix 1 will continue to evolve.

Role of a Space Station

The NASA attitude that led to the Joint Endeavor Arrangements and the research base now being developed in MPS will largely continue to be the thrust of NASA throughout the eighties and on into the space station era and there will be evolutionary changes in the total program. It is important to recognize that just as there needs to be science and applications experimentation in space to develop new products and processes, there must also be institutional experimentation to create innovative arrangements for joint industry-government participation in ventures in space. Thus, the existence of an MPS program and of joint endeavor arrangements are significant steps in the eventual industrialization of space and consequently, the forerunners of things to come.

The period between the present and the advent of a space station will be somewhat analogous to the period when the NACA was assisting a fledgling aircraft industry by providing wind tunnels and technical advice without direct financial assistance. It is envisioned that NASA will have to make available laboratory facilities, drawn selectively from the kinds of processing techniques shown in Table 3, to academic and industrial researchers to further develop the science and technology base of MPS. The basis for use of these facilities (namely, who pays for what and how) has not been determined. However, the motivation is to involve industry during this decade in order to create an awareness of the opportunities of space processing.

On the Space Station it is anticipated that there would be an MPS Processing Laboratory. This would be a large facility performing MPS for science and applications purposes as well as for commercial processing. The key requirements for this laboratory are:

- o Any orbit with low cost, frequent access
- o Micro-g environment (10^{-3} to 10^{-5} g)
- o Short-duration missions (days) for R&D
- o Long-duration (months, years) for commercial production
- o High power and thermal (15 to 25 kw)
- o Back-up power and thermal
- o Manned access
- o Large pressurized volume
- o Isolated, controlled atmosphere
- o Multipurpose laboratory equipment

THE APPROACH

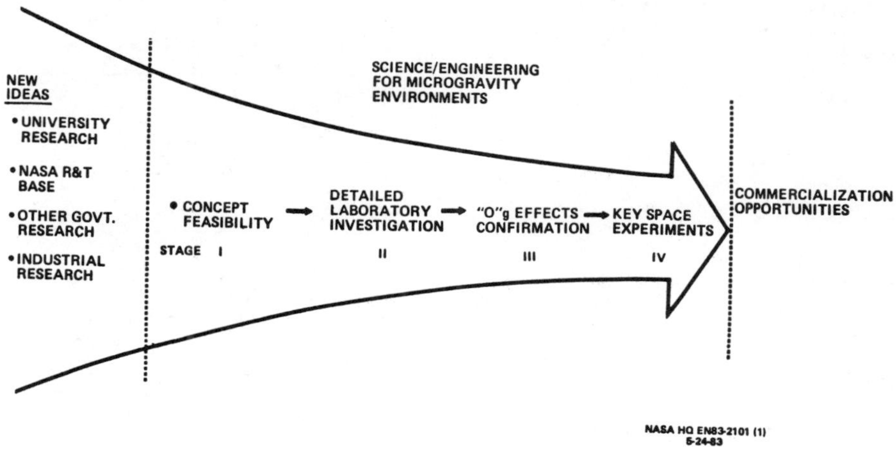

FIGURE 2

- o Low data rate; intermittent video
- o Secure data link materials resupply and product return
- o Protection of proprietary data and products

It is anticipated that the following categories of R&D processing activities will be undertaken:

- o Containerless processing
- o Biological processing
- o Various crystal growth processes

Some commercial missions will require man-tended platforms or free flyers that can be serviced from the space station. This will be necessary for a variety of reasons such as larger power requirements than is available on a space station, undesirable g-jitter, toxicity, etc. Examples of the types of missions that will probably be on a space station are shown in Table 5.

Example Missions	Power kW	Duty Cycle %	Crew Size	Hours/ Day	Pressurized Volume
1. Electrophoresis Pharmaceuticals	15	80	1	2	10-ft Module
2. Materials Research Facility	25	20	2	2	10-ft Module
3. Silicon Ribbon Mfg	15	80	1	2	10-ft Module
4. Melt/Reform Facility	12	20	1	2	10-ft Module
5. Separations Lab	15	20	1	8	10-ft Module
6. Oriented, Composite Mixtures Lab	1.5	20	1	1	10-ft Module
7. Bioprocessing	1.5	100	1	1	10-ft Module
8. Gallium Arsenide Facility	5	20	2	4	2 Racks

TABLE 5

MPS payloads in contrast to remote sensing or communications are power and energy intensive as can be seen from Table 5. Power and energy for MPS research payloads are considerably lower than that required for commercial MPS.

Conclusion

In conclusion it is our belief that the ground work for a solid scientific foundation for MPS and consequently for space industrialization is being constructed and that the institutional and business arrangements originating and centering in the Joint Endeavor Agreement approach will be an ever increasingly attractive feature of the private sector involvement with NASA in the pursuit of joint research and commercial ventures. More specifically, the space stations role in MPS includes opportunities for:

- o Long term "hands on" research which may be viewed as an extension of Skylab capabilities
- o Higher temperature and scaled-up processing concepts via increased power availability
- o Demonstration of NASA's commitment to commercialization feasibility

In summary it is clear that:

- o Certain benefits from current MPS programs will be the improvement of terrestrial materials and material processes through the acquisition of greater knowledge
- o Development in the next 10 years will demonstrate the useful applications for space and Earth
- o Limited by the STS facility, only a few MPS applications have the promise of economic return before 1990
- o Significant commercialization of MPS depends on: (1) success of the MPS development phases in the 80's, and (2) the development of a space platform, power and transport (at <$100/kg) systems to accommodate the phases beyond.

Finally, it should be pointed out that the experiments and processes listed in this paper are based on limited experience gained for the most part through use of ground based facilities on earth. As research is undertaken in the micro-g conditions of a space station who among us earthlings can predict what new processes and techniques will eventually emerge?

Appendix No. 1

Joint Endeavor Arrangements

In 1978-1979 NASA promulgated two policies aimed at creating incentives to encourage the private sector to become involved with its venture capital in space-based ventures. These are the "Early Usage of Space Policy" and "Guidelines for Joint Endeavor Agreements in MPS".

In these joint activities, NASA and interested, qualified commercial organizations enter into "constructive partnerships" as equals who have sufficient motivation toward common objectives to make independent commitments and to share in the risk and benefits. Activities are selected across the spectrum of materials processes to demonstrate that the low-g environment is a valuable tool for isolating and characterizing gravitational effects on ground-based material processes, or for actually producing unique materials in space for commercial application. Since market incentives are presently inadequate to bring about technological innovation based on low-g technology, under its joint activities program, NASA can provide certain tangible incentives such as: (1) providing flight time on the Space Transportation System (Space Shuttle) on appropriate terms and conditions, (2) providing technical advice, consultation, data; and use of facilities and equipment, (3) entering into joint research and development programs where each party funds its own participation and 4) a form of process exclusivity and rights to intellectual property. Joint activities may range from exchanges of technical information and collaboration on low-g ground-based and flight experiments to joint projects (Joint Endeavors) to develop a marketable product.

Terms and Conditions

As stated in the "Guidelines" referenced above, the terms and conditions of joint efforts are negotiable; this includes such factors as the firm's right to proprietary data and/or patent ownership, provisions for a form of exclusivity in special cases, and recoupment of NASA's contribution under appropriate circumstances. Until such time as normal market incentives provide a competitive and self-sustaining condition in the private sector for materials processing in low-g technology, the incentives and negotiability of terms and conditions will be used as a stimulus to encourage the more technologically advanced, entrepreneurally inclined firms to pursue applications of low-g technology.

Since NASA wishes to encourage the private sector in early commercialization efforts and since the regulations for Government procurements are not generally compatible with an entrepreneurial endeavor, NASA does not anticipate sponsorship of commercialization work through procurement type contracts. Rather, NASA is pioneering a concept of negotiating an agreement for joint investigations and joint projects with industry on a case-by-case basis. Thus, the agreement can be tailored to the specific needs of the parties for the specific investigation or project.

NASA has developed three basic levels of working relationships with private organizations.

Technical Exchange Agreement

For companies interested in the application of low-g technology, but not ready to commit to a specific space flight experiment or venture, a Technical Exchange Agreement (TEA) has been developed. Under a TEA, NASA and a company agree to exchange technical information and to cooperate in ongoing ground-based research and analyses. In this way, a firm can become familiar with microgravity technology and its applicability to company needs at minimal expense. Under the TEA, the private company funds its own participation, and derives direct access to and results from NASA facilities and research, with NASA gaining the support and expertise of the industrial research capability.

Industrial Guest Investigator Agreement

Another joint activity is the Industrial Guest Investigator (IGI). The IGI arrangement is applicable to situations where NASA and an industrial firm share sufficient scientific interest such that the company appoints one of its scientists to collaborate, again at company expense, with a NASA-sponsored Principal Investigator on a space flight experiment. Once the parties agree to the contribution to be made to the objectives of the experiment, the IGI becomes a member of the investigation team, thus adding industrial expertise and insight to the experiment.

Joint Endeavor Agreement

Joint Endeavors provide a mechanism for NASA to share in the cost and risk of early space projects. The first Joint Endeavor Agreement between NASA and McDonnell Douglas Astronautics Company was signed in january 1980, and illustrates the key features of this third working relationship. McDonnell Douglas will use the Space Shuttle to develop and demonstrate the technology of continuous flow electrophoresis under low-gravity conditions and to ascertain the applicability of that technology to the production of pharmaceutical products. The agreement requires a substantial investment by McDonnell Douglas and its associates in each of three phases: (1) feasibility studies and planning, (2) flight experimentation and technology development, and (3) applications demonstrations. In return for McDonnell Douglas' promise to make results of the work available to the U.S. public on reasonable terms and conditions, NASA agrees to refrain from entering into similar joint endeavors or international cooperative agreements directly related to the development of commercial devices and processes which would compete with those resulting from the McDonnell Douglas endeavor. However, NASA is not precluded from selling flight time on the Shuttle to any other organization wanting to conduct the same or similar experiments. Significantly, NASA will not acquire rights in inventions made by McDonnell Douglas or its associates in the course of the joint endeavor unless McDonnell Douglas fails to exploit the inventions or terminates the agreement, or unless the NASA Administrator determines that an emergency exists.

SATELLITE/PLATFORM SERVICE AND MAINTENANCE

H. T. Fisher* and K. J. Forsberg**

Lockheed Missiles & Space Co., Inc.
Sunnyvale, California

Abstract

The paper presents the servicing role projected for the Space Station in the 1990 to 2000 year era. Servicing is discussed relative to station based and remote operations as well as orbital mechanics involved in satellite retrieval (for remote servicing) and/or for retrieval and return to the Station for subsequent servicing. Augmenting the Station with a Teleoperator Retrieval System and ultimately an unmanned Orbital Transfer Vehicle is pivotal to capturing the spectra of satellites which can be serviced on-orbit. Although the station offers extensive potential for servicing there are certain instances wherein the Shuttle is a more appropriate service base. Servicing technology is presently available and will be a proven approach which can be expeditiously and readily transferred to the station in an orderly and timely manner upon initial station implementation/activation.

Introduction

The operational support for satellite servicing operations (including equipment changeout, consumable replenishment, and repair) represent one of the major benefits to be provided by the Space Station. Numerous satellites have been or are being designed, and some modified, to permit and facilitate on-orbit servicing from the Shuttle. Important examples include the Multi-Modular Satellite series (e.g., Solar Maximum Mission Satellite), Space Telescope, and ultimately the Advanced X-Ray Astronomical Facility.

It is strongly believed within industry and certainly among many key governmental personnel that the Space Station will offer a more cost-effective base for routine scheduled servicing and maintenance of a large complement of satellites. Major repairs, resupply, reconfiguration, and equipment update may be done at the Station as a more energy-efficient and cost-effective alternative to the present concept of satellite retrieval by the Shuttle and return to earth for major refurbishment.

This paper will focus on two primary subject areas: (1) servicing functions and (2) orbital mechanics associated with servicing. It is patently clear that both the potential and limitations of on-orbit Station-based servicing must be thoroughly understood to facilitate the design and development of the Station and its associated servicing facilities/elements. Thus, the early definition and planning for servicing is critical to the subsequent Station definition and development efforts.

Servicing Functions/Regimes

The operations associated with Station servicing are composed of two primary categories: onboard and remote. It is recognized that servicing also incorporates other Station support activities such as:

- Rendezvous and Docking
- Logistics
- Stage Assembly
- RMS and Crane Operations
- Internal Crew Console Operations

Figure 1 presents an overall synthesized list of crew functions associated with Station operations as indicated above. Within this complement of functions are those associated with servicing either in a direct or indirect relationship.

Servicing categories and the associated functional requirements are presented in Fig. 2. As indicated, three basic ranges of servicing are included; however, only two are nominally planned for EVA support. The servicing categories lend themselves to either Orbiter or Station servicing; however, the more complex repair, reconfiguration, and general service/enhancement functions would be better performed by the Station.

Figure 3 presents the composite spacecraft (satellites) or free-flyer classes used as a basis from which to examine the range of servicing functions. The illustration indicates the use of both the Orbiter (initial servicing platform from the early 1980s to the 1990s) and the Station in a simultaneous role. This dual role will be discussed further under the heading of orbital mechanics.

It is expected that two basic servicing localities will exist. Obviously, the Station provides a self-contained base for the full range of servicing functions envisioned. The second servicing locale is remote. Remote servicing can be conducted in two ways: (1) EVA at the servicing location away from the Orbiter or Station and (2) remote servicing conducted with a "teleoperational presence" away from either the Orbiter or Station. The latter approach necessitates some type of teleoperator with a propulsive stage as well as some type of orbital transfer vehicle. The necessity for both spacecraft to

---

*Supervisor, Crew Systems. Member AIAA.

**Program Manager, Space Station. Member AIAA.

support servicing is obvious when examining the mission model and in determining that the majority of satellites are not in an easy to access low earth orbit (LEO). Thus, an unmanned servicer will be required as an integral element of the station. Whether its operations are limited to simple remote tasks (e.g., refueling of or recovery of satellites) or sophisticated equipment changeout is yet to be determined.

A simplified allocation of crew tasks associated with servicing has been provided in Fig. 4 to include both IVA and EVA locations and the possibility for EVA away from the Station wherein servicing of a Station colocated free-flyer is planned. An unmanned servicer (previously indicated) is also included. Correspondingly, the functional interfaces of the Station with the spacecraft (satellite) to be serviced are portrayed in Fig. 5. It is important to note that the servicing interface is not limited to the Orbiter or Station but rather includes several support functions including several not shown in the Figure (e.g., launch, logistics, training and simulation, ground checkout/test).

As the servicing of satellites evolves during the next 5 to 6 years using the Orbiter as the operations base, it can be expected that the crew will perform an ever-increasing number and variety of tasks. As these tasks increase, so too will the complexity level. Figure 6 presents a simplified list of crew glove related tasks both with and without glove attached augmentation. Accordingly, it is anticipated that the Station will provide an opportunity for greater numbers of actual repair tasks to further extend the range of servicing capabilities.

Pivotal to future servicing is the ability for the crewperson to don a suit (extravehicular mobility unit (EMU)) and exit the Station as rapidly as required. This implies that a no prebreathing requirement exists and that the EMU can be donned and the crewperson can egress on demand. Thus, compatibility between the Station atmosphere and the EMU becomes an important consideration in dealing with the servicing needs. This need is further emphasized when taking into account the necessity for quick reaction for contingency operations associated with the Station, or presently to satisfy the DoD needs relative to Orbiter payload deployment and unplanned EVA support.

Servicing Approaches/Evolution

There are a myriad of servicing approaches applicable and practical for servicing from the Station. Certainly, the Station design approach(es) will take the servicing needs into account very early in the systems requirements definition phase and during the early conceptual design layout. A very simplified example of servicing approaches is shown in Fig. 7. An interesting variation is illustrated wherein the internally positioned crewperson is shown servicing a small experiment or alternatively a satellite.

Other examples in Fig. 7 simply indicate the obvious use both of a teleoperator and an unmanned orbital transfer vehicle (OTV) very early in the Station build-up. The OTV is generally expected to encompass a range of operations exceeding that of the teleoperator (approximately 1000 to 1500 miles) including distances out to high geosynchronous altitudes. The OTV would be outfitted with as a minimum fuel/gaseous transfer capability and a simple satellite grasping technique. The OTV could evolve to a much more sophisticated servicer including a more complete teleoperator front end with substantially enhanced servicing capability. That technology can evolve from the nearer term teleoperator program planned for Orbiter use in the late 1980s.

Both of the aforementioned servicing approaches include the use of the crew as a remote station operator, thus controlling the teleoperator or OTV from the Orbiter and/or Station. This type of remote control is substantially different from that associated with controlling the Orbiter Remote Manipulator System (RMS), for example, where direct control, vision, feedback, etc., is provided the crewperson from the aft flight deck. In the case of the teleoperator or OTV, the crewperson would be operating a console and directing activities in a remote loop configuration including taking into account time delays due to transmission distances.

Servicing approaches between the Orbiter and Station are expected to differ substantially, particularly as the Station evolves and matures. Figure 8 presents a very abbreviated comparison between the two. One of the most significant differences between the two is the continuous on-orbit availability of the Station. This feature permits the servicing functions to be more extensive and allows greater levels of interaction. Thus, the one-week limitation in orbit stay time currently available (up to 3 to 4 weeks maximum potential) in the Orbiter somewhat inhibits servicing flexibility. This is particularly true for servicing on demand wherein the Orbiter may not be on-orbit or available nor configured to conduct the particular servicing needed for that satellite. Another problem arises when considering round-trip sorties undertaken by a teleoperator or OTV. In this case, the time required for launch, retrieval, return, servicing, and relaunch may exceed the allowable on-orbit dwell-time of the Orbiter. Thus, with the Station a totally different servicing approach will be available.

Figures 9 and 10 have been included to portray EVA technology evolution from the Orbiter era to the Station. An example of the EVA servicing hardware technology base is presented in Fig. 11. These equipment items are, for the most part, available with little or no new development required. This is only a partial list; however, it illustrates the fact that considerable equipment exists which can logically transition to the Station era.

Previous maintenance approaches (see Figs. 7 and 8) illustrate a straightforward and somewhat austere set of concepts. A significantly more advanced and ambitious version of a Station attached servicing facility is illustrated in Fig. 12. This facility is designed to permit total on-orbit servicing of an entire spacecraft (satellite) as well as stage assembly, a variety of consumables replenishment, etc. This type of facility and associated servicing approach is envisioned for the mid-to-late 1990s and would represent a mature Station configuration.

Orbit Mechanics

Pivotal to the successful implementation of servicing is the need to gain access to a variety of satellites both in and out of plane with respect to the Station. As indicated earlier, consideration of augmentation methods such as teleoperators and OTVs is necessary to reach the altitude and inclination of a considerable number of the satellites stationed at varying orbits including geosynchronous and high earth orbit. Inherent within this need for servicing at these altitudes is the basic consideration of orbital mechanics which dictates basic transfer requirements, delta-V, and payload size. The following paragraphs will briefly summarize the major factors associated with orbit mechanics and remote servicing.

Basic considerations for servicing at the Station have been briefly synopsized. The next obvious servicing envelope includes the arena of remote operations. These remote operations include servicing and support for four basic categories of free-flying satellites or spacecraft.

The first category is associated with those satellites with the same inclination, orbit plane, and phasing as the Station and within a few miles of the Station altitude. All are readily accessible and provide opportunity for line-of-sight communications and control. This first category includes the concept of free-flying clusters that may contain production facilities for material processing in space or free-flying platforms for various scientific experiments.

The second category includes the support of satellites in nearby inclinations (within 15 degrees from the Station inclination), with the restriction that orbit transfer be made at nodal coincidence. Bounds are provided by the OTV capabilities and payload size. Figure 13 presents the satellite access envelope for servicing at any given instant in time from the Station. Important to note is the rather limited inclination with respect to the Station and the ever increasing limitation as higher altitudes are encountered.

The third category involves orbit transfer from the Station to any satellite in low earth orbit (LEO). The delta V required for a round-trip maneuver will reach about 30,000 ft/sec, which could require a prohibitively large quantity of propellant unless advanced propulsion systems, e.g., electric, are used. Thus, for early Station operations, satellites in this category will likely be serviced by one-way missions only. Also Station-based missions in this category should be restricted to critical activities that warrant the energy expenditure, such as the Space Shuttle rescue mission. Note that even satellites in the same inclination and nearly the same altitude as the space station fall into this category during time periods between nodal coincidences.

Support of geosynchronous earth orbit (GEO) satellites (the fourth category) is also possible from a space station. In fact, the energy to reach GEO is less than that required to support LEO satellites at non-optimum transfer times, since a large plane change from one LEO to another requires a 3 burn maneuver with apogee over twice GEO altitude.

While all of the above categories of space based operations are important to establishing requirements for and configuration of a Space Station, misconceptions seem to have developed about LEO satellite servicing from a Station. Often, it is not recognized that the interval between nodal coincidences for a satellite and a Station in a 28.5-deg orbit will range from 50 days for a high altitude satellite to several years for a satellite at an altitude near that of the Station, thereby significantly limiting opportunities for minimum energy transfers from the Station. Figure 14 presents orbital mechanics considerations for a Station located at 220 nautical miles circular at 60-deg and for a satellite at 1,400 nautical miles circular at 60-deg. Important to note is the time between nodal coincidences when minimum energy transfer can be made; also the time for low energy transfer is restricted to a few days. Conversely, substantial delta V is required when servicing "on demand" may be needed during a period of nearly 3 months between nodal coincidences. Use of aerobraking on the OTV reduces peak energy requirements, but does not change the fundamental problem.

Conclusions

A number of conclusions can be drawn from the extensive ongoing efforts relative to Orbiter-based servicing and the viable thrust to assure the transfer of this technology and operations capability to the Station.

It appears clear that the basis for servicing operations from the Orbiter has been established and that missions in the 1984 time period will verify this capability. Also clear is the very real fact that design of satellites for on-orbit servicing and maintenance (including the associated operations) is generally not considered early enough in the program implementation cycle. The primary concern in the design for servicing/ maintenance is standardization such as:

- Operations
- Procedures - flight and crew
- Airborn support equipment
- Orbital replacement changeout features within/on the satellite and on the unit to be changed-out.

Both operations and hardware technology exist to begin servicing functions now which can lead to important recommendations for design of future satellites for servicing. Nonetheless, little effort has been expended to date on a major effort to define an EVA operations technology evolution and development program for potential transition to the Station.

Obviously, it is not too early to begin developing an orbital operations servicing and maintenance concept for the Station. However, no integrated operational servicing and reliability and maintainability approach (and associated documentation) exists to initiate early Space Station planning and analyses. It also appears appropriate to examine DoD servicing and maintenance operations technology to identify "commonalities" to assist in the overall technology evolution path, and to identify gaps, critical issues, and potential risk areas. This assessment would be highly useful to NASA and aid substantially in the subsequent Orbiter to Station servicing evolution scheme(s).

As indicated earlier, understanding of orbital mechanics and the associated constraints is essential for proper mission planning, and subsequent transfer vehicle development. The Station, in comparison to the Shuttle, appears better for supporting scheduled servicing, maintenance and resupply of:

- Satellites in Station tracking orbits
- Satellites in nearby inclinations at nodal coincidence; to service the majority of satellites requires Stations at 28.5-deg, 60-deg, and 90-deg
- GEO satellites (Station location is not a strong driver)

The Space Shuttle is probably better than the Station for:

- Servicing satellites at non-optimum times
- Emergency resupply
- Highly accessible LEO satellite capture

The Station offers unique capability independent of Station or satellite location for:

- Reconstitution via space-based launch
- Assembly/construction of large space structures
- Associated propellant resupply or scavenging
- Shuttle crew rescue

The final conclusion is becoming patently clear - the Space Station provides a very powerful capability for space-based operations, and one of the most important of these is satellite servicing.

Recommendations

Many of the aforementioned conclusions relate directly into obvious recommendations.

However, there are a number of near-term recommendations which should be stated:

- Require (as appropriate) satellites to be designed for servicing commencing with new program starts
- Establish a responsible NASA focal point for satellite servicing; responsibility includes management, budget, and "clout"
- Initiate an evolutionary EVA servicing technology development program - e.g., equipment as well as procedures
- Initiate planning for development of HEO/GEO servicing capability
- Undertake the development of reliability, maintainability, and logistics requirements for advanced Shuttle payloads including evolution to Station guidelines
- Develop an integrated Station servicing development and interface plan and guideline document
- Prepare a comprehensive and illustrated guideline on orbital mechanics as related to servicing from or at the Station.

These and other previously mentioned areas are strong candidates for study and planning for the next five years. Certainly these efforts necessitate close coordination to be properly in tune with the forthcoming Station studies and subsequent Phase B efforts.

Bibliography

1. SPACE STATION NEEDS, ATTRIBUTES AND ARCHITECTURAL OPTIONS, NAS3684, 22 April 1983, Lockheed, LMSC-D889718

2. SATELLITE SERVICES SYSTEM ANALYSIS STUDY, NAS9-16121, Various Dates (1980 to Present), Lockheed, LMSC-D798839

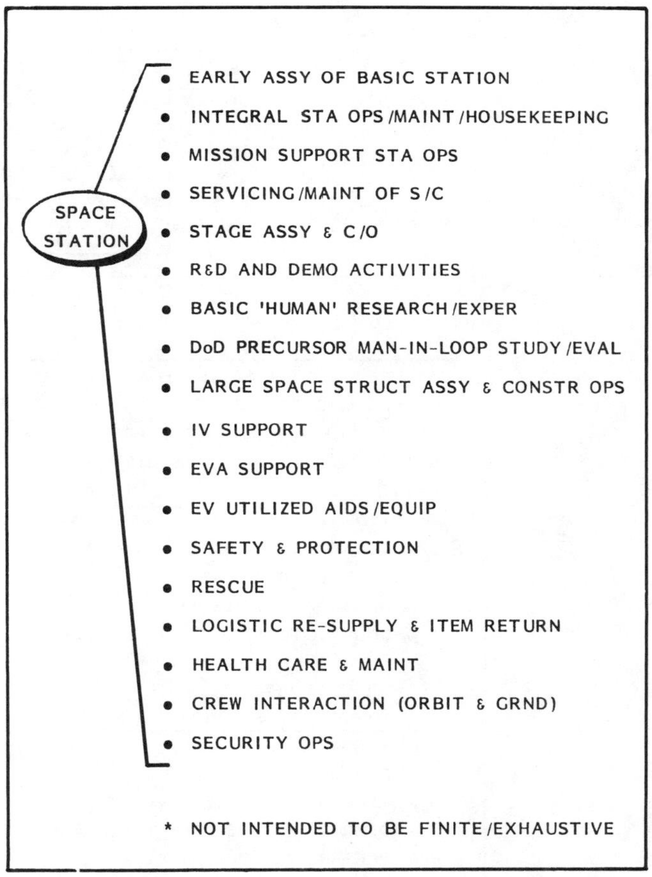

Fig. 1 Selected Crew Operations and Needs

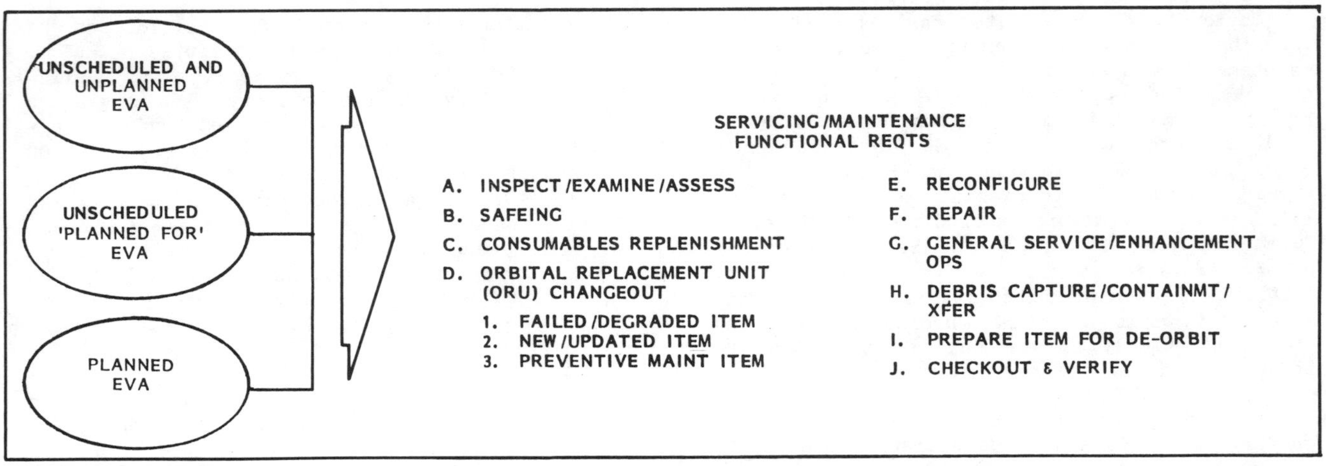

Fig. 2 EVA and Servicing Categories

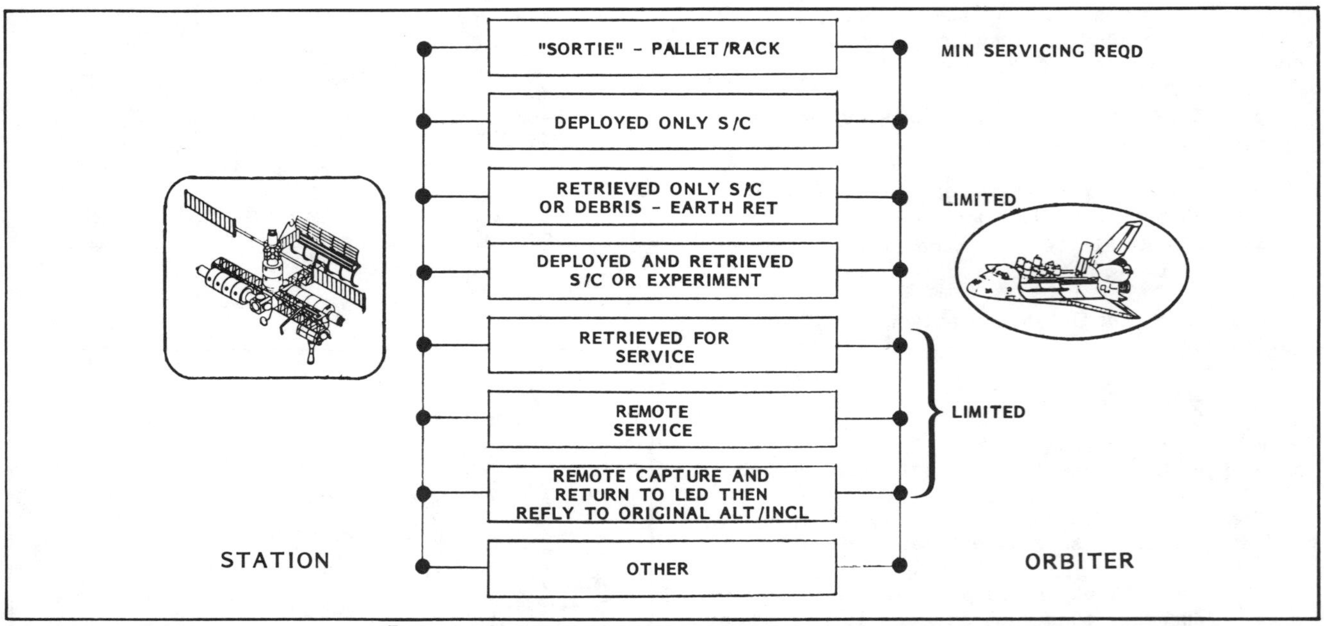

Fig. 3 Classes of Spacecraft or Free Flyers

	STATION IVA	STATION EVA	MAN SER VEH IVA	MAN SER VEH EVA	UNMANNED SERVICER
1. ASSURE ILS & SERVICE SUPPORT READINESS	•		•		
2. ASSURE SERVICING ELEMENT PROPERLY CONFIGURED	•	•	•	•	
3. ASSURE PREPARATION FOR PLUME/CONTAM AVOIDANCE	•		•		
4. SET-UP/ACTIVATE & CLOSE-OUT/DEACTIVATE ASE/FAC	•	•	•	•	
5. CAPTURE/DOCK/HAND-OFF (& REVERSE) S/C FOR SERV	•		•		•
6. OVERRIDE OF 'AUTOMATED' FUNCTION		•		•	L
7. SPACECRAFT (S/C) INSPECTION/EXAMINATION	•	•	•	•	•
8. CONDUCT SERVICING OPS:					
• CONSUMABLES RESUPPLY	•	•	•	•	•
• ORBITAL REPLACEMT UNIT (ORU) CHANGEOUT					
– FAILED/DEGRADED ITEM		•		•	L
– NEW/UPDATED ITEM		•		•	L
– PREVENTIVE REPLACEMT ITEM		•		•	L
• RECONFIGURE		•		•	L
• REPAIR		L		L	
• GENERAL SERVICE/ENHANCEMT OPS		•		•	L
• DEBRIS CAPTURE/XFER & CONTAINMENT		•		•	•
• PREPARE FOR DE-ORBIT		•		•	•
• CHECKOUT & VERIFY	•		•		
9. CONTROL OF REMOTE SERVICER	•				•

L = LIMITED

Fig. 4 Typical On-Orbit Servicing and Maintenance Operations

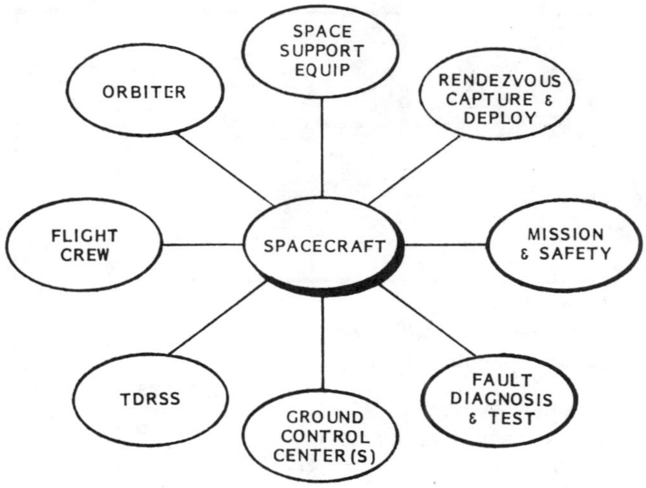

Fig. 5 Typical Servicing Interfaces to Spacecraft and Similarities to Station

EXTENSIVE GLOVE-TOOL/AID I/F
(REPAIR-MAINTENANCE)

• SPLICE	• FASTEN
• SEAL	• CUT
• STRAIGHTEN	• BRUSH
• TRIM/SMOOTH	• SOLDER
• DRILL HOLE	• BEND
• 'SAW'	• SHAPE
• WELD	• SCRAPE
• STRIKE/PUNCH	• FUSE BOND
• REAM	• RIVET
• ETC	

SIMPLIFIED GLOVE-TOOL OR UNIT I/F
(SERVICE)

• RETRIEVE	• STABILIZE
• ADHERE	• DECONTAMINATE
• CALIBRATE	• CLEAN SURFACE
• INITIATE SELF-CHK	• VENT/PURGE
• ISOLATE/DIVERT	• START/SHUT-DWN
• TROUBLE SHOOT	• PERFORM ALIGNMT
• LUBRICATE	• PLACE LABELS
• GAGE/MEASURE	• CLEAN-UP AREA
• CLAMP	• OPERATE D&C PANEL
• HANDLE CABLE	• ETC

Fig. 6 New EVA Tasks Envisioned for Advanced Shuttle Support

Fig. 7 Servicing Approaches

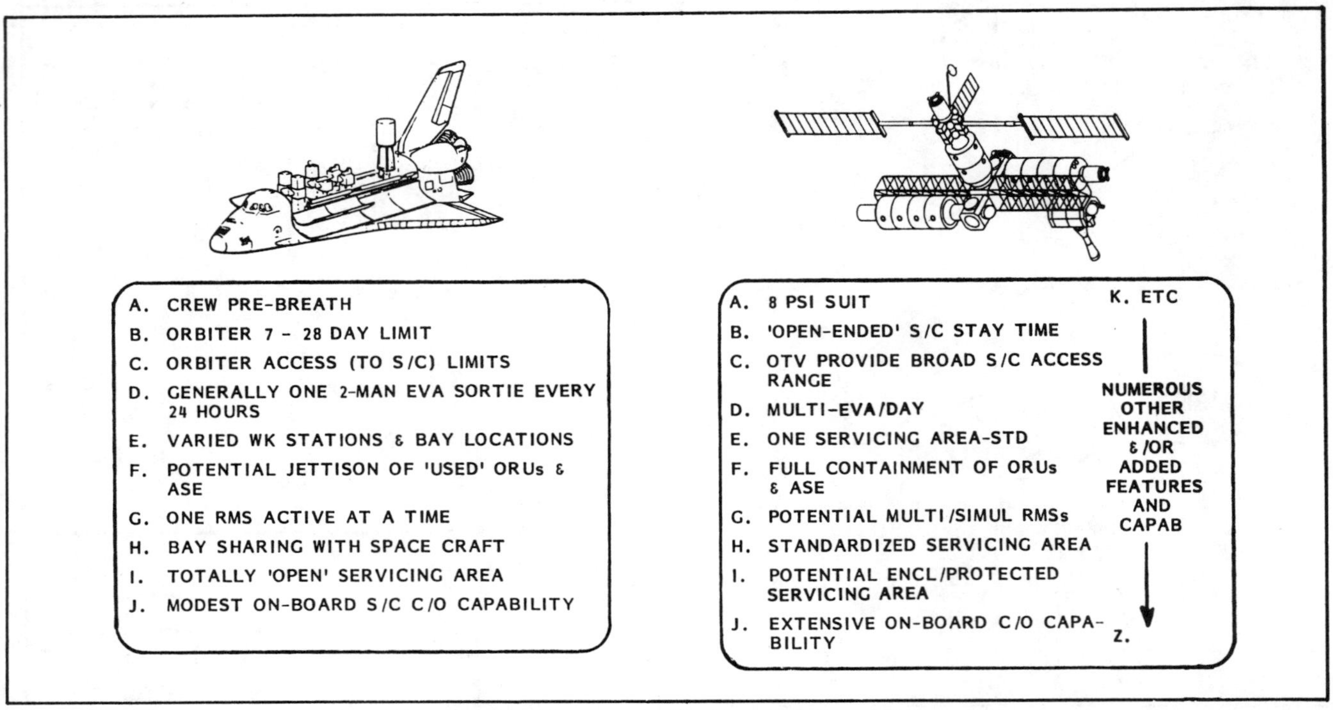

Fig. 8 Shuttle versus Station EVA Servicing Relationships (Partial)

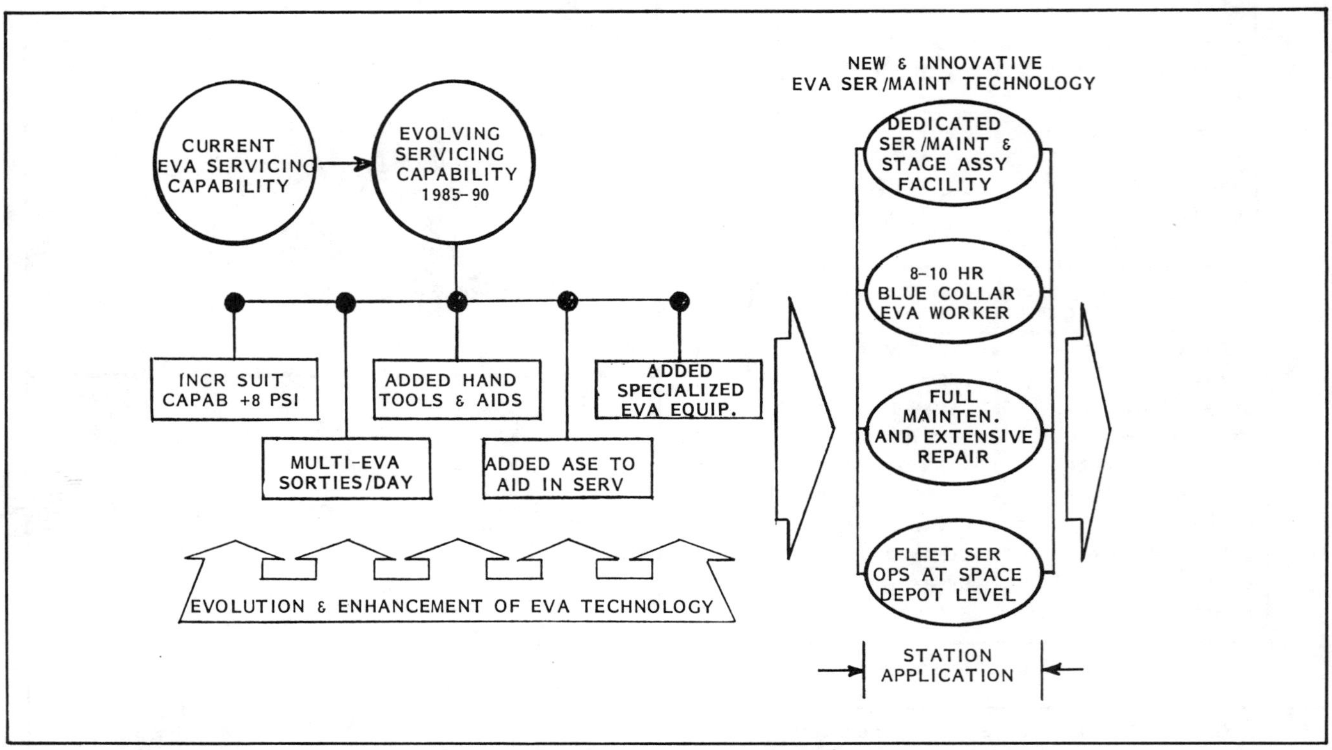

Fig. 9 EVA Technology Evolution--Shuttle to Station

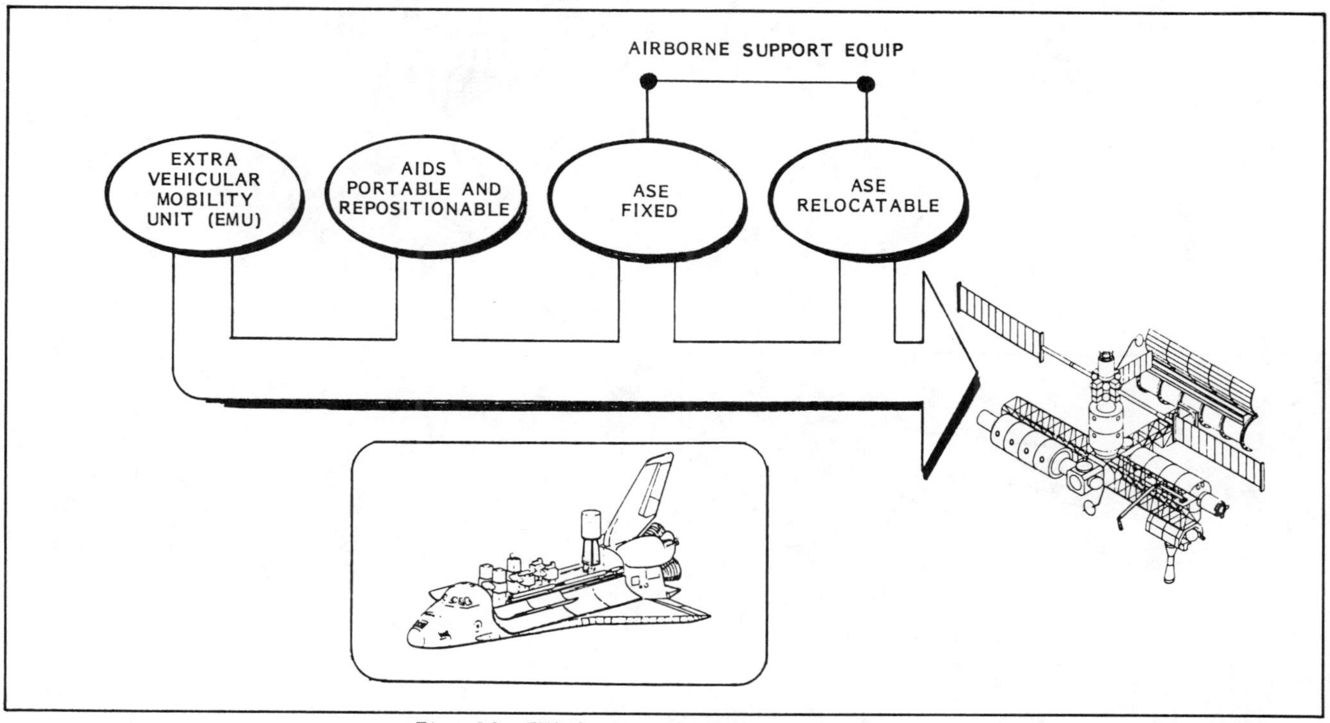

Fig. 10 EVA System and Support Hardware
(Present and Near Term)

CURRENTLY AVAILABLE

DEVELOPMENT NEED / EVA AIDS*	NO NEW DEV REQD	ADDED DEV REQD	EXAMPLE SERVICING BASE USED
1. EMU		●	◀
2. MMU		●	
3. EVA HAND HOLDS	●		◀
4. EVA HANDRAILS	●		◀
5. TETHERS	●		◀
6. SELF-TENDING TETHER	●		◀
7. TETHER RINGS	●		◀
8. FOOT RESTRAINT	●	●	◀
9. FOOT RESTRAINT RECP		●	
10. MINI-WORK STA		●	
11. MESA-MODIFIED	●		
12. TOOL CADDY	●		◀
13. ORBITER HAND TOOLS		●	◀
14. HELMET MTD LIGHTS	◑		◀
15. TILE KIT-STOW ASSY	◑		
16. ETC			

CURRENTLY AVAILABLE

EXAMPLE SERVICING BASE USED	DEVELMT NEED / ORB EVA AIDS*	NO NEW DEV REQD	ADDED DEV REQD
▶	1. RMS		●
▶	2. SLIDE WIRE	●	
▶	3. HAND HOLDS	●	
▶	4. HAND RAILS	●	
▶	5. HATCH MECH	●	
▶	6. CONNECTORS	●	
	7. HNG OVRD MECH		
▶	8. EVA LIGHTS		●
	9. CCTV	●	
	10. ETC		

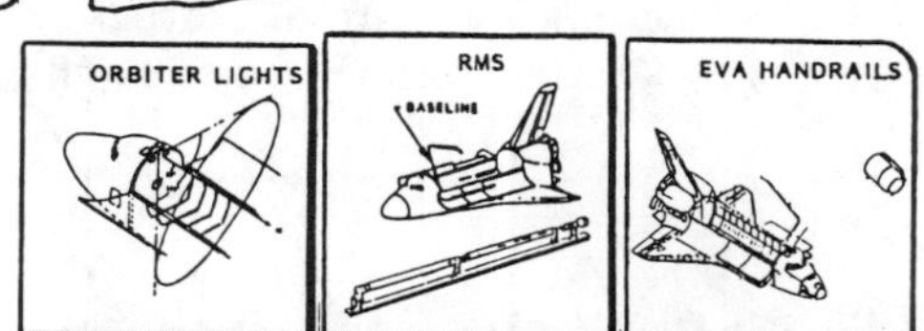

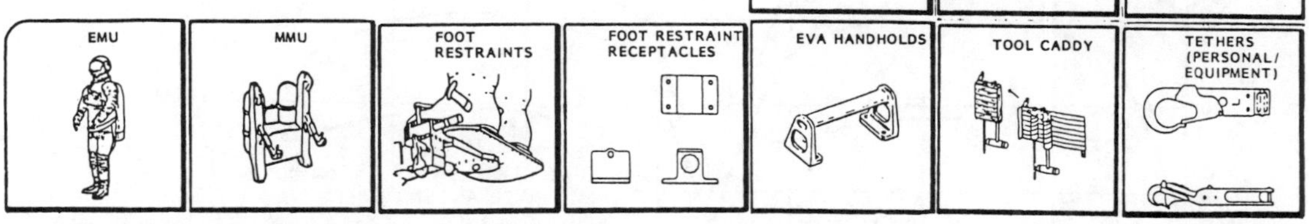

◑ = QUESTIONABLE NEW DEV NEED   * = STATION APPLICABLE

Fig. 11 EVA Servicing Maintenance Hardware
Technology Base

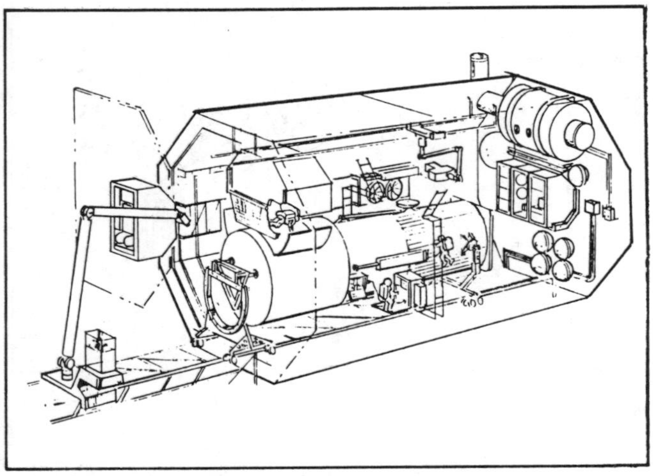

Fig. 12 Advanced Station Servicing Hangar

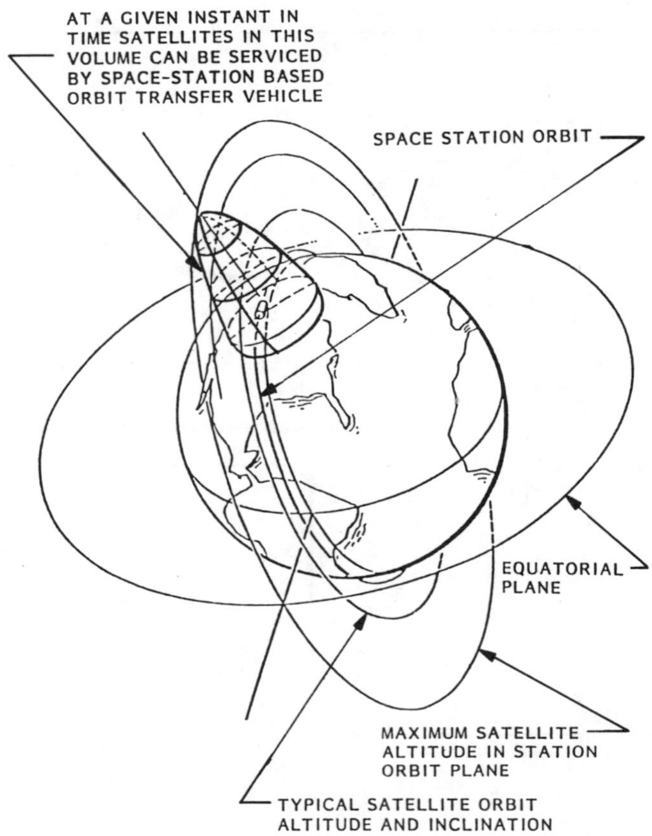

Fig. 13 Space-Based Satellite Servicing Envelope at a Given Instant

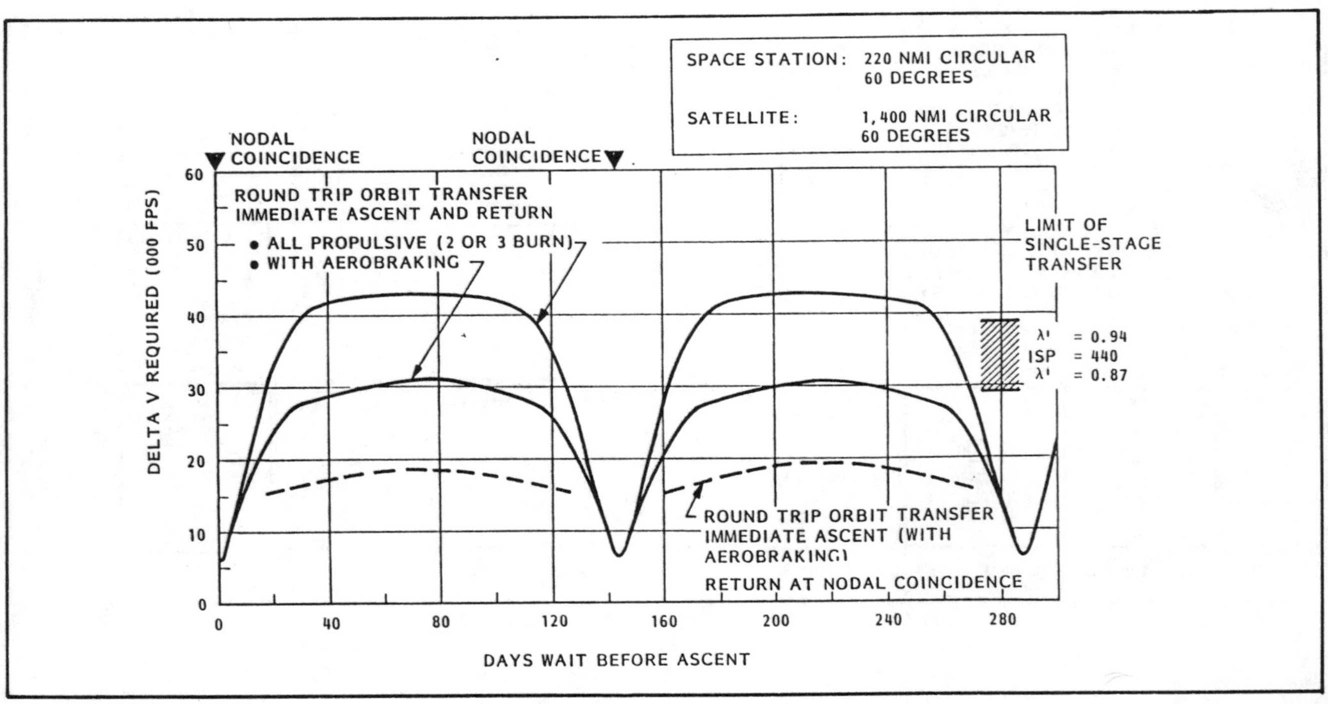

Fig. 14 Delta-V for Transfer at Nonoptimum Time

SERVICE ROLE OF A SPACE STATION: SYSTEM APPROACH TO REQUIREMENTS AT THE TRANSPORTATION NODE

E.G. Cole*

Shuttle Integration and Satellite Systems Division
Rockwell International

Abstract

This paper describes the system analysis approach to defining needs and missions for new space support systems and concentrates on the activities associated with a major transportation node. The process that was used is described briefly and involves considerable interaction with users. Operational locations for payloads were determined and payload size and the growth patterns were determined. Based on a mission model for 1991 to 2000, the distribution of functional support services was determined. A typical communication satellite mission is illustrated and the benefits for all users of the station and the reusable orbit transfer vehicle system are presented.

Introduction

The analysis shows that a space station located at the 28-degree geosynchronous orbit (GEO) transportation node can provide space-basing for a reusable orbit transfer vehicle (ROTV) to transport payloads from the space station to higher energy orbits. In addition to those payloads that are attached to the space station at this location, there are many payloads that will want to be located near the space station and serviced periodically. The synergetic interaction of these different types of payloads going through the station creates a condition in which the Shuttle can be manifested to load factors approaching one and that can exceed one by scavenging residual and reserve propellants prior to orbital injection.

The analysis also investigated orbital transfer vehicles (OTV's) that would be effective in placing payloads in high energy orbits (HEO) in the 1990 through 2000 time period. The most cost effective ROTV system that can handle the distribution of payload sizes in the 1990 time period is a two-stage system: a reusable perigee stage plus expendable storable apogee stages. Although an aerobraker concept gives slightly lower operating costs for payload delivery, it costs considerably more to develop and therefore is less effective for this specific application.

Background

In the late 1970's, Rockwell and many other aerospace companies studied how to construct and assemble large structures in space, specifically in low earth orbit (LEO) and GEO. As the space operation contract (SOC) studies were initiated at Johnson Space Center (JSC) in 1979, a survey was made of potential users to determine their specific needs and the large projects they intended to undertake. In reviewing the various user responses, very few intended to build large structures, but more near-term issues surfaced that were of greater concern to them. The results of these initial contacts is shown in Fig. 1. The major concern to these users was being assured of service availability. There is a need for support-system developers to have firm plans for providing excess capability or expansion to meet the growing demand for new flight support systems.

The second area of great concern was in the cost of services. As the Shuttle comes on line, there is a need to reduce the launch processing cost. In addition, the servicing of spacecraft could provide longer life for critical functions and therefore reduce the costs of doing business in space.

One of the most interesting concerns expressed was that, when asked why they had not transitioned to a more optimum Shuttle utilization, various users answered that they were constrained by the current family of solid upper stages. This meant that they could not grow in size and shape to take full advantage of the Shuttle cargo bay. As shown in Fig. 1, when a payload grows larger than the indicated size for the solid upper stages, it is necessary to jump to the next size solid stage, which represents a major increase in cost. With the advent of the Centaur stage, the size constraints were

CONCERN	POTENTIAL SOLUTION
• ASSURED AVAILABILITY OF SERVICES	• FIRM PLANS FOR EXCESS CAPABILITY OR EXPANSION TO MEET GROWTH
• COST OF SERVICES	
• LAUNCH TO LEO	• LOWER PROCESSING COSTS • FULL MASS LOAD FACTOR
• LOWER HIGH ENERGY ORBIT TRANSFER	• REUSABLE SYSTEM
• SERVICING OF S/C — LONGER LIFE	• CAPABILITY DEMONSTRATION
• PAYLOAD SIZE & SHAPE CONSTRAINTS	• FLEXIBLE SIZE OTV SYSTEM • STAGING AREA IN LEO

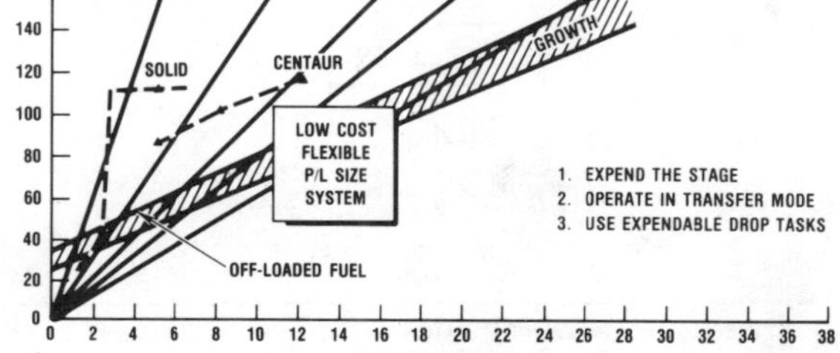

Fig. 1 General user concerns.

*Programs Director,
Space Station Program

reduced and some cost relief was obtained with off-loading; however, there is a need to have a low cost, flexible payload-size system designed for approximately 12,000 pounds, which can be off-loaded to handle small payloads and easily adapted to handle larger payload missions that will be required in the near-term future. This could be achieved by operating in a transfer orbit mode by either expending the vehicle or using expendable drop tanks to increase the delivery capability.

New Space Support System Definition Process

In trying to define new space support systems that fulfill user needs, one encounters the chicken and egg problem. Many users have been reluctant to define new requirements because, in the past, the first user had to pay the initial development cost. On the other side, over zealous developers usually defined optimistic demands to support their specific pet project. It was determined that the best solution to this problem would be a continued exchange between user and developer. This way (Fig. 2) both can develop a better understanding of each other's problems. Each user's requirements can be incorporated into a totally integrated system. The cost of various services are fed back to users who can modify their mission model. Rockwell has utilized this system for approximately two years. The process can be summarized as a progressively coverging iteration between the user and the developer.

Table 1 presents a number of key definitions that were important in setting up the integrated mission analysis presented here. One of the first steps is to develop a definition by each of the users of their mission payload/mission model. Second, as the amount of traffic is observed in each of the functional areas, space support systems options can be developed to accommodate these payloads. The mission model for the space support system must also be defined. The combination of these elements in a time-phased relationship is defined as a mission scenario.

The next step in the analysis was to segregate the various mission payloads into the appropriate functional support areas. The three major areas of LEO, HEO, and payloads attached to the support system are presented in the following.

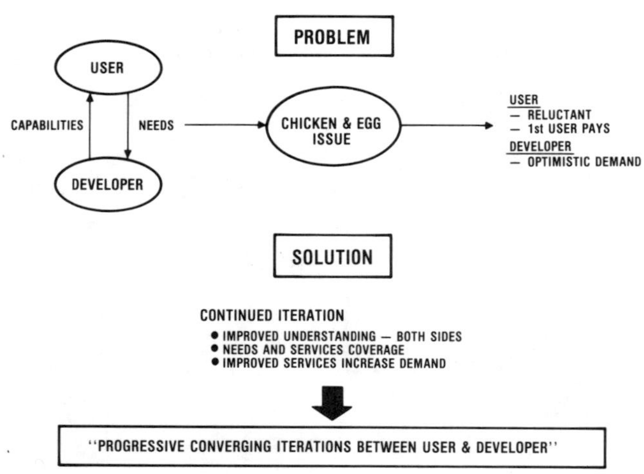

Fig. 2 New space support system definition process.

Table 1 Space operation system element definitions

MISSION PROGRAM —	A SERIES OF INTEGRATED SPACECRAFT AND SERVICE EVENTS (i.e. SPACE TELESCOPE) OR GLOBAL POSITION PROGRAM (i.e. 24 ACTIVE SATELLITES)
MISSION PAYLOAD —	A STS MANIFESTABLE PAYLOAD — • INTEGRATED SPACECRAFT — LOW EARTH ORBIT i.e., LANDSAT — HIGH ENERGY ORBIT (HEO) i.e., TDRSS, GPS SPACECRAFT • ATTACHED MISSION EQUIPMENT — i.e., LIFE SCIENCES, SIRTF CRYSTAL FACTORY, STS SORTIE • SERVICE PAYLOADS — EQUIPMENT OR CONSUMABLE FOR PRIMARY SYSTEM ON SPACECRAFT OR ATTACHED TO STATION OR PLATFORM HEO PAYLOAD CAN INCLUDE AKM FUNCTION & PKM FUNCTION IF MANIFESTED TOGETHER (EXPENDABLE) HEO PAYLOAD WHICH USES REUSABLE ORBIT TRANSFER VEHICLE (ROTV) — SPACECRAFT — MISSION PAYLOAD — FUEL & OTV — SUPPORT PAYLOAD
SPACE SUPPORT SYSTEM PAYLOAD —	ITEMS WHICH PROVIDE SUPPORT DURING SOME MISSION PHASE, i.e., SPACE STATION MODULES, ASE, ROTV, TMS, FUEL, LOGISTICS MODULES
STS CARGO DATA BANK —	LIST OF EXISTING AND PLANNED MISSION PAYLOADS AND SPACE SUPPORT SYSTEM PAYLOADS AND THEIR SUPPORT REQUIREMENTS
MISSION MODEL —	A SPECIFIC TIME-PHASED (YEARLY) SCHEDULE OF MISSION PAYLOADS, LAUNCHES AND SUPPORT FUNCTIONS REQUIRED
MISSION SCENARIO —	A COMBINATION OF A MISSION MODEL MANIFESTED AND ACCOMMODATED BY A SPECIFIC SET (ARCHITECTURE) OF SPACE SUPPORT SYSTEMS

- Low earth orbit (LEO)
  - Spacecraft placement
  - Spacecraft servicing
    - Life extension
    - Space processing supply and return
    - Orbiter, station, or in-situ
- High energy orbit (HEO)
  - Spacecraft placement or escape
    - Expendables
    - Space-based reusable
  - Spacecraft servicing
    - Geosynchronous
    - Space-based TMS
    - Multi-repair carrier
- Attached to support system
  - Space station (manned)
  - Space platform (unmanned-serviced)
  - STS orbiter, sortie
- Support system requirements
  - Crew
  - Power
  - Stable platform—duration
  - Data—communication
  - Stability—g level
  - Fine pointing

In each of these three general areas are additional areas that require different types of support requirements. Almost all of the support requirements can be categorized in these seven specific areas. In addition, this shows the support requirements for those payloads that are attached to a support system. Here, the HEO functional area will be examined.

Studies of larger spacecraft by Rockwell and General Dynamics indicated the efficiency and cost advantages of going to larger spacecraft. So, in the summer of 1981 when Rockwell initiated its space station mission studies, it seemed apparent to a developer that everybody would immediately transfer to larger spacecraft. But after analyzing various mission scenarios and taking these results back to users, insight was obtained into the questions: why and when will those responsible for communications satellites (Comsats) switch to larger spacecraft?, and when will the military want to make their satellites grow? After roughly six iterations (Fig. 3), it can be seen that the current scenario has been established and the high energy histogram of the size of payloads for Mission Scenario 6 is a reasonable prediction of the future.

Mission Payload Locations

In order to understand what kind of space support systems are required and where they should be located, it is necessary to study the spaceography of mission payloads. To develop space support requirements it is necessary to know the payload size and location distribution for a given period of time. Fig. 4 illustrates the results of the medium mission model. Space around earth has been divided into three mission zones: A, B, and C. The shaded areas indicate the regions specified in terms of inclination and altitude where the payloads want to operate. For example, in Zone A many payloads want to come to the low altitude location at 28 degrees, but others pass through this zone going to escape to the planets or on their way out to GEO. At the medium inclination, payloads want to be located at low altitude and a considerable number desire the 12,000 nautical mile region, whereas in Zone C, all of the payloads are at relatively low altitude. The bulk of the mission payloads and the payload mass is concentrated in Zone A with approximately equal distribution in the other two zones. It should be observed at this time that approximately 42 of the payloads in Zone B, which go to high altitude, can also flow very efficiently through the 28-degree node in Zone A. At the bottom of the chart, it shows the total mass flow to the regions in which any propellant or other system mass has been added to the mission payload to make a total cargo mass delivered to low altitude in these regions. Approximately 90 percent of the mass flows through Region A. Distribution of equivalent space transportation system (STS) launches to the various zones are also shown.

Table 2 presents a more detailed distribution of the number of payloads going to each location zone, broken down by user category and also distributed to the functional services to be provided by the space operation system. It is interesting to note that space processing has only 13 large spacecraft factories but requires approximately 395 small servicing packets to complete its missions. Table 3 presents a summary of the functional services in the various zones. Here it is seen that LEO placement missions involve 45 spacecraft with an average size of about 19,000 pounds. Servicing of these spacecraft takes about three times that of many small servicing payloads at the average size of 2,000 pounds. A similar situation is also observed in the attached payloads with regard to the size of initial mission equipment and also the servicing packages. Table 4 shows the percentage distribution across the support functions for the number of payloads, payload mass, total cargo mass, and the man-hours required to support the activities of these various functions.

Space-Based ROTV Concept

After numerous parametric studies, the space-based ROTV concept presented in Fig. 5 was synthesized. It was based on a single-stage space design for direct injection of payloads into GEO and to return the empty stage to LEO. It was sized to deliver a 12,000-pound payload to GEO (Fig. 6). This stage could also be operated as a reusable perigee stage utilizing expendable apogee-stage components,

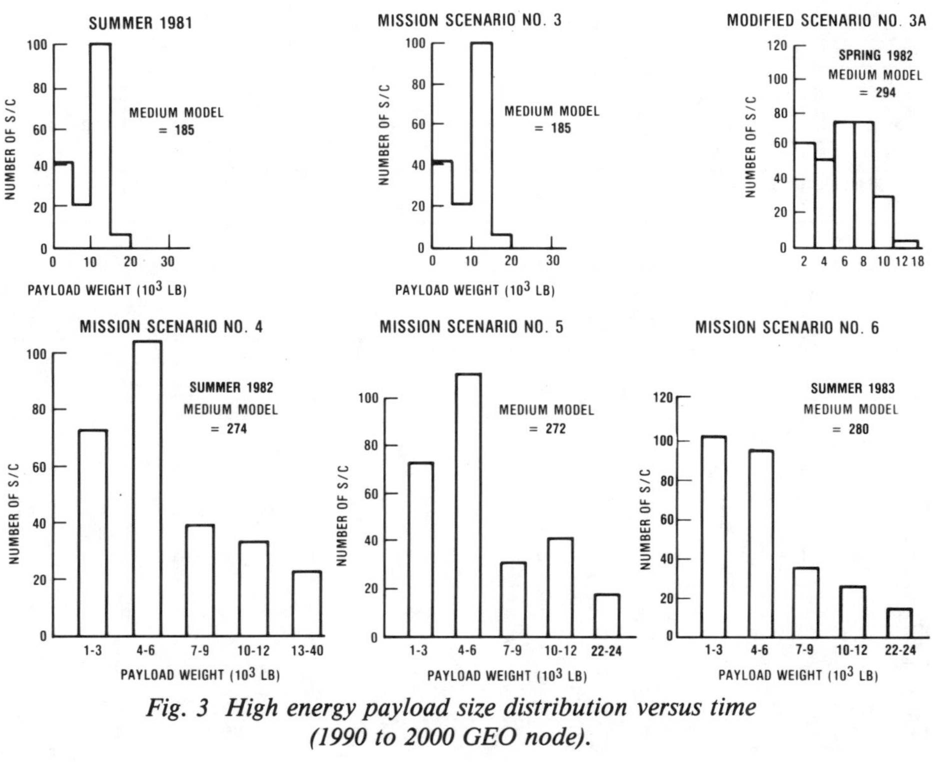

Fig. 3 High energy payload size distribution versus time (1990 to 2000 GEO node).

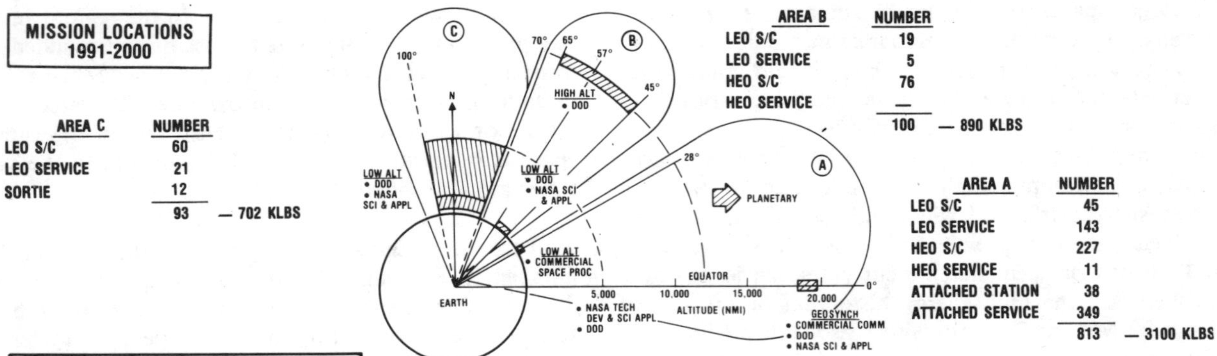

***MASS FLOW THROUGH REGIONS — 1991-2000**

LOCATION	NO. OF MISSIONS	MISSION MASS (MILLIONS OF LB)	PROPELLANT MASS (MILLIONS OF LB)	OTHER SYST MASS (MILLIONS OF LB)	TOTAL CARGO MASS (MILLIONS OF LB)	EQUIVALENT STS FLIGHTS
A	855	3.2	3.1	5.4	11.5	193
B	58	.8	—	0.1	0.9	26
C	93	0.7	—	0.6	1.3	55
TOTAL	1006	4.7	3.1	6.0	13.7	274

*ASSUMES SPACE STATION AT 28.5° INCLINATION

Fig. 4 Where do mission payloads want to be?

Table 2 Space operation system; mission payloads—location and support functions (1991 to 2000 medium model).

LOCATION ZONE USER CATEGORY —	LEO S/C		HEO S/C			ATTACHED P/L			TOTAL
	PLACEMENT	SERVICE P/L	PLACEMENT		SERVICE P/L	STATION/ PLATFORM	P/L SERVICE	SORTIE	
LOCATION ZONE A — INCL 0°-30°			EXP	ROTV	ROTV				
NASA SCIENCE & APPLICATION	11	14	2	3	(INCL BELOW)	19	54	—	103
NASA PLANETARY	—	—	7	—	—	1	—	—	8
NASA TECHNOLOGY DEVELOPMENT	—	—		1	—	10	8	—	19
GOVT ENVIRONMENTAL	—	—	2	3	(INCL BELOW)	—	—	—	.5
COMMERCIAL — COMMUNICATION	—	—	50	114	(INCL BELOW)	—	—	—	164
COMMERCIAL — SPACE PROCESSING	10	117	—	—	—	3	277	—	407
COMMERCIAL — RESOURCE OBSERVATION	—	—	—	1	—	—	—	—	1
GEO SERVICING MISSIONS	—	—	—	—	11	—	—	—	11
DOD	24	12	19	25	(INCL ABOVE)	5	10	—	95
ZONE A — TOTAL	45	143	80	147	11	38	349	—	813
LOCATION ZONE B — INCL 30°-70°				*THRU STATION					
NASA SCIENCE & APPLICATION	3	5	—	1*	—	—	—	—	9
DOD	16	—	34	41*	—	—	—	—	91
ZONE B — TOTAL	19	5	34	42*	—	—	—	—	100
LOCATION ZONE C INCL 70°-100°									
NASA SCIENCE & APPLICATION	8	11	—	—	—	—	—	2	21
COMMERCIAL RESOURCE OBSERVATION	10	—	—	—	—	—	—	—	10
GOVT ENVIRONMENTAL	5	—	—	—	—	—	—	—	5
DOD	37	10	—	—	—	—	—	10	57
ZONE C — TOTAL	60	21	37	—	—	—	—	12	93
TOTAL SPACE OPERATING SYSTEM	124	169	114 (303)	189	11	38	349	12	1006

Table 3 Space operation system; mission locations, mass flow, and average sizes

1991-2000 MEDIUM MODEL									
LOCATION ZONE	MISSION NUMBER & MASS	LEO S/C			HEO S/C		ATTACHED		
			PLACEMENT	SERVICING	PLACEMENT	SERVICING	INITIAL EQUIP.	SERVICE	SORTIE
A 0°-30° INCL	813-P/L's 3100 KLB (3.8 KLB AVG)	NO.	45.	143.	227.	11.	38.	349.	—
		MASS KLB	872.	308.	1091.	117.	186.	526.	—
		AVG KLB	19.4	2.1	4.8	10.6	4.9	1.5	—
B 30°-70°	100-P/L's 890 KLB (8.9 KLB AVG)	NO.	19.	5.	76.	—	—	—	—
		MASS KLB	200.	0.38	673.	—	—	—	—
		AVG KLB	10.5	0.08	8.8	—	—	—	—
C 70°-100°	93-P/L's 702 KLB (7.6 KLB AVG)	NO.	60.	21.	—	—	—	—	12.
		MASS KLB	641.	25.	—	—	—	—	36.
		AVG KLB	10.7	1.2	—	—	—	—	3.
TOTAL NO.:	1006 MISSION PAYLOADS	NO.	97.	169.	303.	11.	38.	349.	12.
MASS:	4675 KLB	MASS KLB	1213.	333 KLB	1764 KLB	117 KLB	186 KLB	526 KLB	216 KLB

Table 4 Distribution among support functions

1991-2000

	NUMBER MISSION P/L's (%)	MISSION P/L MASS (%)	TOTAL MASS (%)	MAN-HRS (STATION) (%)
LEO PLACEMENT	9.6 } 26	25.9 } 33	11 } 16	13.9 } 33
LEO SERVICE	16.8	7.1	5	18.9
HEO PLACEMENT	33.8 } 35	48.3 } 51	76 } 80	21.9 } 25
HEO SERVICE	1.1	2.5	4	2.9
ATTACHED	(38.7)	(16.2)	(5)	(42.4)
• STATION	3.8	4.0	3	16.6
• STATION SERVICE	34.7	11.2	<1	25.8
• SORTIE	0.2	0.8	<1	—
TOTAL	100.0	100.0	100.0	100.0

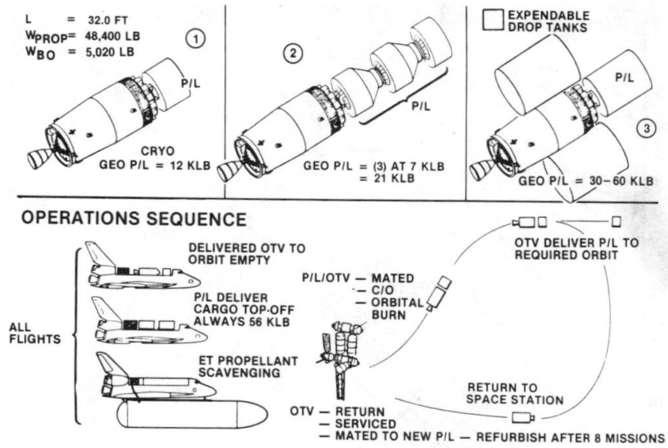

Fig. 5 Space-based OTV concept (phased increased capability—future growth).

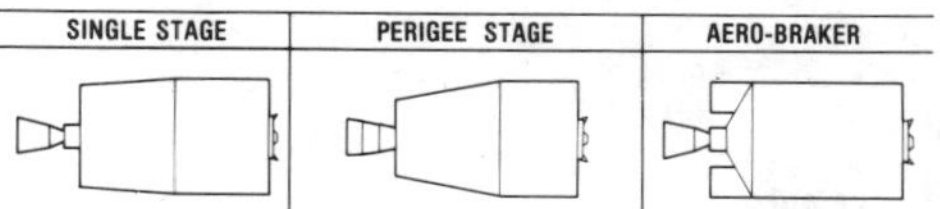

	SINGLE STAGE	PERIGEE STAGE	AERO-BRAKER
ATTRIBUTES:	• DIRECT INJECTION • FUTURE GROWTH	• MULTIPLE PAYLOADS • TWO-STAGE EFFICIENCY • PAYLOAD INTEGRAL PROPULSION	• DIRECT INJECTION • AERO-BRAKING EFFICIENCY • FUTURE PAYLOAD RETURN
CHARACTERISTICS:			
LENGTH — FT	32	26	29
PROPELLANT CAPACITY — LBS	48,400	24,500	35,000
CUTOFF WEIGHT — LBS	5,022	3,554	4,554
PROPELLANT FRACTION	0.906	0.873	0.885
PROPULSION ΔV — FPS	29,000	17,000	21,600

Fig. 6 ROTV concepts; 12,000 lb payload to GEO.

and it could put up the equivalent of about three inertial upper stage (IUS) payloads. It could also be expended at its end of life or expendable drop tanks could be used to increase its performance capability.

The operational concept is presented on the lower portion of the figure. A cargo top-off tank was sized to be used on almost every Shuttle flight. This allows the Shuttle to be filled to its maximum capacity by putting the OTV fuel in the top-off tank. Also, on each mission on the STS performance propellant reserves could be scavenged from lines in the Shuttle and in the external tank prior to dropping it just after main engine cutoff (MECO). The concept is to bring the ROTV's to orbit empty where they would be checked out and prepared for flight. Payloads going to HEO would be brought up on subsequent flights and mated with the OTV and the fuel, sent on their missions to deliver payloads to their orbit, and the stage returned to the station for subsequent flights. Each ROTV would probably fly eight missions, brought to the ground, and refurbished for another eight missions.

As Rockwell's studies progressed and the distribution of the size of payloads was changing with time, an additional concept was studied that was suggested by operational efficiencies and also better matched the histogram of payload sizes (Fig. 3). When the payloads appeared to be growing in mass at a lower rate than was originally anticipated and since many of the today's payloads and those for the planned future have their own apogee motors (attached either externally or integral), it appeared that a perigee stage in transfer orbit mode would provide better performance for a wide range of payload weights. Subsequently, an aerobraker concept was also designed whose design point would be the same as the single-stage direct injection, but it could also be operated in all operational modes both with and without the aerobraker addition. Fig. 7 represents the major operational modes. The single-stage direct injection is a four-burn maneuver that places the payload in its specific orbit without the use of an apogee stage. A stage sized for direct injection could also operate in the perigee-stage mode in which it just goes into the transfer orbit and the payload has its own integral or separable apogee stage. The aerobraker mode is very similar to a direct-injection concept except that it uses the aeroballute, which is the Boeing concept for reducing its velocity on its return to the earth's atmosphere. Fig. 8 shows the performance of the various stages that were studied. Two characteristics are presented, first the total mass in LEO required to put the various payloads in GEO showing the off-loading of the liquid cryo stages. As can be seen, the off-loading slopes for the single stages, such as Concepts 1, 3, 6, and 8, are much the same, whereas the off-design operations of the perigee-stage concepts have improved off-loading in terms of both mass and cost.

There are two major effects in the trade-off between expendable cryo stages and reusable cryo stages. The reusable cryo stages usually require more fuel because of bringing the stage back, but their utilization at the station allows them to operate with an improved Shuttle load factor, gaining a small improvement in this area. In addition, the hardware trade-off between buying an expendable stage for approximately $33 million or the cost of reusing the expendable stage at $3 million per flight constitutes the second cost improvement. From an analysis of mass on-orbit only, it would appear that an aerobraker operating as a direct-injection stage in one payload range and as a perigee motor in an off-loaded range, would produce the lowest cost; however, it can be seen that Concepts 6 and 1 crossover at a payload size of about 8,000 pounds. Since the aerobraker development cost will be about 1.8 times the conventional perigee stage, the perigee-stage concept is more cost effective. Table 5 shows this cost effectiveness where listed are the design, development, test, and evaluation (DDT&E) first unit cost for each system and a comparison of the missions operations cost at a payload size of 5,000 pounds, which is roughly the medium of the payload size histogram. The cost savings of the new systems are then determined relative to using expendable systems across the payload range along with the number of

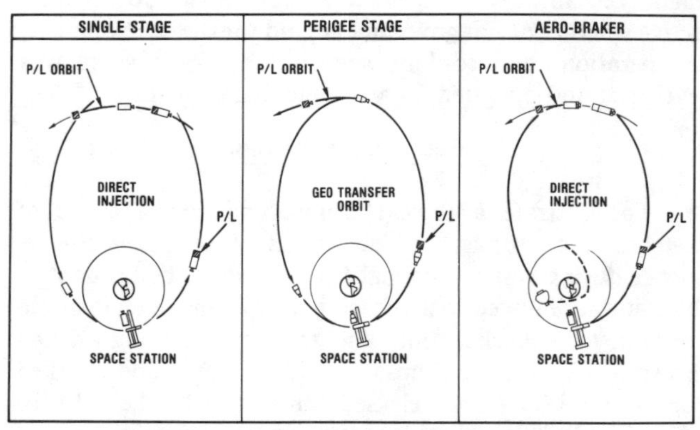

Fig. 7 ROTV operations concepts.

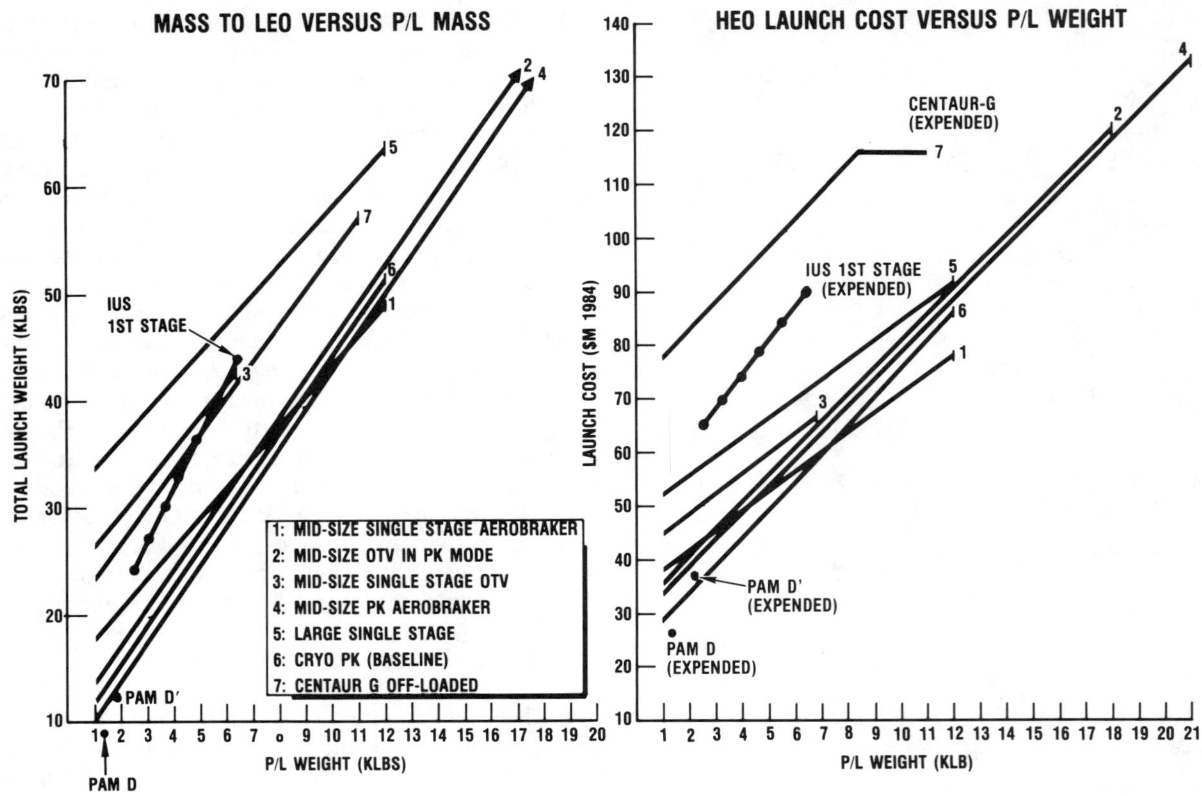

Fig. 8 OTV system comparisons.

Table 5 System return on investment comparisons

SYSTEM	DDT&E/ 1ST UNIT	MISSION COST** AT 5 KLB	BREAK-EVEN*** (NO. OF YEARS)	NET SAVINGS (OVER 10 YRS)
(1) EXPENDABLES	$ — M̄	$72M̄	—	REFERENCE SYSTEM
(2) LARGE SINGLE STAGE — DI	1726	66	7.5	$573M̄
(3) PERIGEE STAGE	1431	48	2.3	4682 M̄
(4) AEROBRAKER	2565	53	4.5	3102
• DI	2565	53		
• PKM	2565	54		
(5) SINGLE DI AB REMOVED	2162	59	*	—
(6) PKM-AB REMOVED	2162	55	4.8	2366

* CANNOT FULLY SUPPORT MISSION SCENARIO
** NO SCAVENGING
*** 50% MULTI-P/L MISSIONS ≤ 4000 LB

flights where the savings just equal the development cost for the new systems. Subsequent to the break even point, the net savings would represent a total operational savings once the development cost had been paid off. On this basis, the smaller reusable cryo perigee stage would be the most cost effective whether or not government paid for the development, its operation, and sold its services, or if the stage was developed and operated by a commercial company.

Typical Comsat Mission

To illustrate a typical Comsat mission a series of sketches (Fig. 9) has been developed, which show the sequence during which a typical Comsat would be brought to the station, removed to a servicing port from which it would be deployed, checked out, and the antennas aligned. As shown in Fig. 10, the Comsat would be mated and checked out on the ROTV and released for launch by the OTV to GEO. The ROTV will return to the station for refueling and subsequent missions.

Benefits of the Station and the ROTV at the GEO Node

When one considers the consequences of having a station located at the GEO node and includes all the payloads that are going to LEO at 28 degrees, a large number come to be attached to the station or are service mission payloads attached to the station. Other earth-orbit payloads want to be placed near and operate close to the station and also be serviced from it. The combination of the payload mass desiring nearby locations with payloads that want to pass through the

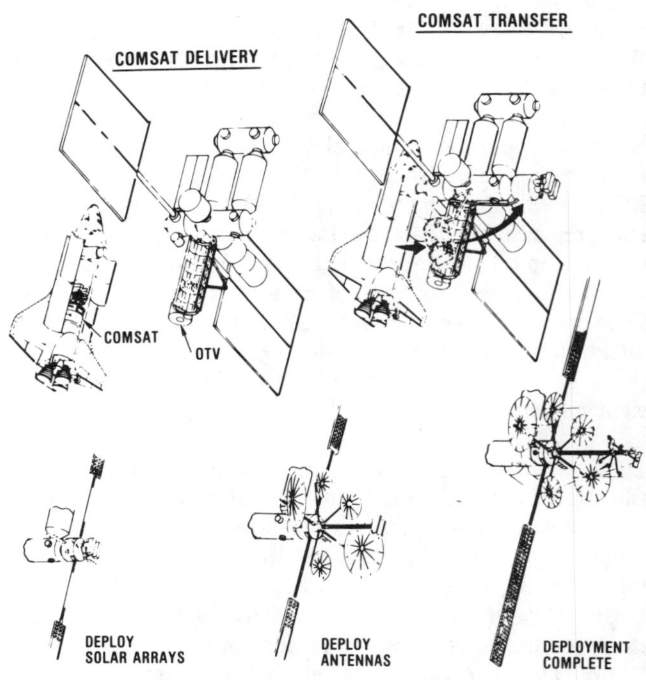

Fig. 9 Typical Comsat mission.

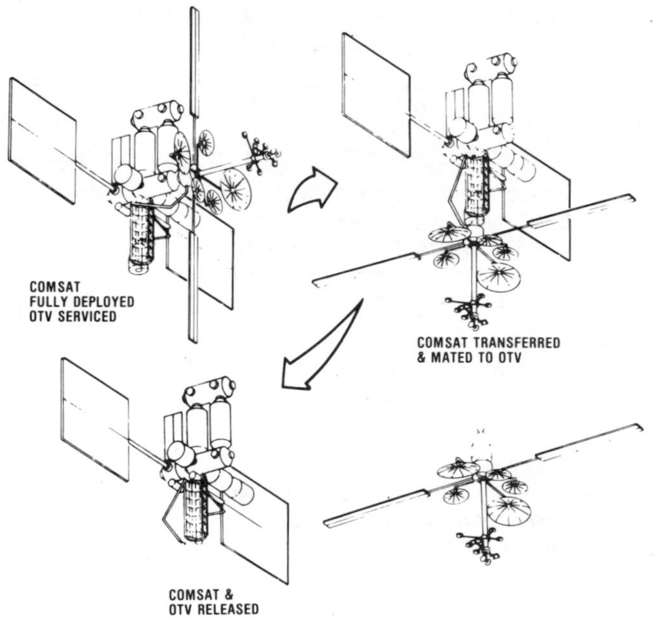

Fig. 10 Comsat/OTV mating and deployment scenario.

station, such as the high-energy payloads and the ROTV's and the fuel, create a considerable mass through one location. All of this mass with its wide distribution of large and small density cargo packets permit the Shuttle to be operated near a load factor of one. The potential for scavenging approximately 8,000 pounds of fuel from the orbiter and external tank results in load factors that could be slightly greater than one. When one looks at manifesting all the payloads utilizing just the Shuttle and without the space station where it is not possible to leave and store cargo packets for subsequent missions on-orbit, a considerably lower Shuttle load factor is obtained. Fig. 11 shows the sources of these savings; therefore, the synergism of the multi-mission service at the GEO node allows approximately a 36-percent increase in the load factor, thereby lowering the transportation cost to LEO.

The second source for savings is the use of ROTV's instead of expendables. The mission cost savings are shown on the right-hand side of the graph. The OTV savings vary over the payload range, but from Fig. 11 it can be seen that the average savings is roughly 33 percent. Therefore, multimission synergism at the GEO node allows maximum STS

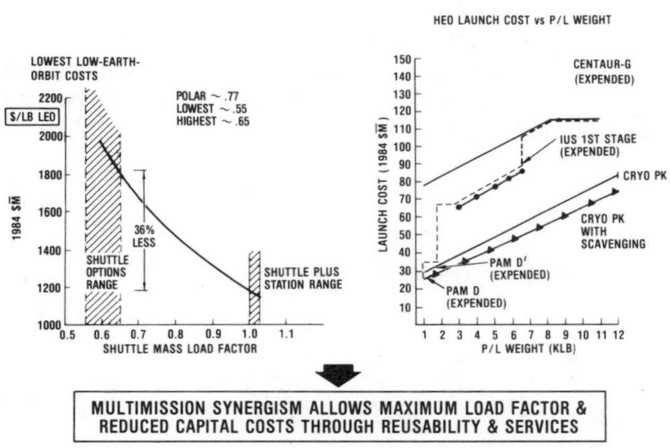

Fig. 11 Source of space station savings.

load factor, reduces capital cost through reusability, provides longer life systems through servicing, and reduces risk and insurance because of checkout.

Insight into why the decoupling of the payload, ROTV, and its fuel provides more efficient manifesting is given in the following and shown in Fig. 12.

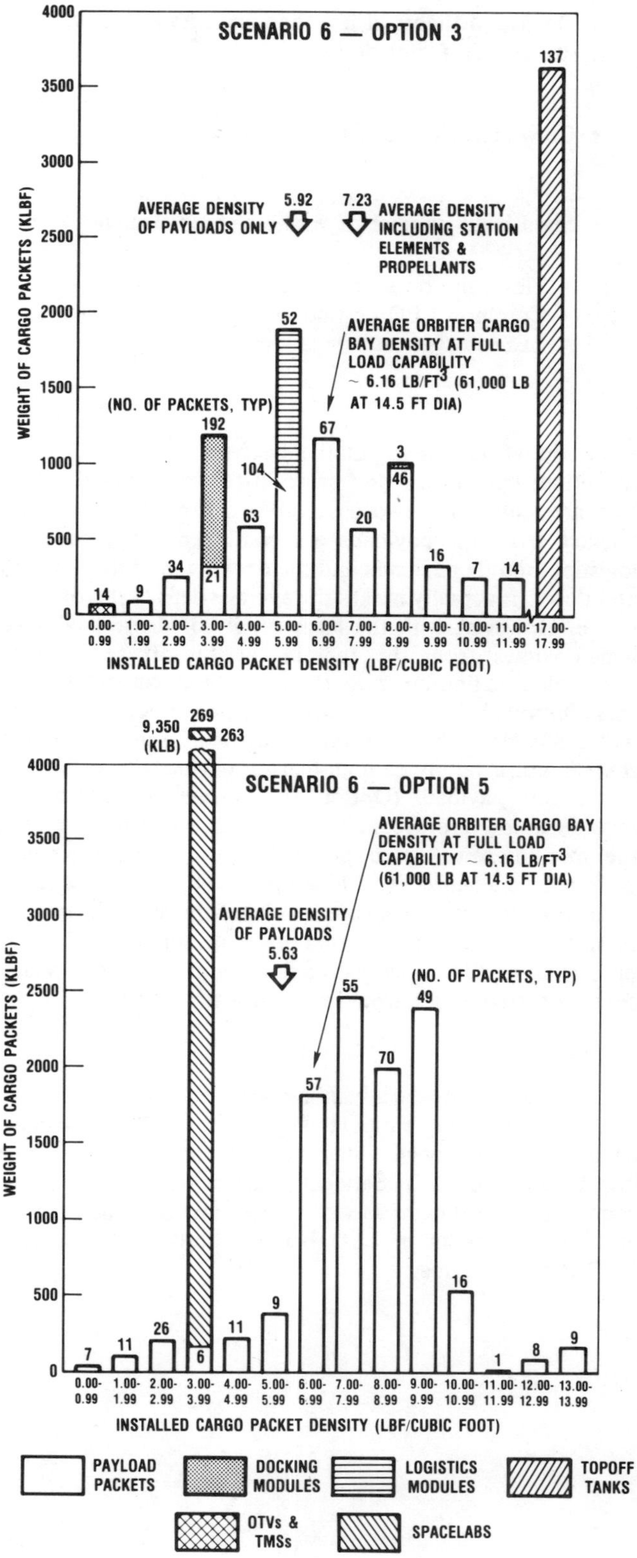

Fig. 12 Installed cargo pocket density plots.

The major influences on Shuttle cargo factor are:

- When and where payloads go—mission scheduling/destination constraints
- Dedicated flights at less than capacity approximately drives load factor down
- Crew intensive—low mass payloads
- Significant number of variable cargo elements
  - Flexibility on specific flights
  - Common LEO destination
  - LEO storage and staging capability

Here, the various items that impact the ability to achieve maximum load factor and cargo-installed density distribution are shown for two cases. When the stable of cargo packets from the payloads are combined along with the logistics module payloads and the docking module going up and down, it actually makes the average density distribution worse. However, with the large family of dense payloads, which comes from the fuel in the top-off tanks, very favorable distribution exists to achieve high cargo load factors (Option 3). If the same mission model is accommodated with a Shuttle-only case, there is a difficult problem because there is not a family of high-density payloads to offset the low-density payloads (Option 5). If some elements of the mission model are accommodated with spacelab missions, the man-intensive space processing and life science experiments encounter a critical problem with these low-density, crew-intensive missions. Therefore, since the cost of these missions would be prohibitive, the number of space processing and life sciences missions in an STS-only, space support system mode would be drastically reduced.

Table 6 presents the advantages of the station and ROTV for providing the support functional areas. Attached payloads have many advantages besides that of the cost savings of mass to LEO. Servicing missions also incur many operational advantages as well as cost savings. The deployment and checkout in LEO benefits both the LEO- and GEO-mission payloads.

Table 6 User benefits—space station and ROTV at transportation node

(INCLINATION = 28°)

SUPPORT FUNCTIONAL AREAS	SAVINGS (%)
• ATTACHED PAYLOADS (STATION)	36%
• CONTINUOUS CREW AVAILABILITY	
• MORE DOWN CARGO OPPORTUNITIES	
• FASTER R&D/SYSTEM DEVELOPMENT	
• ENABLES SPACE PROCESSING	
• S/C SERVICING MISSIONS	36%
• MORE TIMING FLEXIBILITY	
• EVOLUTIONARY TRANS	
• INCREASED HARDWARE LIFE	
• S/C PLACEMENT	
• LEO — C/O & DEPLOYABLE ALIGNMENT	36%
• GEO — CONTINUOUS CREW C/O & REPLACEMENT	0-52%

Summary

In summary, the key points presented here are

- User interaction: user concerns and system requirements
- System analysis: type, location, and quantity — functional services
- Station and ROTV at node: synergistic effects
- Substantial savings to all users
  - Operational
  - Hardware investment

A continuous interplay between the users and developers of the advanced space support system must be maintained so that a timely space support system can be developed that meets the user requirements. System analysis of what is going on in the total system must be performed to find out the degree and amount of functional support required as a function of time. The space station and the ROTV at the transportation node has synergistic effects. As a consequence, substantial savings were provided to all users both from an operational point of view and the cost in investment because their hardware is reduced.

The key to enabling a commercial space processing industry and providing the United States an advantage in achieving the high ground of space is the space station and ROTV at the GEO node.

THE SPACE STATION AS AN ASSEMBLY BASE

Gordon R. Woodcock
Boeing Aerospace Co.

Abstract

The use of a space station as an assembly base is described for four types of assembly operations. Requirements imposed on the space station are summarized and benefits are discussed. Current issues associated with assembly operations technology are described.

Introduction

The Shuttle has reintroduced human presence in space into the U.S. Space program. The fact has renewed interest in space operations that require human presence. One of these operations is assembly.

A space station, through its permanent manned presence on orbit, opens the possibility of conducting complex and difficult assembly operations requiring extensive crew involvement. This paper highlights some current and recent assembly operations studies and some key issues needing near-term consideration.

There are four types of assembly operations, based on their different mission needs. First is the assembly of the space station itself. This process involves deployment, berthing of modules together, and installation and assembly of structural elements.

The second is assembly of payloads to upper stages for transport to mission orbits.

The third is payload deployment assist or final assembly on orbit, to simplify mechanism and structural design.

The fourth is assembly of space vehicles made up of large space structures.

Space Station Assembly

Space stations are conceived as comprised of two to ten or so individual modules brought to low Earth orbit by the space shuttle. Most of these modules will be pressure vessels for which the assembly requirement is simply berthing the module at its intended location. The first module to be launched, however, will be a utilities module that includes a large deployable solar array.

Recent studies of space station mission needs have raised the desired electrical load carrying capability of the initial space station from about 20 kilowatts to 75 or 80 kilowatts. Where earlier studies had assumed that 10 kilowatts to mission users would be sufficient, then recent studies identified user needs in the range from 40 to 60 kilowatts. This high power requirement places an exceptional demand on the early space station for a large solar array.

The large array includes several assembly-related issues. First, the flexibility of the large array may be a limiting factor on space station platform stability. This, in turn, may lead to a preference for a semi-rigid array that requires assembly techniques instead of simple deployment. Secondly, inertia considerations lead to a configuration such as illustrated in Figure 1 to minimize gravity gradient torques on the large array. Thirdly, packaging of large arrays for launch in the space shuttle, especially if semi-rigid, is a challenging issue. Designing the array and its utilities module to survive launch loads may require final installation of the array to the space station as an EVA space assembly task.

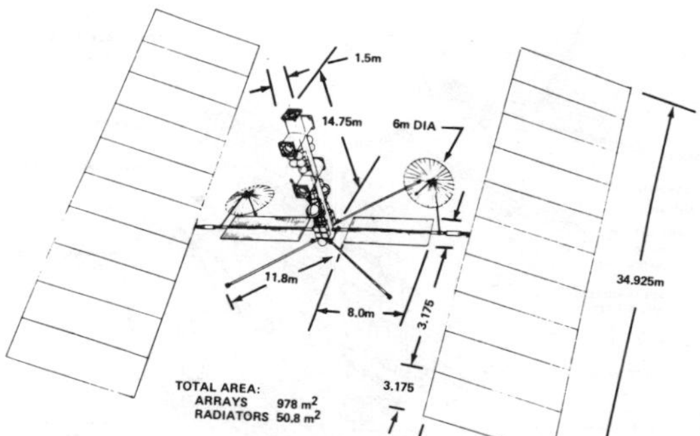

Figure 1. Space Station Power Module

Assembling the space station raises arrangement issues that relate to space station utility as well as to space station assembly procedures. There are three basic candidate arrangement concepts illustrated in Figures 2, 3, and 4. The first is a typical branched arrangement with a central core module sometimes called a multiple berthing adapter (MBA). The second is a more federated architecture that has no MBA, although the function of the work module is very similar. The third arrangement is a side-by-side or "raft" berthing arrangement that minimizes the number of interconnecting berthing fixtures.

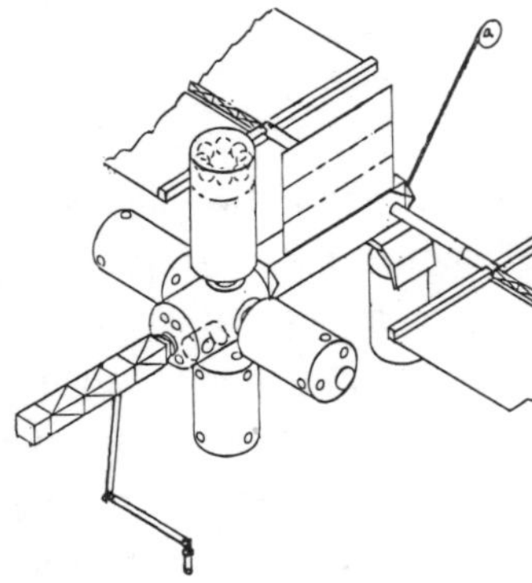

Figure 2. Space Station Arrangement With MBA

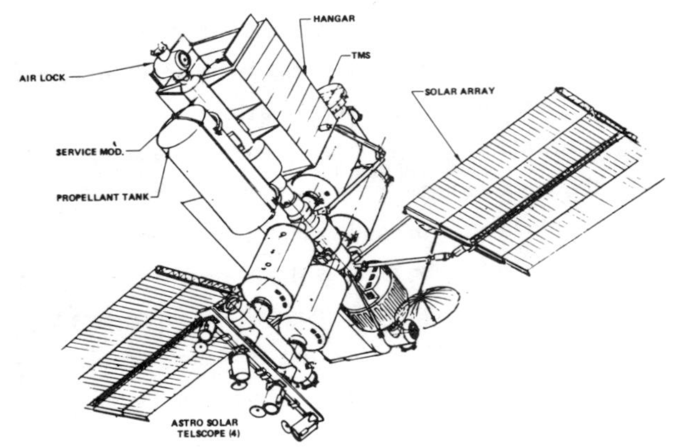

Figure 3. Space Station Arrangement With Utilities Module Core

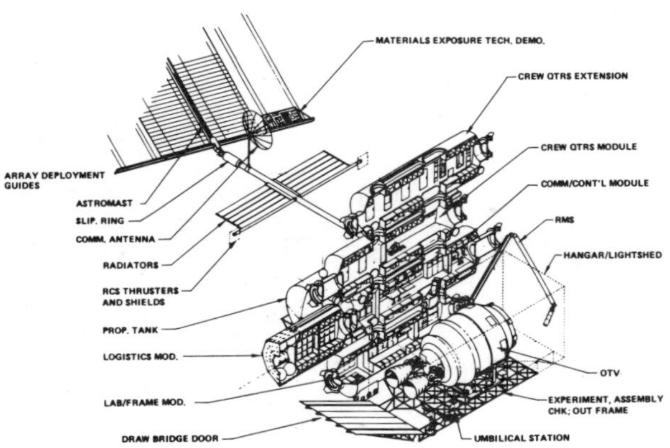

Figure 4. Cutaway View of Raft Arrangement

Arrangement issues center primarily around factors of crew accommodations and crew safety, mission and operational utility, and assemble-ability under the operating constraints anticipated early in a space station flight program.

The first module launched will be the utilities module with solar power, attitude control, orbit makeup propulsion, and basic communications. This initial module must begin its life in space as a freeflyer. It must also serve as the building block to which the later space station modules will be assembled to create a complete space station system. The utilities module will not ordinarily be manned and may or may not be pressurized.

In the multiple berthing adapter design approach, the utilities module is probably unpressurized since the MBA provides the central core to which the other modules are attached. In the federated design concept the utilities module also provides a central tunnel and passageway to facilitate exchange of crew and utilities to the various modules. In the raft concept the utilities module offers pressurized passageways and a standard berthing interface; it is sandwiched between the other modules.

One space station feature that has been debated for years is the need for multiple egress paths from each manned module. This is a safety requirement normally levied on such diverse architectures as buildings, boats and commercial aircraft. Given the small size of the space station modules presently contemplated, the need for multiple egress paths is less clear. One may note, however, that multiple access paths to each module may be necessary to avoid a situation in which the inadvertent loss of a key module such as the MBA makes most of the space station inaccessible to the crew.

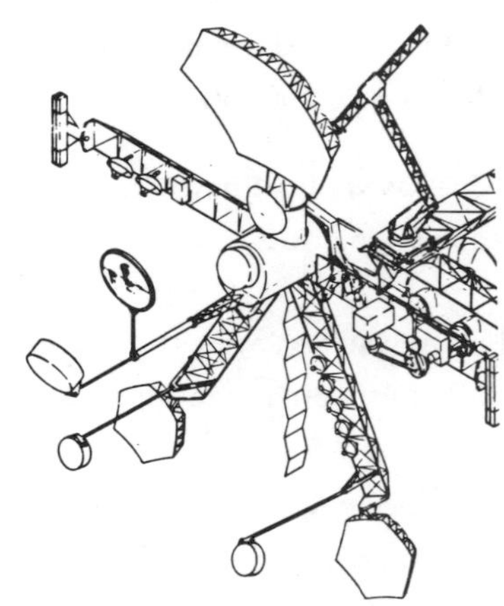

Figure 5. Platform Deployment

If the space station is to be used as an assembly base, a generous structural assembly area is needed. The assembly area must be located where it does not interfere with other space station activities such as science observations or transportation servicing. Figure 5 illustrates the deployment of antenna arms on a large space platform payload illustrating that an extensive unobstructed work space is needed for space structures or payloads assembly.

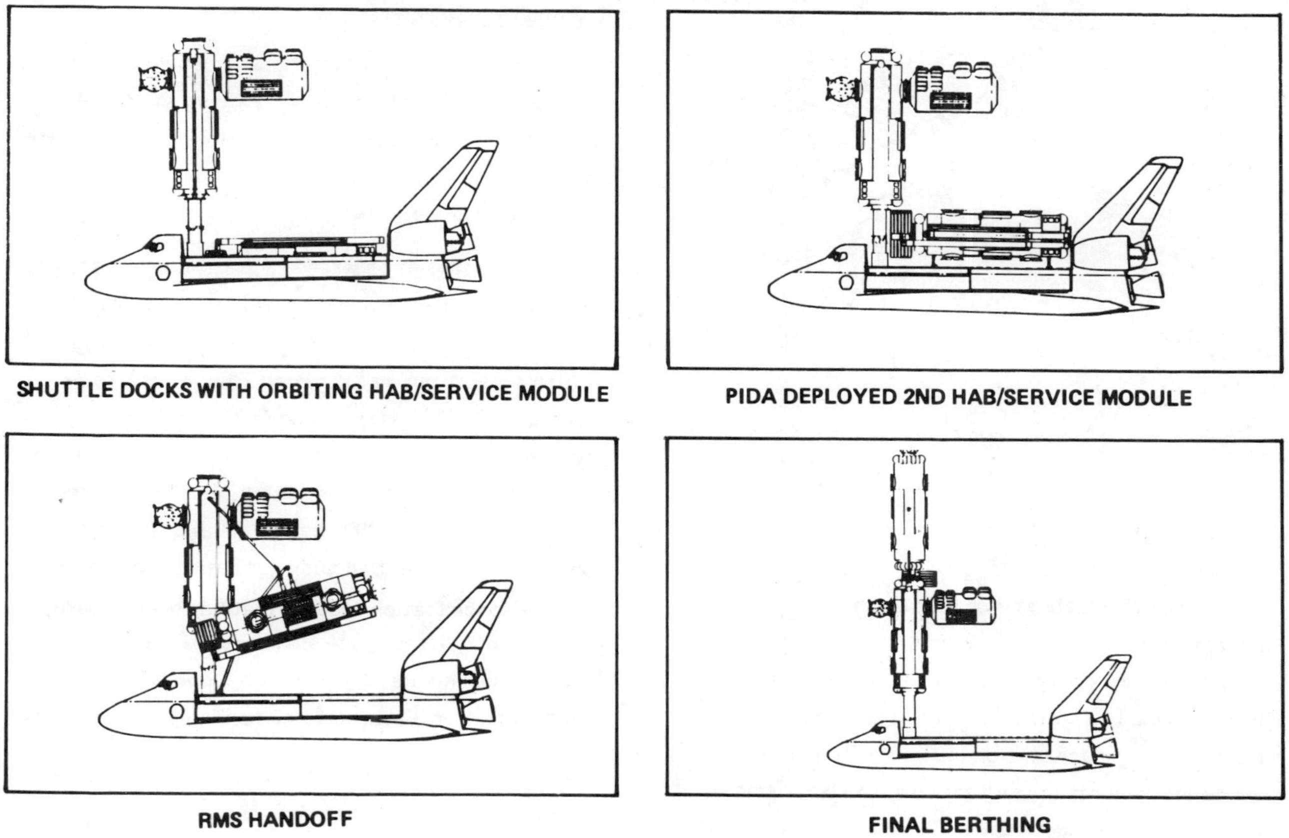

Figure 6. Space Station Assembly

Assembly of the space station modules will be accomplished by use of the shuttle RMS, probably aided with a second RMS installed on one of the initial space station elements, as illustrated in Figure 6. The RMS reach envelope is a strong consideration in determining feasible arrangements of space stations from the assembly standpoint. Some space station concepts have employed the idea of a mobile RMS as on a track to improve its reach envelope. If the space station is to grow beyond a relatively small size, a tracked RMS appears essential.

Some elements of the space station are expected to require assembly operations involving EVA crew similarly to assembly of large space structures. If the RMS is to have tracked mobility, installation of the track systems will probably require EVA activities. A number of space station concepts have employed a so called surrogate payload bay, i.e., a mounting structure that emulates the shuttle payload bay in terms of structural, electrical and data interfaces. These concepts have also included payload hangars which provide a thermal enclosure that also offers some meteoroid shielding for payloads. Figure 7 shows details of a hangar concept. Figure 8 shows an alternate Boeing concept that provides a surrogate payload bay with protective enclosure. It also serves as a large space structures experiment during its construction and initial test.

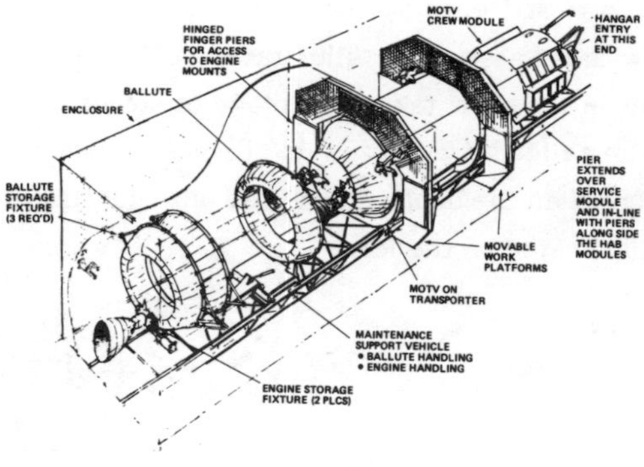

Figure 7. Hangar for Space Based OTV

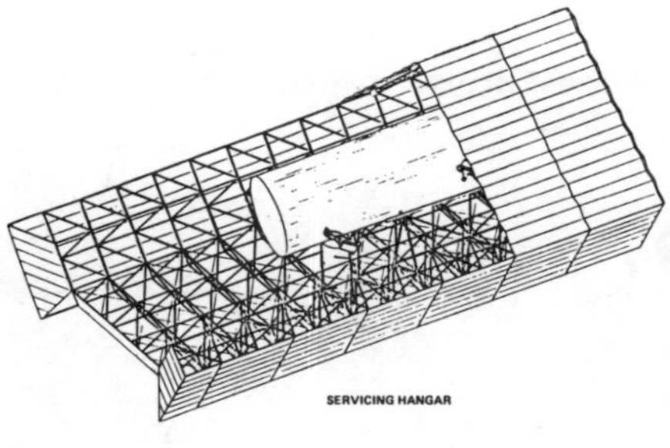

Figure 8. Alternate Hangar Concept

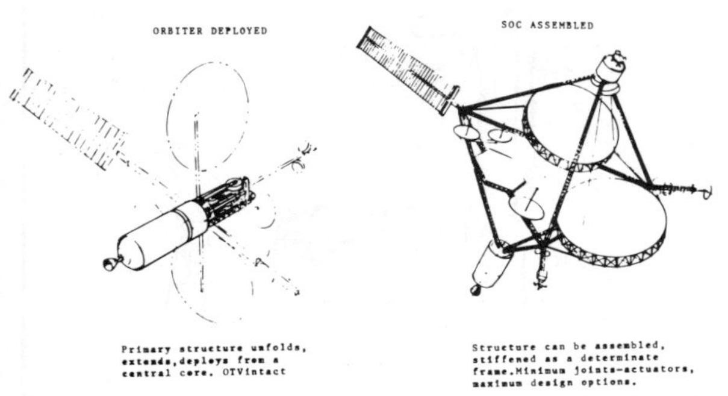

Figure 9. Large Geo Platform Configuration Comparison

Payloads Assembly

Payload assembly at a space station will be a future mission requirement when payloads become larger than sizes conveniently accommodated in the shuttle payload bay. The payloads likely to need this support from a space station initially will be large science instruments in low Earth orbit. Geosynchronous payloads will need such support later on, as spacecraft power and antenna sizes continue to grow.

In addition to accommodating large sizes the assembly of payloads at a space station can simplify deployment and deployment mechanism design as well as provide for increased stiffness of large payloads. Evolution to stiffened payload structures made possible by crew-aided assembly is illustrated in Figure 9. The payload on the right takes advantage of structural braces installed by EVA crew to provide a much stiffer structure than the deployed payload with cantilevered arms illustrated on the left.

Large Space Structures

"Large space structures" have been described as a concept in search of a mission. Indeed, some of the early large space structures technology concepts were intended simply as demonstrations that large space structures could be built.

Recent studies of large space structures have concentrated on concepts that provide mission utility. I will briefly describe four. Power needs for the space station range from an estimated 75 kilowatts to as much as 200 kilowatts, depending on the level of mission activity devoted to commercial materials processing production. The associated solar array sizes range from about 15,000 sq ft to about 50,000 sq ft. A football playing field, including the end zones, is 54,000 sq ft. The solar array sizes now projected for space stations imply large space structures. (These high power requirements may be met by more than one space station or platform or more than one solar array.)

The hangar and work shed illustrated is a large space structure of modest size, but one that can provide very useful experimental results and structural dynamics, stiffness, damping and assembly methods as well as serve a useful function once completed.

Various studies have characterized large antennas for a number of purposes including very large microwave radiometer spacecraft, as well as radar systems. Some of these may be deployable, but in the larger sizes a construction effort appears to be required. A recent Boeing study for Marshall characterized a large microwave radiometer antenna that could be assembled at a space station. This is illustrated in Figure 10.

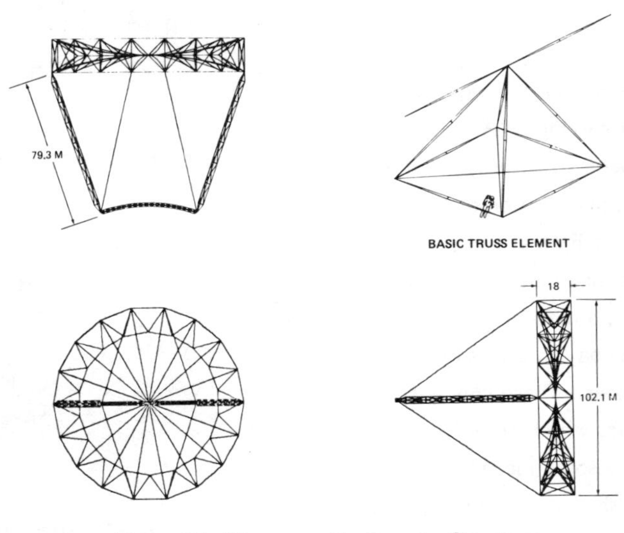

Figure 10. Microwave Radiometer Structure

One of the most interesting potential applications of large space structures and assembly operations is for large optical or infrared systems such as the telescope illustrated in Figure 11. An alternative version of this telescope, using hexagonal sub-reflectors is illustrated in Figure 12. This telescope had a 10 meter aperture with the primary mirror composed of 7 subunits. Figure 13 shows its assembly timeline.

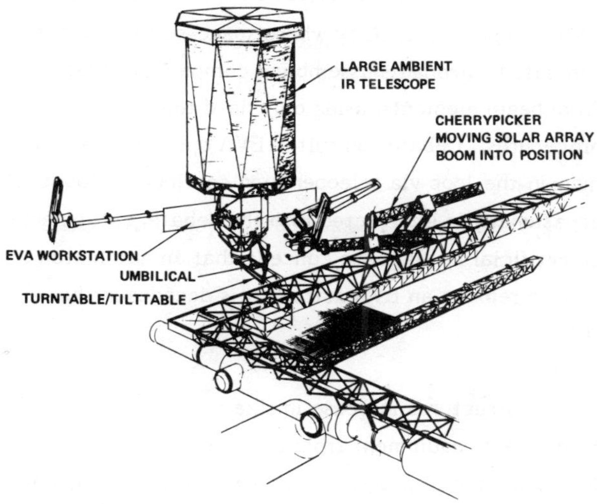

Figure 11. Large IR Telescope Construction

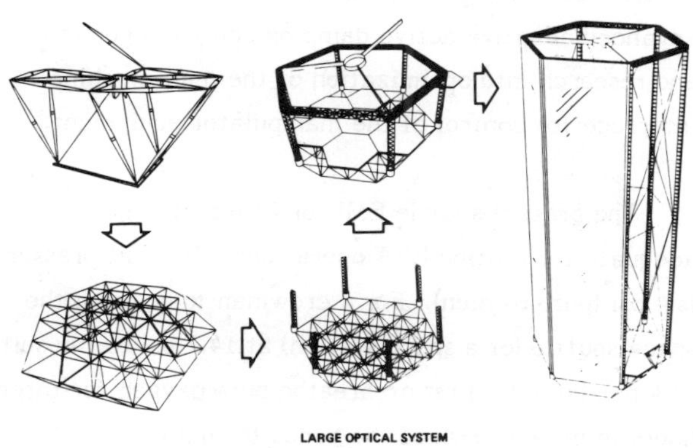

Figure 12. IR Telescope Construction Sequence

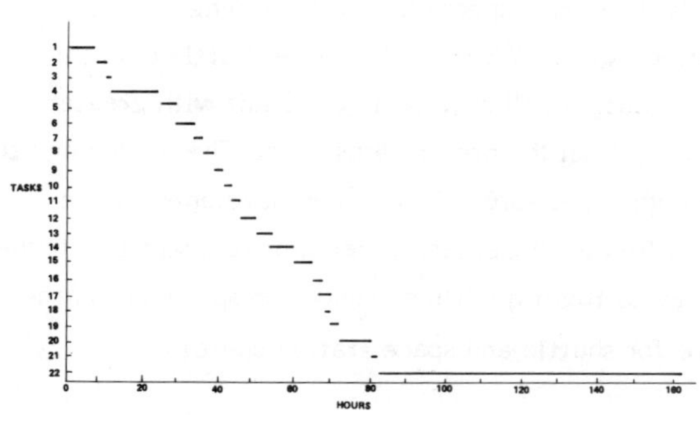

Figure 13. Timeline for Optical Telescope Assembly

Ames has studied a large IR optical system called "Large Deployable Reflector", with a 20-meter aperture that might be comprised of 37 or 61 hexagonal subunits from 2.2 to 2.8 meters in size. Such systems are highly complex and will definitely require assembly involving EVA crews with a variety of construction aids, as illustrated in Table 1.

- PERSONNEL
  - ENGINEERING
  - S/C SYSTEMS OPERATIONS—DATA
  - S/C SYSTEMS OPERATIONS—ELECTRONICS
  - S/C SYSTEMS OPERATIONS—MECHANISMS
  - SPACE STA. SUBSYSTEM OPERATIONS & MAINTENANCE
  - EVA CHERRYPICKER OPERATIONS
  - EVA WORKSTATION OPERATIONS
- SUPPORT FIXTURES
  - STORAGE FACILITY
  - CONSTRUCTION FIXTURES
  - MISC. CONSTRAINTS AND HOLD-DOWNS
  - S/C ORIENTATION FIXTURE
  - ARTICULATED HOLDING FIXTURE (LASER MEAS.)
  - DOCKING/BERTHING PORT
  - STRUT ALIGNMENT AND ASSEMBLY FIXTURE
- INSTRUMENTATION
  - STRUCT. DYNAMICS (ACCEL., STRAIN, LOADS, ETC.)
  - THERMAL RESPONSE (THERMOCOUPLES)
  - POSITION/DEFLECTION (PRECISION LASER RANGING, CORNER REFLECTORS)
- DATA SYSTEMS
  - RECORDING
  - STORAGE & RETRIEVAL
  - MANIPULATION (EDP)
  - TRANSMISSION (UPLINK & DOWNLINK)
- UTILITIES
  - ELECTRICAL POWER
  - LIGHTING
  - REMOTE TV
  - OXYGEN (AIR)
- HANDLING EQUIPMENT
  - RMS (CHERRYPICKER)
- SMALL TOOLS
  - MAINTENANCE
  - CONSTRUCTION

Table 1. Crew and Equipment Needs for Construction

Structures of this nature, during the construction phase and in some instances when completed, exhibit very low ballistic coefficients. Orbit lifetimes may be as short as a few days as illustrated in Figure 14. Consequently, use of a space station as an assembly base is essential to avoid inadvertent reentry and loss of the payload in the event a shuttle revisit were delayed.

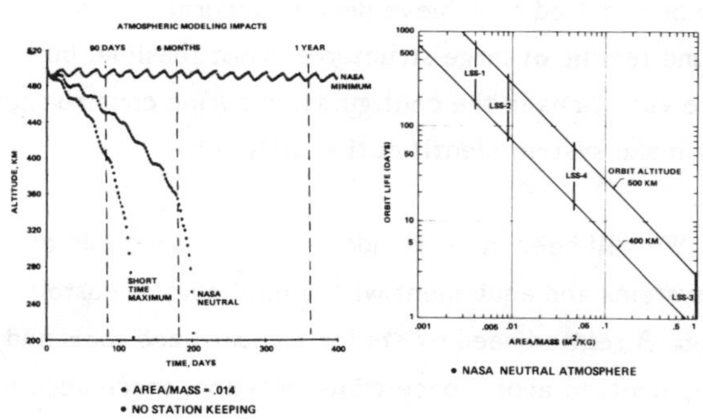

Figure 14. Orbital Lifetimes for Large Space Structures

Technical Issues

Space construction will require more than just fabricating large space structures in orbit. The central issue is that we must carry out final assembly, test, and checkout operations in space that have heretofore been done only in factory assembly areas employing far more labor content than one can realistically plan to use in space. The methods, tool, equipment, crew skills, and times required for test and checkout in space have not been well-defined. Better understanding is needed.

It will be necessary to predict accurately through analyses and simulation the flight and structural dynamics of these systems, both as final configurations, and in all phases of construction and assembly while mated with the space station and construction fixtures used in the construction process. These predictions must be made without benefit of full-size structural dynamic test articles and must be sufficiently accurate to ensure dynamical stability of flight control systems.

Large structures will be dynamically excited by thermal inputs such as emerging from shadow into sunlight if their thermal and structural responses have comparable time constants. Thus the control system may have to actively remove dynamical energy from the system. These dynamic characteristics along with control/structural frequency overlap dictated by control bandwidth requirements and low densely packed structural frequencies demand state-of-the-art advances in large space structure control and dynamic model identification to insure stability and performance.

These control techniques are dependent on good models of the structure with frequencies in or near the control bandwidth. Advances in on-orbit identification may be required to achieve desired accuracy because ground testing of large structures is not feasible, but large variations in the configuration during construction may make system identification difficult.

We will need more standardization, especially of subsystems and equipment with a minimum of customizing. A related need exists for standardized tools and equipment, to avoid space transportation, space accommodations, and crew training costs for specialized tools and equipment.

Along with standardization of subsystems, equipment, and tools, we will need to develop methods and technologies for in-space maintenance of large complex spacecraft. Initial steps have already been taken. The multi-mission spacecraft (MMS) modules developed by Goddard Space Flight Center are designed to be removed and replaced in space. The Space Telescope, under development by MSFC, is designed to be serviced in space. As development of techniques and equipment for in-space maintenance (often called "satellite servicing") continues, it will be important to develop a set of design standards for widespread use in the industry. These standards should enable spacecraft designers to confidently design for in-space maintenance.

The installation of subsystems and equipment as well as the assembly of complex structures from structural beam elements is not easily automated. MIT, under contract to the Marshall Space Flight Center, has conducted an extensive series of simulations in the water immersion facility where space-suited students simulated astronauts assembling complex structures from beam elements, using only hand tools and EVA procedures. Debates on suited EVA operations versus man-in-the-loop via teleoperation continue. "Robotics" occasionally is interjected into the debate, but experts on artificial intelligence concede that an automaton able to replace an EVA astronaut is decades in the future.

As structures increase in size the scale-ups of manipulator equipment from the size of the shuttle RMS will lead to progressively lower natural frequencies of the assembly equipment and correspondingly longer times to accomplish even simple tasks. This problem could be greatly mitigated by the use of advanced adaptive active damping control algorithms and research into optimization of the man-machine interface for control of the manipulator equipment.

The present shuttle EMU or EVA suit is not adequate for routine EVA operations. The suit pressure is 4 psi (pure oxygen). For a crewman to go from the space shuttle (or a space station) at 14.7 psi to the suit at 4 psi, he must first prebreathe pure oxygen for three hours to prevent risk of the bends from pressure reduction. Experiments with reduced cabin pressure at 10 to 12 psi have proven less efficacious in preventing bends than was expected. It is becoming clear that routine use of EVA from the space shuttle or from a space station will require an 8 psi suit with greater mobility than the present 4 psi suit. The technology for the higher pressure suit has been developed by the Ames Research Center. It remains to adapt this technology to flight qualified production space suit hardware for shuttle and space station operations.

Improvements in glove tactility and mobility as well as elimination of the water boiler suit cooling equipment have been studied and technology work is underway to effect these improvements.

Conclusions

A space station will be enabling for advanced space assembly operations. Benefits include the cost advantages of space-based upper stages and the ability to assembly and test large space systems and instruments not presently practical. Utilization of these technologies can make a strong contribution to continued U.S. leadership on the space frontier.

PANEL ON INTERNATIONAL UTILIZATION

Moderator: Kenneth S. Pedersen
Director of International Affairs
National Aeronautics and Space
Administration, USA

Horst Anton
Deputy Head
Department of Shuttle/Spacelab Utilization
DFVLR, FRG

Karl Doetsch
Assistant Director
National Aeronautics Establishment
National Reasearch Council of Canada

Luciano Guerriero
Director
National Space Plan
National Research Council of Italy

Georg Peters
Space Transportation Systems
European Space Agency

Masatoshi Saito
Assistant Manager
Space Shuttle Utilization Department
National Space Development Agency of Japan

Kenneth S. Pedersen: The international aspects of space station planning are very important in NASA's viewpoint, although they may not have received the visibility that some of the other aspects have. It has been obvious to those in NASA who have been actively engaged in this planning effort that international interest in NASA's space station planning has been extremely high. I think it's fair to say that representatives from most of the free world's spacefaring nations -- those countries that have cooperated with the United States in a variety of ways through the years -- have visited NASA to familiarize themselves with the current space station planning efforts.

A number of countries, including the countries and organizations represented on this panel have gone beyond that. They have determined that they have a sufficient degree of interest in participating in this planning activity to take the important step of investing their own money in similar and parallel activities, to look at how space station might fit into their planning, how it might relate to the objectives of their particular country or organization.

NASA has encouraged this activity, and we think that it has benefited us substantially to have the valuable contributions they have made to our planning.

The fact that they have funded these studies on their own has indicated a degree of interest and commitment that's also been important to us in demonstrating to people in the United States and elsewhere that this is a very serious planning effort and one that we intend to do in a thoroughly professional manner.

To some degree, the way we've approached the space station internationally has been just a little different from NASA's traditional procedure. Historically I think it's fair to say that in most cases NASA has done much of the early planning on major projects internally. Only after we had the concept quite well refined did we go out and begin to discuss with potential foreign cooperating partners what interest they might have in the project. In the space station we've attempted to involve other countries and international organizations that might be interested very early in the process, so that they could be aware of what our interests were, what our objectives were, and, of equal importance, so that we could benefit from their contributions, thoughts, concerns, and questions.

When we had our first meeting with some of these people more than a year ago, I said to them, "This is going to be for you an exciting and a frustrating process: exciting because I think you will see just how a program like this gets put together from the nuts and bolts stage; frustrating because you're going to suffer the stops and starts and uncertainties that all programs like this go through in the early stages."

I think almost everyone that's gone through it agrees that they have indeed enjoyed the thrill of the chase and the frustrations of the uncertainties that we've encountered. However, only they could tell you whether in their judgment it's all been worth it up to this point.

We have attempted to involve them in our planning very early in large measure out of self-interest. I don't think they're spending their money for philanthropic or charitable purposes, and we aren't either. We look at this process as a means of meeting some of our critical objectives. For example, we do not view the space station as a space spectacular. We view it as a working orbital facility. As such it is in our interest to have as much information as possible about user requirements, how space station customers and potential users will want to use the station, and how it can best meet their needs and their objectives. In a way this is helping us do something of a market survey: finding and talking to the people who in the future will find use for a space station and getting that information into our planning process.

We all recognize that the very scope and complexity of the space station process tends to suggest that foreign participation, if it takes place, is going to entail fairly sizeable financial and

---

From edited transcript.

political commitments on their part. Thus, it's only fair that they have the opportunity to be involved early so that they can judge in the most fully informed manner possible just what the risks and the opportunities available to them are. While we may have had some stops and starts and some uncertainties, we have also -- I think they would agree -- honored our commitment to keep them informed. I believe that when and if the time comes that we have the opportunity to proceed "full speed ahead" with the space station, they and their countries are going to be in a better position as a result of this activity to know what their interests are, to know what their level of participation might be, and in what areas that participation might be most mutually beneficial.

Horst Anton: I want to emphasize some ideas that I would like to call ground rules for participation in the space station program.

The first, and in my opinion the most important one, is that the element of our participation has to be an integral part of the German/European space program. The technical layout of this element should allow for independent use and operation of the system. By independent, I mean independent of the space station architecture.

We would like to have continuous use and operation of this German/European element of the NASA space station architecture. There is clearly no intention of single-shot participation.

We offer fair partnership and expect the same from NASA. Here, fair partnership means a fair balance of services and a fair balance of money flow, cash flow, bartering, etc.

We offer access to all relevant information during the preparatory and operational phases of the space station program. By this, I mean the software or paperwork. In principle, we want to have access to the entire space station architecture for scientific and commercial use.

We want to see true competition for German/European industry to design/build those elements or hardware developed for existing and future programs.

We would like to see clear technical and organizational interfaces. However, for the present, the cooperation should be on a governmental level only.

The last ground rule is also quite important: the cooperation should be based on a long-term intergovernmental agreement, similar to that we already have for the Space Transportation System. A similar agreement should be extended for the space station utilization.

Regarding German/European utilization of the space station, I have one basic assumption: Europe will not only stay in but also increase its engagement in manned space activities. With this presumption, I will try to outline what this participation might look like:

1) Europe has to participate in the space station for political and user-oriented reasons, the political reasons being "transatlantic cooperation"; the user-oriented reasons being that experimental programs are already underway in Germany and Europe which clearly show an interest in utilization of the space station.

2) Germany puts special emphasis on the lead function for these activities, not only during development but also during the utilization phase.

3) We are interested in being an attractive partner to NASA--attractive from both the financial and technical points of view. I think we have already proved our technical capabilities and potential to be an attractive partner.

4) The participation should evolve from existing programs and hardware, e.g., Spacelab, the free-flying system SPAS, and the European retrieval carrier Eureca.

5) You may know that we have made extensive investment in the infrastructure of the German Spacelab mission D-1. We would like to see this investment utilized in, for example, integration, operation, crew training, and mission management of our own elements in the space station architecture.

6) The elements of our participation should fit into our own program, but our program must also be of interest to NASA.

7) European investments should be an integral part of the space station architecture. These investments will be made available to NASA and we would like to see fair compensation for the use of this investment by NASA for their own purposes.

We have identified several possible elements of this cooperation/utilization of the space station architecture, fulfilling these ground rules, which in my opinion, seem to fit into the overall NASA planning. Without going into great detail, I mention only four projects: Columbus (a manned payload-dedicated module), free-flying platforms, orbital transfer and logistical vehicle (a modular system), and modular solar array (a double-folded + 15 kW system).

DFVLR will have a detailed proposal regarding international cooperation/utilization by the fall of 1983.

To summarize: the good cooperative relationship already existing between Germany and NASA in the Shuttle utilization program will be continued. I'm quite convinced that in a few years there will be no symposium on the possible international utilization but rather one on the results of international utilization of the space station.

Karl Doetsch: Canada is a vast country. It has a sparse population that is not uniformly distributed, and it's blessed with natural resources that are the envy of many. For these reasons, we have been involved in space activity for a long time, both in terms of talking to ourselves and in terms of looking at ourselves and our resources. It was quite natural, then, that when the invitation came along to participate in looking into the benefits of a space station to Canada, we should grasp this opportunity to investigate how indeed this future development could aid us.

James Beggs[1] indicated that NASA was interested in international participation, that one of the reasons was competition, and that cooperation would be a better means of focussing the activity than perhaps straight competition with individually limited resources would be.

A country such as Canada cannot compete across the board. We must look very carefully at those aspects that are of use to us. We have done this through user surveys. We have gone to our industry, universities, and the public sector, to see what use a station might be to us.

Education of the potential users is important. There's a good feeling that comes to the fore immediately, which says the space station is great, we want to be part of it, we want to be part of the whole development. However, as has been pointed out several times, one needs a little more than that. One needs to find good reasons for it.

The interest that has developed has been extensive and in our view surprising when you consider the time scale before such a space station will become a reality. Not surprisingly, the key interest shown in Canada has been in the area of remote sensing, because we do need to do a lot of that, and also in technology development.

We're currently evaluating those responses in terms of their strategic benefits, and our technological readiness to support them.

The results, as we see them at the moment, are that the space station is necessary, but not sufficient as a single unit. A space platform will be required--in fact, more than one. At the very least, it needs to have a high inclination. We will not define at this point exactly what that inclination should be, because there are options. You can have other platforms at higher inclinations to give the same frequency of coverage demanded by our users.

The other outcome of our study so far is that the technology development to support our users is very closely related to the technology development required in the design of a space station or a space infrastructure. Many of the sensors now being proposed for remote sensing do require higher power, better data handling, and better turnaround than those currently existing. These will feed very significantly into the basic development of the space infrastructure.

On the technological side of things, our perceived need at this time is that a research and development laboratory should be an appropriate and important part of the first space station.

Also, man will be an important part of that first space station, for reasons that really follow what Jack Lousma was saying--that man adds flexibility.[2] We are talking about an evolutionary system, which means that we need to build in flexibility because we don't know quite what we want to end up with. This is not to say that there is going to be a requirement for an exponential growth in the number of people in space. It is saying that as you develop and automate some functions, there will be new functions requiring man's direct participation. We may even level off in the number of people required.

All of this leads to what we see as the space infrastructure. Figure 1 shows that the space station is more than a single entity. The whole infrastructure, which involves not only the station but the platforms and how you get from one to the other, needs to be reviewed in the same way that one reviews matrix management vs vertical structure management. In general, as the projects become larger, there are advantages of flexibility which come about with matrix management, and we feel that those aspects are lessons that probably will be applied in developing an infrastructure such as shown in Fig. 1.

How much autonomy, how much flexibility, how much control on a project basis, what the single elements are, how much sharing of data, how is control at the top exercised, and how can one maintain a focus of activity -- these are the sort of questions being addressed at this time, although we're not yet ready to lay out exactly what we would expect of an agreement.

We also think that the space station as a development and as a technological stimulant has strong justification in its own right. This is coupled with the needs of the users, but it musn't be forgotten.

The rate of return on investment is important, but the strategic benefits are also important.

When man does go into space, we need to work very carefully to make sure that the proper symbiosis between man and machine is arrived at--and not only between man and machine, but also between the serviced and unserviced elements that will form part of the infrastructure.

We're positive about the steps which are being taken at this time. We anticipate going to our government in the same time frame that NASA is going to its for support and funding. We look forward to participating in a development of this excitement.

Georg Peters: I will briefly summarize where the European Space Agency (ESA) stands in the planning and preparation for utilization of a space station. However, let me briefly recall that the first step in US/European transatlantic cooperation is now at the point of successful termination with the Spacelab development, funded and developed in Europe, and with the preparation for the first Spacelab flight. Also, the utilization of the Shuttle Spacelab by European organizations beyond Spacelab I is in preparation, as has been mentioned. The most important development is the D-1 mission, with the considerable contribution of ESA experiments. The International Microgravity Laboratory is under discussion and early planning and the European retrieval carrier (Eureca) which will make use of Shuttle's deployment and retrieval capability is in development. The first mission is planned for 1987.

However, the next major opportunity for international cooperation is the space station program.

1 See Keynote Address.
2 See Jack R. Lousma, "The Role of Man in Space."

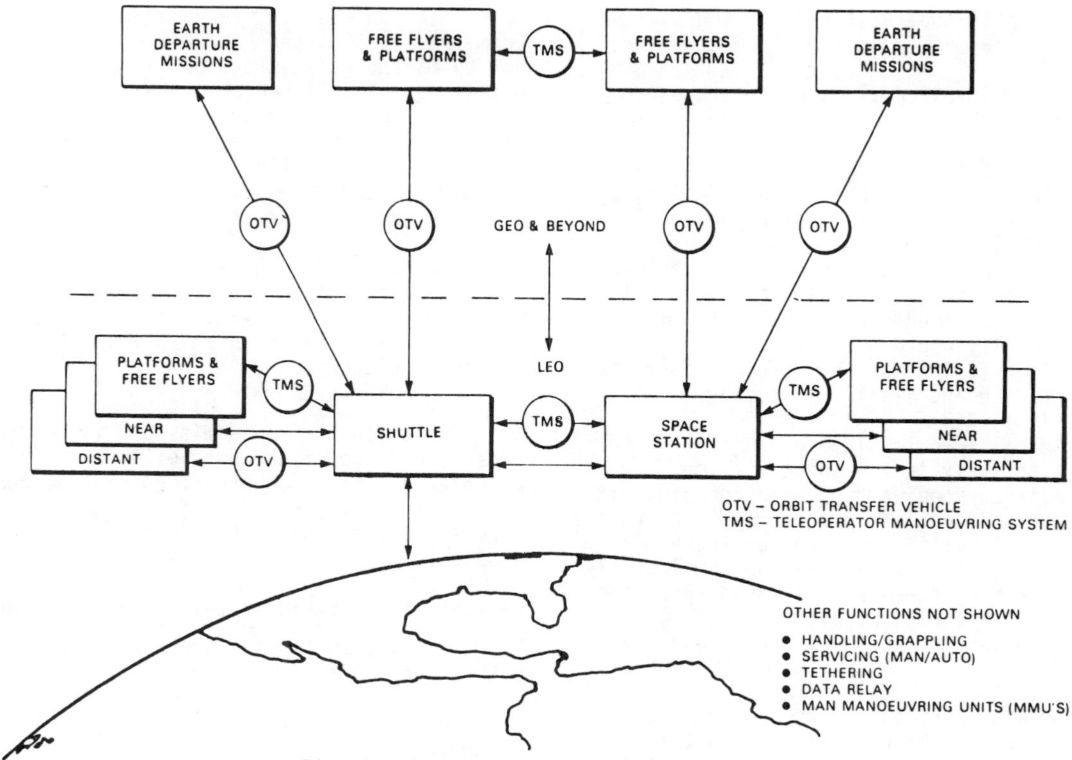

Fig. 1 Elements of space infrastructure.

The member states[3] have asked ESA to prepare and plan for a possible European participation in a space station program. This activity is part of an evaluation of ESA's long-term participation in the STS program which comprises scientific trans-atlantic cooperation, definition of the future European launcher, and the so-called space in-orbit infrastructure (IOI), including automated platforms. Decisions on the major new European STS elements can be expected in the 1985-86 time period. This date seems to fit comfortably into the present NASA space station program planning.

The results of the European-user survey on space station utilization are only preliminary, but contact is being maintained and several updates are planned.

The results have been analyzed and a utilization study will be sent to DFVLR and other ESA advisory user groups.

The greatest support has been in the micro-gravity research area, particularly in the life sciences, where whole new areas of investigation can be opened, and in materials processing, where results of the experiments in space may lead to a better understanding of ground-based processing and preparation for commercial processing on a space station.

The attitude of European space scientists has been more conservative. Those developing programs for the late 1980s or early 1990s are not comfortable with the proposed space station orbit. However, there have been proposals and ideas mentioned for instrument packages that could be directly attached to a space station.

[3] ESA members are: France, West Germany, United Kingdom, The Netherlands, Belgium, Luxembourg, Ireland, Denmark, Sweden, Italy, Spain.

In the area of Earth observation, there has been a limited interest in support of operational missions, in particular the use of the metric camera, which will fly on Spacelab I. However, there is considerable interest in advanced instrument testing, for example, an infrared radiometer and calibration of laser operations.

In the area of telecommunications, the present baseline is the direct launch of Ariane into geostationary orbit. However, a space station could be of use in advanced payload testing.

The area of commercial space station utilization is one that in Europe still needs a lot of work. The motivation of private industry to invest is still in the embryo stage, particularly for pharmaceutical or materials processing in space. However, profitable demonstrations of the type proposed by McDonnell Douglas and Johnson & Johnson, described earlier, could act as a catalyst and accelerate the process.

We are planning to intensify the contacts with commerce and industry and will concentrate such contacts on the most promising possibilities, since the use of a space station for commercial applications may be a major motivation for Europe to participate in this program.

I should briefly summarize the elements necessary for European participation in the space station. ESA has conducted, with the help of the European aerospace industry, four studies on what potential elements are wanted. The Spacelab-related applications include the pressurized module, the unmanned pallet, and instrument-pointing system. In the area of Eureca-derived applications, an evolution to automated free-flying platforms appears promising. A great interest has been expressed in the teleoperator maneuvering system, particularly in the European aerospace industry.

Selection criteria will have to be applied to reduce this still large number of possible elements.

Finally, I want to comment on the operational schemes for the space station. Europe definitely looks for some kind of operational responsibilities, which will depend on the level of utilization and on the kinds of elements to be provided. As an example, an attached research module could lend itself to very good operational autonomy. This responsibility would be different for different levels of participation in different elements in a space station and its infrastructure.

Summarizing, I would like to say that Europe participates actively in the space station planning. It has established good working relationships with the NASA Space Station Task Force. It is currently spending about $1 million per year on these planning studies and this amount could be considerably increased. ESA has requested increased funds for the considerable effort that will be required for the Phase B studies in case NASA goes ahead next year.

Dialogue with European users has been established on space station utilization and these contacts, particularly for commercial utilization, will be intensified. The European aerospace industry has started planning and definition efforts for the space station elements. A tentative data exchange with potential American partners has been started. This data exchange will be improved should we go into Phase B. ESA continues space station planning with NASA and with its European member states.

Masatoshi Saito: The Japanese Space Activity Commission (SAC), the national commission for policy making and overseeing all Japanese space activities established an ad hoc committee on the space station program to study the potential utilization of the space station and related candidate development items. This ad hoc committee issued an interim report on June 15, 1983, which I will briefly summarize here.

Figure 2 shows the current thinking on the Japanese utilization of a space station. Astronomy and astrophysics observation includes typical missions such as a large infrared telescope and x-ray telescope clusters.

The timing of these missions is not decided at present because of the possibility of international cooperation, in which case the international science community will decide more precise time frames. At present we can just show the candidate missions.

In the early 1990s, we think that the conventional satellite is enough for Earth observation. But in the 1990s, new Earth observation missions might come up, such as the very large aperture radiometer antenna. So early in the 1990s, a technology will be developed in advanced space technology experiments

In the area of communications, until the mid-1990s, conventional satellites will be sufficient for Japanese purposes. But beyond that time a large communications satellite may be needed, its assembly and launching made possible using a space station.

		Candidate Development Themes	
		MSS Attached Module	Subsatellite Type Platform
Candidate Space Station Utilization	Science Observation	•Astronomy & Astrophysics Observation (High energy particle, etc)	•Astronomy & Astrophysics Observation (Infrared ray, visible light, X-ray, submilimeter wave, etc)
	Earth Observation	Ocean, Land, Atmosphere Observation (which require the analysis or judgment by crew)	Ocean, Land, Atmosphere Observation (which abhor the waste gas pollution, the disturbance of pointing, etc)
	Communication	Large Communication Satellite Assembly, Launching	
	Material Processing	•Material Processing Experiment Commercial Material Processing (which require the operation and the continuous monitoring by crew)	•Material Processing Experiment Commercial Material Processing (which require the high quality micro gravity)
	Life Science	•Life Science Experiment Commercial Bio-processing (which require the operation and the continuous monitoring by crew)	•Life Science Experiment Commercial Bio-processing (which require the high quality micro gravity)
	Advanced Technology Development	•Advanced Space Technology Experiment	•Space Energy Experiment

• Shows the candidates for the early phase of '90's

Fig. 2 Japanese proposals for space station utilization.

Japan is just beginning a space experiment program using the Space Shuttle. Our first experiment has a launching time frame of 1988. So in the early 1990s, materials processing experiments will still be in the research and development phase. A manned laboratory is most efficient for us in that time period.

But later in the 1990s, some materials processing such as compound semiconductive single-crystal growth, electrical device manufacturing, and heat-resistant alloy processing may be investigated for future commercial applications.

The life science area is somewhat similar to the materials processing situation. In the early 1990s, a manned research laboratory will be necessary for life science experiments. Commercial bioprocessing, such as separation and cultivation, will come up in the late 1990s.

In the area of advanced space technology experiments, a manned facility is needed. These experiments include the assembly and testing of large space structures, such as antenna and solar array panels, and the development of service technologies such as teleoperations.

Another advanced technology development area--space energy--is high-power and distribution. The investigations should include such things as a space plasma experiment, electric propulsion tests, and microwave energy transmission tests.

Figure 3 shows the relationship between the elements of the space station and possible mission objectives. We have made a preliminary study of the modules and platforms needed to accomplish our suggested missions. The question is--what kind of functions should these modules and platforms fulfill? We will continue this effort in the coming year.

To summarize Japan's thinking on the space platform:

1) Science observation: most observations will be made using a space platform, but some will be done with modules.

2) Earth observation: this is still not clearly defined, but, for example, our ocean observations require trained personnel in the loop of these systems.

3) Communications: the large communication satellite assembly and launching will be done on the module or in the vicinity of an attached module.

4) Materials processing: in the early 1990s most experiments will be done in attached modules. But some experiments, such as chemical vapor deposition, will be done on a platform. Later, commercial materials processing will be done primarily in a more automated space platform.

5) Life sciences: a scenario similar to materials processing.

Before closing I would like to quote Japan's interim report on the benefits and impact of a space station. It said that: 1) space processing and commercialization can be realized; 2) a stable and high level of science and Earth observation can be realized; 3) a very large space system can be built; 4) the risk of losing all satellite communication as a result of a partial failure can

	Early Phase of '90s	Later Phase of '90s
Science Observation	Astronomy & Astrophysics Observation	Astronomy & Astrophysics Observation
Earth Observation	Technology development at "Advanced space technology experiment"	Ocean, land, atmosphere obervation
Communication	Technology development at "Advanced space technology experiment"	Large communication satellite assembly, launching
Material Processing	Material processing experiment	Commercial material processing
Life Science	Life science experiment	Life science experiment; Commercial bio-processing
Advanced Technology Development	Advanced space technology experiment	Advanced space technology experiment
	Space energy experiment	Space energy experiment

Fig. 3 Space station utilization.

be minimized; and 5) the space station is the first step of the enlargement of the living space of human beings.

Luciano Guerriero: Italy's interest in the space station derives from the fact that besides our participation in the ESA activity, we have in the past few years been developing a national space program that considers NASA's Space Transportation System as a key element in its future activities.

This program has several projects, such as the interim upper stage system and a data satellite, both of which would be operated in conjunction with the Space Shuttle.

The results of the studies presented at this symposium seem to confirm substantially our previous basic understanding of a properly designed space station, which will support both science and application developments in the medium and long term.

We fully agree with the NASA evaluations of the key role that a space station will play, together with the Shuttle, to obtain real cost effectiveness for the NASA Space Transportation System. We consider this the most fundamental factor in the development of all future activity in space.

This conclusion remains valid in our opinion in spite of the understandably cautious position of the scientific and national security communities. We deem that these positions are mainly dictated by short-term budgetary constraints.

Moreover, we fully share the point of view so clearly expressed by NASA Administrator James Beggs that the most important economical benefits will probably be those that we have not yet dreamed of. In this sense, we fear delays might come from preconditioning the real launching of this program to an a priori guarantee of commercial return.

With respect to the space station utilization, Italy has contributed to the ESA mission analysis study, which indicates several potential uses in the space science and application areas and in which the Italian community is also planning to operate.

In addition, in the framework of the ESA programs, Italy is very interested in activities based on the utilization of Spacelab and Eureca, which can be employed only in conjunction with the NASA Space Transportation System.

We have been listening with great attention to the renewed NASA interest for international cooperation in the space station program. We agree that cooperation is the best way to reduce duplication of effort and to optimize the economic efficiency of such an enterprise. Cooperation, not competition, should be favored among Western countries, establishing rules that will not endanger the success of the program and not be a prejudice for the legitimate interests of the partners.

This cooperation cannot be limited to the utilization of the future space station, but requires some form of involvement of the partners in the design and the development of the system. In this perspective, the National Science Council and NASA cooperation program for the joint development of the tethered satellite system contains all of the basic ingredients for the type of cooperation we have in mind. In fact, it will produce a well-defined system. It will maintain a clear interface in the definition of roles and responsibilities, and it does not require any exchange of money between the participating countries.

The tethered satellite system can also be considered an ideal candidate passenger for the space station, where it could benefit from the longer mission duration to perform all of the experiments now planned in connection with the Shuttle.

We'd like to take this opportunity to suggest to the NASA Space Station Task Force consideration of the tethered system as a permanent operational facility to be installed on the space station to extend its capabilities, no matter what architecture is chosen for the station itself. In this regard, the Italian program and NASA are already carrying out a joint study to identify the possible applications of the tethered concept to the space station.

The Italian space program is organized in terms of five-year plans: however, we have provisions for updating and reorienting it every two years to take advantage of any evolutions in the field. Therefore we can quickly adopt appropriate action to stay in step with our cooperation program with NASA, including the space station development.

For this cooperation to continue, we need a clear understanding of the NASA policy on international participation in the space station program. We hope that this will soon be available.

Pedersen: My experience suggests that the United Nations could take this panel as an example on how to keep remarks short and to the point and leave some time for some--hopefully--relevant questions.

I would briefly summarize the panel's comments by saying that I think what you've heard is indicative of both the interest that I spoke of in my opening remarks, as well as the candor and frankness with which this whole international dialogue has been conducted.

NASA has been aware throughout this space station study that we do not have an approved program and, are therefore, not in a position to begin discussing agreements or specific formal arrangements. Thus, we have not wanted to create unnecessary and unwarranted expectations that would come back to haunt us, or to find this consulative process we started in an open and positive manner backfiring on us. To date that has not happened, and I think the reason it hasn't happened is that we at NASA and the countries that have been working with us have tried to be as open and candid with one another as we could in terms of the state of play, the current situation with respect to the decision-making process, and what we believed a realistic schedule would be. Some of that candor and concern has been voiced here today. I take heart from it.

Roy Gibson, GTS (United Kingdom): I'm not sure whether I'm in order, because the title of this panel is international utilization. But participation is also mentioned, which is quite different from utilization.

Several of the speakers, particularly from Germany and ESA, mentioned the possibility of hardware contributions from Europe. This seems to set up the need for a whole different examination or a whole different policy statement, which goes beyond NASA into the U.S. State Department or even DOD realms. It is pretty hard to see how this can be done on the time scale that's being talked about now.

Is consideration being given to this, and are we learning from the Spacelab/Shuttle discussions that went on about 10 years ago, where Europe was always behind the ball because the political issues were never faced until much too late. Thus, what came to be a free choice from the large washing list of possibles was scrubbed down to the inevitable Spacelab. Is it going to be the same in the space station era, or are we going to have an earlier start at these necessary, very boring, political discussions?

Pedersen: You sound like someone who knows of what he speaks, having been an active participant in the Spacelab negotiations.

We are trying very hard to learn from the Spacelab experience and from all of the discussions that surrounded it in the early 1970s. And I mean learn in both the positive and negative senses, because I think there were some good lessons in that experience.

Of course, to be realistic, we have to recognize that there are certain lag times in political decisions that are difficult to force in any circumstance. Yes, it would certainly be a very good thing right now to be in a position to say that when the space station is approved, here is the policy that will be followed, here are the ground rules under which international cooperation will take place, here are the financial arrangements, here are the quid pro quos, and so on. We would very much like to be in a position to do that now, instead of just sitting on the sidelines, ready to trot in when the game begins.

Realistically, I think that we have to take a step-by-step approach. What we are doing in this regard is:

1) Trying to get individual countries and international organizations involved very early in the program so that they can voice to NASA, the U.S. State Department, and other interested participants (including American industry) their concerns, the things they feel they need, the kinds of guidelines and policy they need, etc. If we are aware of their concerns, hopefully we can be more sensitive in developing that policy.

2) A part of NASA's efforts, as well as in other U.S. government agencies and departments, is a major current interagency study that culminates in a senior interagency group, a so-called SIG, that brings all of the various agencies together, including the ones you mentioned. International participation and involvement is an issue specifically on the agenda of that group. So the problem is being addressed on an interagency basis. I would be the last to say that having such a process in place guarantees the proper outcome in the proper time period, but it is the way that I think you can maximize the chances that it will happen.

I think the two toughest questions we're going to have to deal with are, assuming we get approval of a space station, what systems, components, or pieces of hardware are best provided internationally? I say "best" in terms both of NASA and U.S. interests and of those who might contribute or provide the hardware, components, etc. As both Horst Anton and Georg Peters indicated in their presentations, there have already been significant investments in Europe and elsewhere in hardware that might be well suited to a space station. These things have to be taken into account.

My hope is that this issue will be resolved in productive discussions and that we will not be in a situation is which at the last minute we are all dealing from a very rapidly diminishing list that leaves everybody on both sides unhappy and dissatisfied. We're working very hard for that not to happen.

Gibson: I'd like to encourage you to follow up on the ideas that you have just expounded. There's a considerable concern in the United Kingdom at the moment over, shall I say, the ambivalence between policy at the most senior levels and the actual policy that has been implemented at the working level. One can see that in the debate on the commercialization of satellites, where the official view is that agreements entered into a number of years ago are now being interpreted as mere understandings. That causes considerable concern.

On the operational side, when information and technology need to be transferred, one sees the attempt to apply the Munitions Control Act. This is causing considerable concern, and I'm certain that such factors have to be fully tackled and resolved to give the programs you anticipate following the best chance for success.

Pedersen: I agree. The technology transfer question is one that I think NASA has dealt with very openly in the past and it is still under discussion with regard to the space station. In its cooperative programs through the years, NASA has had to face this question of technology transfer many times. I think we have arranged our programs in ways that minimize that kind of transfer through providing for clean interfaces and requiring that the contributing partner assume full responsibility for the particular component being developed.

I might add that even if technology transfer were a major concern in an earlier period (that is, of American technology being transferred abroad)--and maybe it was, and perhaps legitimately so--the rapid growth and increase in capabilities abroad tends to suggest that we're now facing a much more balanced situation than existed in the early 1990s when we were looking at the Space Shuttle. I think that while technology transfer has to be a concern, it's also got to be a concern for the parties we're talking about cooperating with. I think this is an area where good solid self-interest tends to be the best regulator. If there's anything I'm sure of, it is that we are all approaching this from a sound basis of self-interest. I hope that none of us here are looking at the space station any other way. I assure you that NASA and the United States are considering the space station primarily in terms of self-interest.

Question: You mentioned that two tough questions were on the agenda. The first was what components or pieces of hardware were to be considered for international participation. But, I did not get the second tough question in my notes -- What was it?

Pedersen: I was referring to the SIG agenda, which includes international cooperation. In the early 1970s European organizations were invited to look at the Shuttle and a number of possible cooperative ventures were raised. However, when it came down to the final moments, a lot of those suggestions began disappearing as candidates for international cooperation. The Space Tug was perhaps the most famous casualty.

We are trying very hard to avoid a situation in which the opportunities to cooperate become vanishingly small and the vital and important systems of the space station are all reserved for the United States. If that were the case, our question about international cooperation would be solved: we just wouldn't have any. We might have utilization, but we wouldn't have cooperation.

Question: There is a technical answer to the policy question raised by Roy Gibson. From the very beginning, we envisioned the space station as a distributed architecture type of device, as has been described by all of the speakers. Because of that distributed architecture approach, NASA thinks that the problems of coexisting with various constituency users will not be anywhere near as difficult as they have been on a highly integrated system like the Shuttle. The bottom line is we feel there's an architectural solution to that problem.

Pedersen: I appreciate that remark. The very nature of a dispersed system tends to make some of the technology transfer concerns a little easier to deal with -- hopefully.

Question: What tax advantages are being considered by foreign governments for private domestic concerns in their countries as an incentive for private space investment. The reason I ask is that one of my company's major clients is a certain Canadian oil concern that is very interested in the private development of Earth-sensing technology. I would suggest that this is one area that I believe most nations would find a very useful mechanism for stimulating their participation in space and that we have found a surprising interest in space by private enterprise overseas. I believe that will eventually be the major means of foreign participation.

Doetsch: That is a good point. Certainly as the users emerge and the Canadian government's relationship with those users is developed, it would be a factor to be considered in our space program. But it is too early at this point to make any specific statement as to which way those discussions might go.

Question: There has been a lot of discussion of international cooperation on a space station. The question is to what extent will there be international cooperation and coutilization of the facilities. There has been a lot of discussion about separating the German facility, the ESA facility, and the American facility. Will there be any effort to allow American scientists to use German facilities, German scientists use American facilities, etc.? What are the political and policy aspects of that question?

Pedersen: The opportunity to have a variety of facilities provided by a number of countries that are then available to all participants strikes me as a very efficient way of operating. I think we want to avoid an excessive duplication of facilities, i.e., where each country had their own particular laboratory. I see nothing right now that would rule out joint facilities.

Anton: Such cooperative facilities already exist in the D-1 mission and the planned follow-on missions. There is also a program called the International Microgravity Laboratory in which we are trying to incorporate this philosophy. Half of the elements are provided by NASA and half by Germany. There will be common use of this equipment. We could just extend those procedures to the space station later.

Pedersen: We should not confuse the philosophy of distributed architecture, which is the basis for solid international cooperation, as also applying to international utilization. International utilization would certainly mean cooperative use of those various facilities. Even though the program talks about international utilization, we are also talking about international cooperation, and the two are quite different. Each might very well move forward using different approaches and philosophies.

Question: There is a major element of space station capability missing from the program of this symposium -- the Soviet program. There has been considerable cooperation on the part of the French in recent months with the Soviet Union. Are there any prospects as the Soviets progress with their space station activity (which they say will expand considerably during this decade) of any cooperation, coordination, or interaction in any way with their work?

Is there a prospect for using nongovernmental organizations, such as the IAF or the International Academy of Astronautics to further this kind of east-west activity.

In other words, is this something we're going to continue doing separately for the rest of time, or do you see any situations in which there might be some sort of coordination?

Pedersen: I think that the situation with regard to the US-USSR cooperation in space has always been ruled by political realities and political issues that transcend space per se. It has tended to reflect rather than affect the climate of relations between the two countries. So one had a situation like Apollo/Saylut, which occurred at a time when detente or rapprochement was the order of the day.

I am not a foreign policy expert, but the current situation strikes me as not being particularly conducive to large-scale joint American-Soviet undertakings in space, which at this time would have to include space station activities. That is not to say that the situation could not change, but my guess is that it would be as a result of political changes and decisions that will be made for outside of, above, or beyond NASA.

NASA has had experience working with the Soviets through the years. That body of experience and knowledge still resides in NASA. There is also the acquaintanceship of people in both countries. But, I do not think the political will is there right now.

Peters: I have to say that I am not authorized to give any official statement on this topic, but my personal feeling is that there will be no cooperation with the USSR on the space station. From the Russian side there is some inclination to get us involved. They have offered to use their rockets to put European satellites into orbit. But my feeling is that it won't work this way.

I stated earlier that at present cooperation must be only on the government level. However, in my opinion, that restriction applies only to the present starting phase of the space station. Later on, there will be a whole lot of cooperative projects, governmental and nongovernmental.

Doetsch: I think that the infrastructure in general will be developed by specific funding from various governments, and they will probably maintain control over how those funds are spent. So I think it quite unlikely, in the early stages of the development of the infrastructure itself, that that authority would be passed over to nongovernmental parties.

LUNCHEON SPEECH, JULY 19, 1983

The Honorable Don Fuqua*
Committee on Science and Technology
U.S. House of Representatives
Washington, D.C.

In the enumerated objectives of the National Aeronautics and Space Act of 1958, which we celebrate 25 years of this year, one objective is specifically devoted to the preservation of the role of the United States as a leader, and I underline the word leader to you, in aeronautical and space science and technology. That leadership shines through more than 25 years of our country's space history and its diverse accomplishments in space activities. These activities have not only carried us beyond the confines of our own planet, but they have brought us knowledge of Mars, Venus, Saturn, Jupiter, and, through brilliant achievement, in planetary exploration.

We've expanded as well as reduced the cost of world communications and, through remote-sensing techniques, we've been able to monitor the planet's weather and agricultural patterns and better locate our own resources. We've developed and operated with extreme success a Space Transportation System whose potential uses and scientific achievements and benefits are just beginning to be appreciated.

So, we've done very well. We've been very fortunate. But the challenge of leadership is always to do better, and a firm commitment on our next major space goal will help us to meet that challenge by continuing leadership. Through that leadership we'll be better able to foster the cooperation of all the free-world nations to keep space an open and accessible domain to the nations and citizens of the world.

In planning America's future space activities, we must allow for international cooperation. Over the last few years, we have increasingly seen the development of the space station concept as being particularly adaptable to an international cooperative effort.

We know that the lead times for major space systems are long. The cycle for a space station could approach 10 years. This, coupled with the Soviet commitment to a permanent space station necessitates more than just studies and reviews of the space station possibilities on the part of the United States. We need a firm presidential commitment to the space station goal to protect against any short-term political and economic events that might cause deflection of such a decision.

This does not mean that the specifics of a program plan cannot and will not change or evolve as we develop the concept further. It means rather that the plans can remain flexible while the commitment remains firm.

It's important to view the idea of a space station as a concept, rather than as a physical outpost in space. The phrase "space operations center" has been used frequently and I think it conveys a broad meaning: a center around which independent platforms could offer both diversity and flexibility. In this way the process of development could be evolutionary, as functions and uses are needed and components can be added. In fact, we might well begin with an unmanned platform and evolve into manned stations.

In our consideration of plans for a space station, we should not make the mistake of concerning ourselves solely with the engineering and technical developments. It is also important to plan the institutional mechanism to present optimum opportunities for the private sector's participation. These opportunities should not be limited to the manufacture of products and materials in space, but should also include identifying with parts of the overall infrastructure, and how that could be provided to the private sector.

As we begin to define the potential user community of the space station as a conglomerate, we should also make a concerted effort to include that community in the planning process. They will not only be able to identify their own needs and capabilities, but also provide innovative suggestions for the overall space station scheme and consequently a unique perspective.

The development of technology for a proposed U.S. space station could begin before the system requirements are established. The technologists should anticipate requirements. Although we are able to build a space station with Skylab-level technology now, it would not allow much potential for growth and it certainly would not enable us to benefit from the advantages and the advancement that innovation can bring.

What has also characterized America's space program to date is the overriding creativity of effort. I think that a major new product, such as the space station, offers us the opportunity to continue pushing the frontiers of our technology.

As the Space Transportation System moves us headlong into routine space travel, we need to focus on how each sector of our economy can use the new access to space for its own commercial benefit. This will be the responsibility of business and industry in particular. They will need to develop their own ideas and make plans. The government cannot and should not be dictating where and what the commercial opportunities are. And the borders of business and industry do not stop with those of the aerospace community. If there's any barrier that separates aerospace from nonaerospace, as far

From taped transcript.
*Chairman.

as the commercialization of space is concerned, it does not exist in practicality, but rather in the minds of those who perceive it that way.

A few years ago in the House Science Committee, we introduced legislation known as the Space Industrialization Act, the purpose of which was to stimulate thinking in the business/industrial sector about the possible uses of space and its unique characteristics to promote product and process development. As business begins to operate in the space environment, they will no doubt see opportunities emerge in directions that were previously unimagined.

As we considered then, when we undertake a space station, we must remain steadfast in our commitment to continue a healthy space science and applications program. We know that some tasks are best accomplished by free-flying unmanned satellites, while others must be done by man in space.

A major factor in our continuing strength and success in space will be the diversity of our efforts. Space science and space stations are certainly not mutually exclusive. It is now the time to move from the contemplation of a space station to a commitment. I cannot think of a more exciting era in which to live or a more exciting endeavor with which to be associated. We are indeed fortunate to have a part in such plans. And if our efforts equal the inspiration of our task, we will surely continue our leadership and success in space development.

SPACE STATION ARCHITECTURE

Chairman: Lt. General J. A. Abrahamson
Associate Administrator for Space Flight
National Aeronautics and Space Administration
Washington, D.C.

In January 1972, citing national priorities, President Nixon gave the go-ahead for the Space Shuttle, but vetoed the combined development of the Shuttle and a space station. Obviously, the Russians made an opposite decision at about the same time. However, NASA has quietly kept space station studies and activities alive in the intervening years. These efforts have intensified since James Beggs brought up the matter during his confirmation hearing before the Senate in 1981.

Although it is fun to design space stations—what modules to have, even what color they should be—the effort of NASA's studies to date has been to answer the question: Why should the United States build a space station? Hence, the numerous studies of requirements for a space station made to date, aimed at generating new ideas and possibilities as part of the revolution in space that we all hope is and will be taking place now and in the future. The Shuttle is the current agent of that revolution, proving the capabilities of a work horse vehicle and human beings working in space. In addition, as time goes on, the limitation of the Shuttle will become obvious as it goes back time and time again to the same region in space. This limitation will lead to a second step that will further this revolution or evolution—and that is what the space station is all about.

The presentations at this session will not detail the nuts and bolts of the space station—nor its color. They will be describing the architectural approach to a space station. By architectural approach we mean the entire space infrastructure with which NASA and others are trying to proceed in an orderly manner. Such an infrastructure includes manned elements, unmanned elements, elements in one initial orbital orientation but with the possibility of adding others later and at different altitudes. It may include remote vehicles, such as the teleoperator maneuvering system, and may even include an orbital transfer vehicle.

The point of these presentations is to emphasize the diversity of the space station and the consequent fact that the station is not a singular element. Rather, it is the next step in the full infrastructure for near-Earth operations, which will, hopefully, in turn be the springboard to the numerous possibilities beyond near-Earth orbit. What are those possibilities? Well, how about going back to the moon, or to Mars, or even to the stars beyond?

So, the space station is the second step and as the requirements and some details of a station structure are discussed in the following papers, keep in mind such long-term possibilities.

ARCHITECTURAL OPTIONS AND DEVELOPMENT ISSUES*

Luther E. Powell**
National Aeronautics and Space Administration
Washington, D.C.

First, I want to clarify the definition of a space station. When we say space station, most of us automatically lock on visually to this thing that we're now calling the manned core. But if you look at what NASA has tried to do with the mission analysis contracts -- to search out the requirements -- you can see that the view has broadened to the point that everything possible is included to satisfy the mission requirements being developed.

When considering the space station, keep in mind that what we're talking about are those elements that will be used to conduct the missions of the space station, which means both manned and unmanned capabilities. Thus, the space station will have many elements. The first is the core element, which is the habitat, the utility module, the operations module, and the docking hub where everything is brought together. Here also are those things that will enhance the space station's capability -- TMS, EVA, the tether, etc.

Other elements to be brought into the realm of operations with the space station are a space-based OTV, the Earth ops, the microgravity, all materials processing operations, any astrophysics studies and all of the science and applications payloads associated with that. Other things will also enhance the growth of the space station, making it more usable and increasing its capabilities -- i.e., the laboratory modules. And, naturally when the capability increases the logistic supply to it must increase.

In general, that's the kind of overall system meant by the term "space station concept."

The topic of this paper is architectural options and development issues of the space station. I will discuss this from the perspective that nothing has been designed yet. What we have tried to do is look at what Brian Pritchard gave us in terms of the requirements developed by the Space Station Mission Requirements Workshop held at Langley in May 1983. We've assessed those results from a standpoint of what's required in the architecture. Although we have not necessarily tried to broaden it too far, we did want to make certain that we tapped the sources of the NASA's current capabilities and all of the things that we will probably be able to call upon in the next 10-20 years. With that, let me outline NASA's current thinking. The initial approach has been to set up a Concept Development Group. We have tried not to call it systems engineering too early -- in my opinion a wise move because we want people to realize that we are not designing anything. We're going to keep on doing that until there's a real reason to start talking about design.

I am probably the only speaker you'll hear at this Symposium who doesn't have a pretty colored viewgraph of a concept. One interesting comment I would make that is that we've had several significant meetings recently with all of the NASA Centers and contractors. At each, as we talk about our concept, they all conceptually visualize in their minds what the station looks like. These visualizations come back to my office. I have about a half a dozen viewgraphs that I could include in this presentation, but then I realized that it would just be confusing. Interestingly enough, every one of them is different. So thus far, we've been able to keep the thinking about the space station fluid and open to all suggestions and interpretations.

Thus, I'd like to talk about what the capabilities of the architecture really means and what options we're looking for. Then I'd like to talk about the development issues, which I call challenges. It is in that area that you can take a really hard look at some of the design problems we think are facing us down the road.

There has been a running discussion between certain elements in the Administration about the cost of a space station. One position talks about the cost of a large space station and the other talks about the cost of one not so large. So we took on the task of defining the "bolt of the cloth" from which we could pull the space station, yard by yard. We will look at this on an incremental basis.

What we have tried to do is take the requirements of the overall mission developed by the analysis made at the May 1983 Langley workshop and put together what we think is a total ultimate space station. Then, we have backed away from that ultimate and picked the elements we thought could be the stepping stones with which to initiate the program and from which orderly and logical evolution could be made to the ultimate space station we want to have by the year 2000.

Once this perspective of the overall total space station architecture was established, we began to look at the available options. We then set these in order and began to evaluate them. What we've tried to do in each case is to take each option and look further downstream to develop its architecture to the point that we could satisfy ourselves that it would fully, partially or not satisfy the mission requirements. So, if you will, what we have come out with is in essence a listing of capabilities. When you look at Table 1 you can probably conjure up particular elements that would satisfy these requirements. Granted, there is an awful lot of vagueness about all of this at the

*From taped transcript.
**Concept Development Manager, Space Station Task Force.

present time. However, we have tried to keep this at the point where there are a number of capabilities at each level where any options can be exercised to the fullest. Until such time that NASA feels it must narrow this to some kind of perspective for the initiation of a particular effort, we are going to leave it in this nebulous vein.

Naturally, any kind of space station must be capable of supplying all of the required resources of power. I might say that from what we have seen the spectrum of the power requirements in the space station are power rich. Also, there is the constant necessity of being able to control the dissipated heat -- which is generated by that rich power utilization.

We foresee tremendous amounts of data being generated as a result of experimentation on the space station. That volume will tax our capability to process it in an orderly fashion and to provide it to the user (if he can use it, make the most of it, and, in most cases, try to react with it on a near real-time basis).

There has always been the requirement from the science world for stability in pointing, but we're uncovering more now: the necessity to be more conscious of the stability requirements for the MPS potential users. The problem we see of microgravity research and microgravity product production is going to cause us to be more sensitive to the need for a controlled environment. Back in Skylab days we had an experiment that tried to find out what kind of forces we could expect from crew motions. As you may remember, when Conrad decided he was going to use the watering lockers around the workshop as a racetrack, it drove the scientists on the Apollo telescope mount wild.

So we have had to learn more about what happens in space -- and we're still in that game today. We are trying to understand more of the kinds of forces experienced in space and how they affect not only the scientific endeavor, but also materials processing.

At the same time, we recognize the problem that has been with us for some time (even back to Skylab days) in a controlled environment -- contamination, particularly EMI. This is the sort of thing that causes us concern about the capabilities of operating the necessary payloads on the whole architecture of the space station.

Such capabilities are what any architecture must have to be attractive to the user. The term "user friendly" has been used often, but it is important that all of the capabilities must be of such an attraction that the users will have the impression they are truly welcome and on friendly ground.

Facilities designed to take advantage of such resources are going to be those necessary to do research and development, particularly in the science and application area. Consequently, all of the experimentation has been in that direction. Potentially, there will be unmanned platforms; there will be payloads in the science and application area attached to the space station. There will be different needs for each one of these applications, and those needs will dictate the actual location of those particular payloads.

A number of times we have tried to coin the phrase that we're going to have a national research facility in space. By this we mean a laboratory that has the transparency, if you will, of a general laboratory here on Earth. We are anxious to make that laboratory really inviting to the science community and to industry as one in which to conduct R&D operations.

Table 1 Architectural capabilities

Utilities
- Power
- Thermal control
- Data processing and transmission (with security)
- Stability and pointing
- Controlled environment

Facilities
- Science applications experimentation
- Laboratory (R&D)
- Pilot production
- Production

Functional support
- Man interaction
- Servicing
- Maintenance
- Logistics

Table 2 Architectural options

Present STS

Extended STS
- With unmanned space platforms

Limited duration manned transportation node
- With unmanned space platforms

Manned space station
- With unmanned space platform(s)
- With transportation node capability

Table 3 Preliminary results

Manned "core" element
- Habitat
- Laboratory facilities (R&D)
- Pilot and production capabilities (MPS)
- Science and applications (S&A) experimentation
- Servicing (satellite, TMS, etc.)
- Logistics support
- Transportation node capability

Unmanned space platforms
- S&A experimentation
- Production capabilities

Space-station-based TMS

Space-based OTV (reusable)

Launched and serviced by STS

Ground support capabilities

There are people in the commercial world who see a potential for space industrialization, but who must have more tangible proof that they can utilize this space station to the ultimate to make a return on the old dollar. We think that this is going to happen and, that when it does, some kind of pilot production capability ought to be developed before they go into full production.

Then there are those who are going to need, shortly after the initial activation of a satellite, capabilities to increase the throughput and therefore increase their production yield.[1] We also want to be able to provide that.

Although these things are not that definitive, they are very broad in nature. There are a number of ways to satisfy those needs. And when we talk about the manned space station architecture, man can be involved, whether the space station is the actual center of activity for that particular payload or whether it's on an unmanned platform where the crew can actually maneuver to it on a required or regular basis.

Functional support is one of the key things that can be offered through manned facilities. Jack Lousma has mentioned a number of things that the crew can do.[2] I thought Jack was going to state the way he told it to me at the first CDG meeting: when talking about the potentials of man in space he said, "Luther, I can do anything up there I can do in my garage, if you just give me the tools and the restraints. Give me the foot restraints and the tool kit, and I can do anything up there I can do in my garage." I believe that. I saw Pete Conrad do mundane things with a hammer on Skylab. In that incident a charger battery regulator module had gone out -- at a time we were power hungry because of the famous loss of one of the solar arrays. We worked out on the ground how to make a possible fix. We told Pete the exact spot on one of those boxes to rap one time very strongly with his hammer. Ground tests said there was a good possibility that might fix the problem. We stood there holding our breath while Pete hauled off and whacked it -- and bingo, she worked! Now, that's a fact and it's just the kind of innovativeness that's available to us when man is there.

There is another story I thought was very interesting. We had a problem with the suit loop, that is, the loop that keeps the suit cool. A Sears and Roebuck saddle valve was sent up with the second crew to recharge that loop. Thus, the recharging of that loop was performed on orbit with a simple and cheap Sears and Roebuck saddle valve. There is no limit to what we can do if we tap the sophistication that NASA has the potential of supplying.

Servicing is one of the potential future functions that no one has a handle on yet. I think it's just one of those things out there waiting for us. If we have enough vision and intelligence to approach this thing logically, I think we can realize a tremendous advantage from the servicing aspects of a space station. I think it will be one of the functions the space station is really going to offer to the community: servicing payloads, repair, maintainability, etc. For example, maintenance of the space station itself, the maintenance of those co-orbiting platforms (whether they be large or small), and maintenance of the experiments on the space station itself are going to give us a tremendous enhancement of the capability of the space station.

The logistics function is one that first supplies the needs of the crew. But it will also have the capability of bringing up the everyday needs of the payload, such as raw materials for the materials processing capability.

However, no matter how good the potential capability might be, we have to be frugal as we go about putting the space station together. So what we've done is to go back to the options and look at them in light of the desired capabilities, based on the integrated mission requirements that Pritchard outlined.[3] Particularly, we have looked at how we can best utilize the present STS. What can it offer us in its present form and how can such things as the present EOVs augment as well?

We also looked at how the STS could be extended by such things as the power extension package (PEP), which will increase the capability of the Orbiter in terms of mission time by augmenting the fuel cells. We've looked at how we could just add the cryocapability with the cryowafer. In the past we had a program called the 25 kW power module that was designed to put the STS in orbit for better than 30 days.

We've been looking at how unmanned platforms, along with the extended STS, could satisfy the spectrum of requirements we have from the mission and requirements communities (Table 2).

So, rather than trying to go into a full blown space station concept immediately, we have considered just taking a manned transportation node and developing it for a limited duration. We have also added to that the augmentation with unmanned space platforms, thus bringing into focus all the requirements in the mission community.

Another option is a manned space station beginning with just the manned core, expanding it to include unmanned space platforms, and then adding the transportation node capability -- a slightly different approach than the limited duration mode, but still one that gets it all together.

In each case, those options were evaluated against the set of requirements shown in Table 1 and what that meant was trying to determine the capability of these options to satisfy the requirements of the space station in terms of the coupling of a long duration on orbit with a high manned involvement. That is the primary question, although there are a number of other things that go along with it in terms of resources and refinements.

We also wanted to make certain that any options that we considered were not "dead ended." Above all, when we initiate the space station, we

1 See James T. Rose, "Role of a Space Station in Pharmaceutical Manufacturing."
2 See Jack R. Lousma, "The Role of Man in Space."

3 See E. B. Pritchard, "Integrated Requirements for a Space Station."

do not want one that is functional for only a few years and then having to start over again convincing people that now we need a bigger and better space station. We want to start with a station that has a logical beginning and a logical evolutionary process.

Another consideration is the space station's capability to stimulate technology. Rather than just putting the technology into the space station initially placed in orbit, we want to have the capability to drag technology with us for the next 15-20 years -- that is, the space station should be the thing that motivates the technology. This is called technology transparency.

The results of the preliminary evaluation are shown in Table 3. This table summarizes months of thinking by NASA and eight contractors trying to hone in on the question: What should the architecture of a space station be? In very broad and general terms, Table 3 lists some of our conclusions as of July 1983. They will be refined in the next few weeks and months in terms of what's necessary for any particular gate we try to go through during the ensuing decision process.

There has to be a habitat, naturally, if we're going to have a crew. We're going to need laboratory facilities and power production capabilities. We must offer the science and applications community the wherewithal to conduct their experimentation, whether it be on the space station or an unmanned space platform. We will offer satellite servicing, in the beginning to be done primarily at the station. However, as the TMS evolves there will be more capabilities of doing situ servicing of satellites -- a function that will grow as time goes on. Logistic support will be provided to the payloads on orbit, whether it be in the space station or on unmanned space platforms. And we will evolve transportation out of the initial space station that services the TMS and then supplies an OTV at the proper time.

The unmanned space platforms will be oriented toward the science and applications experimentation, primarily astrophysics. This will also be potentially useful to the materials processing people if for any reason they want to move away from the space station -- maybe they've got specific requirements that the space station cannot meet in the right perspective, maybe they've got a large power requirement, or they've got a much lower gravity requirement.

And we will have space-based TMS eventually. This will evolve from this present STS capability to the point that a space-based TMS can be serviced and supplied from the space station. We also want to evolve a space-based reusable OTV. That's probably our biggest challenge as far as the future is concerned.

This space station architecture will be launched and serviced by the STS. In my opinion, that is a very good attribute. It gives us a tremendous amount of flexiblity, utilizing the STS to its fullest to assemble and service the space station.

Ground support capabilities will be needed to keep the space station on orbit and to support the payload community. Such support will also be needed for the OTVs and TMSs as they move away from the space station on any kind of mission.

Table 4 Development challenges

Modularity Proper level Launch, assembly and activation Cost (commonality)	Ground support vs control
Initial IOC configuration Level of resources Crew size Servicing Number of elements in architecture	Automation/autonomy
Initial IOC configuration as beginning for ultimate configuration	Servicing (satellite, platform, OTV, TMS, etc.) EVA Robotics
Transportation node for reusable space-based orbit transit vehicle Cryo-fluid handling and storage RSBOTV assembly, C/O and launch	Rendevouz, docking vs berthing Orbiter to space station Space station element to element TMS to space station OTV to space station
Technology transparency	Flight orientation
On-orbit maintainability How? What level? Spares? Where stowed?	Power generation
Safety Dual ingress/egress Safe haven Rescue	Controllability
	Thermal control
	Contamination control Particulate EMI
	Stability (gravity-level)
	Pointing base
	Habitability

Those are the architectural options, as envisioned in the summer of 1983. However, I do not want to leave the impression that things are all that vague. As can be imagined if you look at the requirements, potentials, and options in detail, there is a lot of activity going on trying to sort out what the first decisions should be in terms of architecture, and what impact they will have with regard to such mundane things as procurement.

Every facet of the space station architecture has some permutation that ricochets throughout the whole Space Station Task Force. At this time, we are merely trying to sort these out and find the right path to take until the President makes a decision regarding the whole project.

Now, let me give you the opportunity to play designer. Evaluate the problems facing us and offer suggestions that you may have for their solution. Table 4 summarizes what we think the problems are. I will highlight some of this list below.

The STS (Shuttle) will be used to launch and assemble the space station. Thus, the right level of modularity must be chosen carefully in order to optimally assemble it for future expansion. Much of this decision will be dictated by cost.

Which leads to the next point -- the initial IOC configuration. To most people, IOC means a date on which the "initial operational capability" is going to be on line. In my opinion, it will occur over a span of time because one launch is not going to build a space station -- it will require multiple launches. Initially, a large part of the activity up there will be putting together the elements to make a space station. Thus, the IOC will be the point at which this creation starts some sort of mission.

Technology transparency is essential. Over the next 15-20 years, we want to be sure we have the capability to incorporate better and cheaper technology as it becomes available and to do it through the maintainability aspects of the space station.

There has been a lot of discussion just on the question: What is maintainability? How do we do it? What are the tools needed? It is a large area that needs to be solved early in the program.

Ground support vs ground control is a constant question. Many people have noted the "standing army" we now have for ground control of the STS, although it is decreasing. Our approach is to try to make the space station autonomous, which leads to the next question: How much autonomy should we have? Is it dependent on how much automation is available? I'm not sure that there is a whole lot of differentiation in most people's minds between automation, autonomy, and robotics. But those differences must be sorted out: what will be done by robotics and what by automation; how autonomous the station will be from the crew and how autonomous the whole station and crew will be from the ground.

What kind of servicing capabilities are we going to have? How far can we take EVA? How much development do we need from now until the time the space station comes into fruition? How far must we push development of the suit and the tools necessary to increase the productivity of man in orbit.

Most of us are familiar with past operations that brought two elements together with a pretty good energy dock. With the space station we need to bring these elements together in a much more gentle fashion. So, how do we rendezvous with the elements of the space station, how do we construct a space station with berthing capabilities as opposed to docking?

All of the particular elements listed in Table 4 are the kinds of things normally faced in design. What is the flight orientation? That is dictated a lot by the kinds of appendages inserted for power generation and heat rejection. How do we get the stability in terms of low gravity level for the MPS people? Do we have isolators? Do we have soft mechanisms? Most of us are probably looking at the old pointing bases used in the past. Probably the IPS will need to be updated significantly to handle future requirements.

And what about habitability? Lousma said that seven days in the Shuttle is like a camping trip. In Skylab days, they were up for 54-89 days and there were some habitability problems. I don't think we know all we should about the requirements for crew habitability for long-duration space flight or service on the space station.

This has been a brief insight into some of the problems NASA has tried to address in developing a space station. NASA welcomes all of you to contribute your expertise to help us in the solutions.

SPACE STATION ARCHITECTURAL ISSUES
THE VIEW FROM SCIENCE

Frank B. McDonald, Chief Scientist
Kenneth J. Frost, Space Station Task Force

National Aeronautics and Space Administration
Washington, D.C.

INTRODUCTION

The architecture of the space station is a very dynamic quantity which will go through three more iterations. Thus it is necessary to take a very broad look at the science impact on the architecture issue. As you have heard in this symposium the needs of Space Sciences and Applications have been characterized by the Office of Space Sciences and Applications current mission plan. The plan has been generated in a rigorous fashion with the assistance and concurrence of various discipline and agency advisory groups and has been reviewed by the Space Science Board. It genuinely represents the long term interests of the Science and Applications communities.

I would like to sketch very briefly the current state of Space Sciences and Applications and what I think will be its near term evolutionary path. In the last 25 years we have completed a remarkably successful exploratory phase.

In the planetary area there are detailed analyses of Apollo lunar samples, the Viking landing on Mars, probes into the atmosphere of Venus and close-up looks at the magnetospheres, atmospheres and surfaces of Mercury and of Jupiter and Saturn and their vast systems of moons and rings. In the near future there are Uranus and Neptune encounters, the Galileo Orbiter and Probe to Jupiter, the Venus Radar-Mapper with the Planetary Observer are on the horizon.

In Astrophysics the OAO's and IUE with responses extending into the far ultraviolet have completely changed our view of stellar atmospheres and the interstellar medium. IRAS is giving us an unprecedented survey in the far infra-red. X-ray Astronomy has provided a new and unexpected window on high energy processes and Gamma-ray Astronomy offers a new way to study the structure of our galaxy and energetic extra-galactic objects. After IUE and IRAS there are ST, COBE, GRO and EUVE with AXAF, XTE, SIRTF, FUSE and STARLAB waiting in the wings.

In the Solar-Terrestrial Sciences, the studies of the solar wind, the structure of the solar coronae and the impulsive release of large amounts of energy by solar flares have greatly extended our knowledge of our own local star and its extended heliosphere. This system interacts in a complex way with the magnetosphere of the earth and a larger number of plasma processes have been identified than could only be observed by in-situ measurements. We have studied the dynamics and chemical constituents of the earth's outer atmosphere and established the great value of oceanographic observations from space. DE, IMP, ISEE, SMM, SME, and Landsat will be joined by ERBE and AMPTE with UARS, TOPEX and OPEN in the queue.

The more than 15 active U.S. scientific satellites are a very important element in maintaining the scientific vitality of our space program. We are clearly moving toward an era of long lived observatories and planetary orbiters. In the future long lived will mean 1 to 2 decades. The 200 inch telescope on Mount Palomar is still very useful after thirty five years as its detector systems are continually upgraded. The Space Telescope, the Advanced X-ray Astronomy Facility, the Gamma-ray Observatory and the space Infra-Red Telescope Facility - the Astrophysical Quartet - will be retrieveable and serviceable in orbit with long term detector refurbishment planned for ST and AXAF.

This long term observing capability will be important for other areas particularly the global habitability studies of the earth, its atmospheres and oceans and its biosphere. It is important that we understand such transients as the " el Nino currents " which are having such effects on our climate.

The concepts of on-orbit maintenance and repair represent one of the great advantages that the NASA Space Science Program will realise from the space shuttle. Five years ago few of us could have envisioned this great change . In the same spirit, we do not believe that we fully envision the new opportunities that will be offered by the Space Station.

To meet the future requirements of this plan the Space Station will provide facilities within and attached to the station for the conduct of science; support the maintenance and repair of spacecraft and provide platforms , some in polar orbit for earth observations and some in low inclination orbit for astrophysics.The polar platform will be serviced by the STS and the co-orbiting platform by the station itself.

The Science and Applications discipline offices at NASA headquarters have been actively included in shaping the current architectural concepts of the Space Station. The Space Sciences Board and the Space Applications Board of the National Academy of Sciences has received regular briefings on the Space Station developments. Since the first version of Station architecture is very recent nothing that could be called a community opinion is available now. However, a number of common issues have been raised by individuals in the briefings and some of these are addressed in the following discussion.

What we would like to cover in this discussion is the following;

a) the status of implementing the current mission plan using the shuttle and free flyers

b) the space station viewed solely in the context of the current mission plan

c) the future interests in space sciences and applications that require a Space Station-like system

d) how these needs should drive the architecture

e) the need for a smooth transition from shuttle missions to Space station missions

THE CURRENT MISSION PLAN

The mission model that was used to drive the first version of the station architecture was described in detail in Brian Pritchard's paper in this symposium.This model was not devised in response to the question "what would science and applicatins do if it had a Space Station ?". It was generated by the science community and NASA Headquarters as described in the introduction. Furthermore it is responsive to and consistent with the recommendations such as that of the Astronomy Survey Committee report titled " Astronomy and Astrophysics for the 1980's ". That the Space Station be capable of supporting these missions is a necessary requirement if the Station is to meet the needs of Space Sciences. Obviously the science mission model is only one of a number of different justifications for a Space Station.

It is important to note that several of the free flyers and attached payloads have been shifted from the late 80's to the early 90's as a result of budget limitations. It would be of no benefit to the science community if bringing the Station on line further delayed these missions. Such delays would also impede the bold, new innovative steps that the Space Station will offer. A way to minimize if not eliminate this problem is to accelerate and even enhance the level of science that can be done with the shuttle in the remaining years of the 80's. Of great importance is finding new ways to do the missions in the late 80's and early 90's in a far more cost effective manner. Now that the shuttle is operational and there is some hope of the budget restrictions easing, an intense effort should be made to utilize the cargo bay to the maximum extent possible both in the interest of accomplishing the goals of Space Sciences and to prepare for the Space Station opportunity.

As emphasised previously, repair and refurbishment are key elements in NASA's current planning. A little less than a year from now the first activity in the repair of a free flying spacecraft will be conducted with the Solar Maximum Mission. One of the most dramatic moments in space will occurr when the untethered mission specialist leaves the orbiter in the Manned Maneuvering Unit and goes out to bring the Solar Maximum Mission back to the cargo bay. This will be a very important step in developing techniques to support large observatories permanently in orbit. The first extensive use of this capability in the future is the maintenance of the Space Telescope and the Gamma Ray Observatory. These requirements will persist and grow to a higher level in the Space Station era using the Tele-operator Manipulator System and spacecraft designs incorporating higher levels of maintenance and reconfiguration characteristics.

THE SPACE STATION AND THE CURRENT MISSION MODEL

As discussed earlier, the current science mission model was conceived without a space station in mind. Hence it does not take full advantage of the potential possible in a Space Station concept.A great deal of effort is now being given to taking maximum advantage of the shuttle in its current operational mode. Of necessity this must be the first priority for science in the remaining years of the 80's.

If we think no further and do not put ourselves in the frame of reference of what one could do with a Space Station beyond the current interests of the science program we would indeed be missing a great opportunity.

We must require that the Space Station accelerate and expand our opportunity to do science in space. We cannot allow our requirements to rest at their current state of development. We must take advantage of the enormous new potential that the Space Station will offer.

In the coming year, as the architecture is moved through the additional cycles we must begin to address long range scientific needs and interests. These are not currently in the mission plan primarily because they cannot be implemented with the launch capabilities we now possess. It is in considering these objectives that the Space Station will acquire its true character as the new high ground for Science and Applications.

FUTURE REQUIREMENTS FOR A SPACE STATION

The challenge of the Space Station, its great opportunity, is the ability to go beyond our present visions of how we do science in space. There is already some consideration in the literature (1) of the opportunities that lay ahead of us if we could but coalesce our technology in structures, data systems, detectors and optical devices into an activity of assembling large observatories in space. There are a number of such observatories in the very early concept stage that could capitalize immensely on this capability.

The following serve as illustrative examples just from the field of Astronomy and Astrophysics:

a) Orbiting Very Long Baseline Inteferometer

b) Infrared Submillimeter Telescope

c) Infrared Inteferometer

d) Ambient Deployable Infrared Telescope

e) Hundred Meter Thinned Aperture Telescope

f) Very Large Space Telescope

g) Pin-hole Occulter Facility

h) Coherent Optical System of Modular Imaging Collectors

These instruments are but a sample of what the future, that is the 1990's, holds for Astrophysics. There are similiar requirements or interests in the areas of earth observatioons, communications and experiments to detect gravity waves and verify aspects of the general and special theories of relativity.All of these payloads are beyond the launching capability of any known booster and all press the capability of the Shuttle itself. The STS will be able to carry into orbit the materials needed to assemble one of these observatories. However, the stay time, crew complement and assembly facilities are inadequate for the level of assembly and in-orbit verification that these systems require. Commitment to a Space Station with a facility of sufficient capability and magnitude to asssemble the instruments cited above would provide Space Sciences with an opportunity to advance our knowledge of the universe in an awesome way. Indeed it has been claimed in one of the study documents which describes these instruments that " Today, Astrophysics has within its reach the ability to bring about one of the greatest scientific achievements ever -- a unified understanding of the total evolutionary scheme of the universe ".

THE TRANSITION FROM THE SHUTTLE TO THE SPACE STATION

I believe that we can now describe what the priorities for developing the Space Station architecture for science should be. The Space Station for the mid-90's to the year 2000 should be the prime focus. If that architecture incorporates the facilities just described it would provide to Science and Applications a spectrum of capabilities that it currently does not possess. Concurrent with developing the Space Station architecture we will seek to enhance the level of science done with the STS for the remainder of the 80's. As these two activities mature we can look foward in the next year to defining that transition process that will move us smoothly from research on the STS to research on the Space Station. This effort will define the initial configuration of the Space Station and the manner in which science should be done in the STS in the late 80's in preparation for the Space Station. I believe that this approach will move us into the era of the Space Station with optimal advantage to both the Science community and the Space Station itself.

REFERENCES

1) GURSKY, H., " Future Space Telescopes " in Technology for Space Astrophysics Conference p. 1 - 5 American Institute for Aeronautics and Astronautics, New York, 1982

SPACE STATION ARCHITECTURAL ISSUES AS VIEWED BY THE USER COMMUNITY: APPLICATIONS*

William P. Raney**
National Aeronautics and Space Administration
Washington, D.C.

In talking about the space station and applications the question is always: "What's an application?" It is debated constantly, and has been since the beginning of the space program.

There are those who measure the worth of applications only in terms of whether somebody can make a profit from them. There is a large school of people that say we should constantly test our program by whether or not it can lead to commercial profit. But there is another set of people who say it's much broader than that. There is a large group of governmental or public service users who might also be able to gain practical benefit from space activities. And I'd like to include a whole class of work, which I would describe as applied research, aimed at future applications that are now only imperfectly perceived.

Those applications we traditionally think about in NASA are only a few in number. Although the program sometimes tends to look complex, I've been impressed by the descriptions of the complexities in space station architecture and the complexities of the science program made at this Symposium. In this presentation, I would like to try to examine what are the fundamental facets of the applications we want to pursue. I do this partly as a means of reassuring those folks who are just as vocal in the applications community as the pure scientists are in the science community, wondering whether or not the heavy investment the government will have to make in a space station will scuttle activities near and dear to their hearts.

Let me start out by noting, as we all realize full well, that the applications we do in the space program have all grown up over the last 25 years in an era when what people had to work with was expendable launch vehicles -- which meant applications that were consistent with free flyers. It is only recently that we have been able to modify that approach with the Shuttle capability. But still the things in the pipeline and the things we have been doing have all grown out of a particular capability.

That's where the dreams begin coming, because we're now at a point where a whole new plateau of capability is at least within grasp. Whether or not we will really manage to grasp it is a current question. But we've been thinking about it. It is clear that there is a great deal of potential, although not for all of the applications -- unless we look at the whole concept of a space station in the larger context, that is as a whole complex of capabilities in the space station concept, rather than as simply the specific manned element.

What are the current applications? The first is communications. This area has been very successful financially, not necessarily for the spacecraft builders, but surely for the transponder brokers, those who sell programs, and those who provide bulk communications. Yet those who have thought hard about the future of space communications realize that a great deal of it has had to do with only two classes of service: the handling of bulk traffic, and the beginning of a direct broadcast capability.

The direct broadcast industry is going to have technical limits because of the permissible field strengths. The bulk traffic communicators are going to discover that there are other technologies coming along that are going to shift the balance of economic advantage of doing communications by satellite. And so I foresee a shift in the sort of things that satellite communications will be called upon to do. In other words, the people who say that the terrestrial trunk communication will be done economically in the future with fiber optics probably have a case to make.

So there are going to be changes. Thus, what does the communications community expect to be able to do and how will a space station or associated capability help that community?

An approximate answer is that a space station won't help very much right away, because almost all satellite communication is done from geostationary orbit, and it's going to be a while in the space station era before we have the capability to use the space station as a waypoint on the path to geostationary orbit. But what it does offer is the future capability to assemble and then get into geostationary orbit large structures, which will be needed to open new frequency regimes to satellite communication. And that, we have not been able to do in the past. As a matter of current economics, a space station is not very interesting. For the future where new services can be developed, the space station is extraordinarily interesting. As soon as people get over wanting instant payback for that particular application of a space station, they'll realize just what potential is there, and I think the enthusiasm will grow. It's the same story over and over again as I talk to applications people. Once they begin to think about it, they begin to realize just how many things in the applications world can be assisted by a space station.

Consider another area, the classical one: remote sensing of the Earth. I have been amused

*From edited transcript.
**Assistant Associate Administrator for Space Science and Applications (Programs).

when talking to people who somehow assume that the only group seriously interested in looking at the Earth is the Defense Department. In fact, the whole remote-sensing program that we've called applications concentrates on looking toward the Earth.

The requirements differ a little in each of the three regimes usually covered by remote sensing, that is, the atmosphere and the environment, the land, and the ocean. Yet they all have something in common: the need to survey the whole Earth. So the fundamental thing here is that for Earth remote sensing, a polar orbit is needed. Continuity is also essential. That is the cry in each of the discipline areas of Earth remote sensing. For an operational application, it's not terribly interesting to make an occasional observation of the Earth, except for things that never change. But the meteorologists are constantly looking at change, and the land people -- at least the people who look at growing things -- must constantly make detections of change. Obviously, the people studying land use must constantly look at change detection all over the world. That means uniform access to the whole surface of the globe, repetitive coverage, high precision, and a broad spectral content. Such systems get expensive, and therein lies one of the more interesting things that can come out of the space station -- namely, the capability to repair rather expensive missions to provide exactly the continuity that has oftentimes been missing in the past.

There is a tendency to say that our program is just R&D, and therefore continuity isn't important -- it's only the operators that really need continuity. That assertion just is not so. The R&D people need continuity for long-time series observations just as much as the operators do. The problem is that the former have never been able to make a cost-effective case, whereas the operators make a dollars-and-cents argument that says continuity is important and, therefore, it's important to be able to extend the total service life of such satellites.

Then we begin the tradeoff arguments about how much it costs to go and repair the thing vs the cost of putting up a brand new one. Those arguments will never stop, and they will provide a very healthy pressure on the space station to continue to make more economic provisions for the maintenance, repair, and refurbishment of missions that need to be long-lived.

There is no question that this sort of application will continue. There are applications of high social value in understanding the global environment in all sorts of ways: food supply, contamination of the environment, economic effect of weather, etc. In all of these things, it is absolutely impossible to do a decent job of predicting what comes next without global observations.

Local observations are not enough; spot observations scattered sparsely throughout the rest of the Earth's surface are not enough. There is only one way to get the necessary data and that is through satellite systems.

At the moment, remote-sensing interests are settling for being part of the space station community in that we need to have polar platforms for Earth observations. The capability to service equipment easily on polar platforms from the space station will come. It's part of the vision. And as we learn how to do it, and how to do it economically, there will be new applications conceived and perfected.

The next area is materials processing. Other presentations at this Symposium have mentioned the very substantial capabilities and the very likely possibilities that profits can be made in this area. The materials work NASA has sponsored has lately been concentrating on the underlying science.

We used to tell a lot of funny stories about the generous supply of vacuum, but it's really cheaper to build big vacuum pumps. And we said we could make good use of zero gravity, but zero gravity doesn't exist very often or in many places as an engineering reality. The real question that besets us in discussing the sort of facilities we need with the space station complex is how close to zero gravity must we be and how much crew attendance must there be for the experiments and the required material functions.

We can argue forever about the details. The point is that the space station architecture we've been discussing with the Space Station Task Force seems to be able to do a great deal to help us. It's a major step forward. As we learn more about the business and learn more about the power required, the level of gravity required, the contamination requirements, and the associated communication requirements, we will be able to improve the system.

At the moment I think we shouldn't spend much more time arguing endlessly about the details. In point of fact, the plan is moving forward in a helpful direction. I think rather than try to plan for every eventuality from the beginning, we'd better decide quite soon just what it is that we're going to do as a first step and set about doing it.

Finally, the point I'd like to emphasize is that in the applied research area there is an area not normally treated as an application, and yet most of our advisors in the business of trying to learn to do practical things in space have emphasized it heavily. That is the whole question of how to maximize the benefit of man's presence in space.

We have not thought about it hard enough and we have not worked at it systematically enough. But we have the makings of a very good application because we've been doing some of the fundamental physiology. The business of habitability means something more than the restful "government green" with which many of us are familiar. But it's hard to come to grips with exactly what will enhance people's productivity and/or their pleasure in their jobs. We do not want a set of US astronauts to complain, as some of the Soviet astronauts apparently do, that they were very excited about going, but they would not necessarily want to go again after they've been on station for a long time.

If we are to succeed in making the best out of this matchless opportunity to combine the capability of sophisticated machines with the capability of sophisticated astronauts, I think we had better tackle this as a real application and a real area of fundamental importance to NASA and its future. I want to mark that as something we do not traditionally think of as one of the applications, and to suggest that it is one. The imminent presence of a space station gives us the best possible chance to move forward in this supremely important area.

SPACE STATION ARCHITECTURAL ISSUES AS VIEWED BY THE USER COMMUNITY: COMMERCIAL USER MISSION CONCERNS

Peter Wright Wood, Senior Vice President, Booz, Allen & Hamilton Inc.
Myron Weinberg, Ph.D, Weinberg Consulting Group Inc.

Abstract

The paper discusses the place in Manned Space Station development of the non-aerospace user firm, the identification of the function the firm plans to perform on orbit, and the need to include these inputs in planning and configuring the proposed space station complex. Techniques used in stimulating and assisting candidate users are described. User concerns and other obstacles faced by users are discussed on the basis of the authors' experience in commercial user development for the space station, and several key steps to allaying concerns are suggested. The authors carried out their work at Booz, Allen & Hamilton Inc., international management consultants, first under subcontract to MacDonnell Douglas Astronautics Co., and now under direct contract to NASA, supporting the Space Station Task Force.

Background

There will be a Manned Space Station -- and it will fly the Stars and Stripes.

That sounds like an article of faith. In fact, most of NASA's best known projects proceed on a certain amount of faith. Jim Beggs, NASA's tough-minded Administrator, quoted in the Harvard Business School Bulletin last December, said:

> "Sometimes you've got to go out and do things without knowing what the results are going to be; otherwise you'll never make progress."

So we don't pretend to know the full range of rewards this daring project will return to the United States, its people, its industry, its international partners.

But to define the building blocks, set the time table, and allocate resources of the country, industry, and investors, we need to know all we can about the early users, and what they will want to accomplish. It's clear, even at this early stage, that important drivers of Space Station Architecture will be the commercial users and their orbital ventures.

Why is NASA interested in commercial ventures in space? President Reagan announced his Space Policy Objectives last July, and NASA's response was to set a new goal -- to expand opportunities for United States private sector involvement in civil space and space-related activities. The Space Station is the most dramatic initiative toward meeting that goal.

Finding and Stimulating Commercial Users

NASA, and those who recognize the value and potential of the Space Station concept, know that there are many forward-looking business men and scientists who can turn Space Station to economic advantage, and meet many national needs.

The immediate problem was one of identifying the early users, and bringing definition and focus to their activities. Booz, Allen & Hamilton, having served clients throughout both U.S. industry and the space community, became an active player in solving this problem. Our efforts have taken two forms:

- Finding the promising commercial uses and probable users, informing the user community of the opportunity to benefit from space, and stimulating users to plan and to make a commitment to participate
- Gaining a clear picture of the concerns and obstacles that might deter such private sector commitment, and finding avenues to overcome such concerns and obstacles.

Aerospace industry, like NASA, has invested years of analysis and lots of ingenuity toward planning a permanent presence in space. But non-aerospace industry is nowhere near as well-informed or able to deal with the specifics of space phenomena. We were all hoping this community would:

- Find a reason to move into space
- Commit to serious planning, research, and concept development
- Then participate actively in orbital proof of concept, and in the commercial development of Space Station

But a catalyst was needed to bring about this sequence. Booz, Allen treated the problem as a marketing strategy in a brand new market.

- We segmented the market, sketching out the broad areas that could use attributes of space
- We developed seed concepts in each area - ideas that would communicate what could be done and would stimulate other ideas
- For each set of concepts, we targetted qualified firms -- those

who would understand, might react, could take risks and make long term investments

- Then we set up meetings - often through key executive acquaintances, sometimes in the shelter of confidentiality agreements, with the blessing of decision makers, involving the pathfinders in research.

Booz, Allen's job was to inform, to stimulate, to listen, to answer questions, to help, to simplify access to a dimly perceived Government agency, and perhaps to aerospace. We had to deal with technology in diverse fields, with risk, and with large investments. Most of the work has been carried out on a one-to-one basis, with approaches tailored to each firm, and with great care in selecting points of contact, building confidence, working patiently to fan sparks of interest, and following up.

This approach has provided successful, and a number of firms are moving toward commitment of resources and planning joint agreements with NASA. Interest in notable in such fields as chemicals, metals, glass, communications, crystals. Most of the interested users are well known U.S. industrial firms, but several European firms and several relatively small or new businesses are included. There are also projects such as a commercial fee-for-service laboratory in space, which are attracting interest from venture capital.

Awareness

Because the tailored approach, while bearing fruit, limits the range of users contacted, an awareness program is also being carried out, including articles, interviews, press releases, platform appearances, and frequent discussions with corporate executives in the framework of other assignments. A number of newspaper and periodical articles have resulted, such as the recent cover story in Business Week (June 20, 1983), and a flow of inquiries has been generated from firms not previously contacted, some with well-thought-out ideas for material processing in space. These efforts are helping to identify those most likely to benefit from NASA's forward looking planning. Momentum and excitement are building.

Commercial User Concerns

As user company staff interest grew, and promising applications began to emerge from the seed concepts and brain storming, we had to remain continuously aware that commitment on the part of a user firm meant risking large amounts of money, with return lagging perhaps ten years, and with alternative investments beckoning. Numerous other concerns made themselves known, many related to dealing with the Government, a new experience for some. Such concerns, operating in what is already seen as a high risk environment, can rapidly become barriers to serious consideration and to commitment.

Experience with industry, and the contact with potential user firms, as well as extensive work with NASA and other federal agencies provided a basis for identifying concerns and obstacles to industry commitment to space station participation. Whether the sources of concern are real or perceived, they need to be dealt with. They can be viewed as three groups: technical matters, economic matters, and government-related areas, although some partake of all three areas. Let's look at the more prominent ones.

Technical Barriers

The principal barrier encountered in early contacts with non-aerospace firms willing to discuss space station is typically a fairly broad lack of knowledge about space and the impact of space attributes -- especially vacuum and microgravity -- on physical, chemical, and biological phenomena. It is a mistake to assume intuitive knowledge of the effects of such attributes.

Closely related is the difficulty otherwise competent research managers and scientists often have in conceptualizing a process in microgravity. Having developed life-long instincts in a one-g environment, it takes a significant and deliberate effort to set gravity aside.

Aside from the need to acclimate our thinking, there are simply too few answers about how such familiar evolutions as cell growth and multiplication, behavior of molten metal, and heat transfer are influenced by the environment of space. More science is needed, and managers need more confidence in how processes will work in space.

Technology issues are the key element in risk identification and risk management. (Other elements of risk relate to staff, property, investment, schedule, the market for product produced, and the possibility of competing technologies nullifying investment before pay off can begin.) A special risk issue flows from uncertainty that space transportation and the space station will in fact be accessible when needed.

A special technology issue about which more needs to be known is the matter of human endurance on orbit, and the ability to staff the orbital facility and keep staff motivated and productive.

Finally, the protection of intellectual property and proprietary processes is a frequently mentioned concern, relative both to the development and demonstration phases and to the space station operations period where competing processes might operate in adjacent modules.

Economic Barriers

Cost and cost risk, compounded by technical uncertainty, constitute a formidable barrier to all but the strongest, most confident firms.

Cost factors include research, development, prototyping, data, tests and trials, system materials, process raw materials, transportation, insurance, and all related burden costs. The size of the expenditures must further be viewed in the context of a long (9 to 12 year) payback period, high risk of failure, uncertain return in a relatively unexplored market, the possibility of being undercut by an as yet unidentified competing process, and alternative opportunities for investment. Timing of investment raises questions that may stall a decision. (Is it too early for the space station? Shall we wait for someone else to learn some lessons or make a breakthrough?)

Government-Related Concerns

Some of the firms contacted during recent studies had not done business with the Federal government, or at least with NASA, and were reluctant to "take on the bureaucracy." Many perceived an unfriendly interface, or were unwilling to take a chance or blaze a trail. Some saw the path to a Joint Endeavor Agreement as frought with hazards.

Concerns about stability of Government programs (will there be a Space Station when our process is ready?) or tax policy or trust policy all call into questions long term investments, especially if alternative investments must be foregone. Access to space, where size and weight require use of space shuttle, depends upon Government determination, and may soon be subject to some form of regulation, possibly oppressive.

Finally, some people are concerned that military shipments on shuttle or Space Station may somehow have priority, interrupting planned sequences and transfers.

Some Solutions

While it is not appropriate here to address each concern or obstacle listed, a number of measures offer considerable promise for encouragement to would-be users.

- An outreach program can provide specific information on space phenomena, space transportation, and related procedures and costs

- Assistance can be offered to candidate user firms in forming concepts, developing designs and interfaces, and planning commercial orbital programs. NASA can call on space knowledge and technical expertise at all of its centers, and can thereby promote the commercialization of space.

- The Government can simplify procedures, focus interface points (one stop service), provide a warmer, friendlier aspect to approaching users, and find ways to be more positive and responsive to legitimate trail blazers. This should continue to include economic assistance including space transportation for R&D and proof-of-concept efforts under appropriate conditions

  Most importantly, the Government should try to establish a formal program for Space Station, with sound supporting policy, and build confidence and stability into it.

Finally, a well executed program of communications, making known technical and financial success stories, will be a most welcome and supportive measure when it has participation from both Government and industry. One special measure which is being considered is the establishment of a permanent intermediary. It would be well versed in procedures, motivations, and constraints of both government and industry, fully accessible, prepared to help with inquiries, contacts, and authoritative answers, help in forming teams and entering into agreements, and be responsible for discreet assistance in getting new concepts into development.

* * * *

The dramatic achievements of our space pioneers make a permanent manned presence in space easy to believe in. Now we need to cooperate in creating the most supportive possible environment for the commercialization of space.

NATIONAL SECURITY IMPLICATIONS OF A U.S. SPACE STATION

Charles W. Cook*
U.S. Department of the Air Force
Washington, D.C.

July 20th we celebrated the 14th anniversary of one of man's greatest achievements--in 1969, a human being set foot on the moon for the first time and walked on its surface. That "one small step for man, one giant leap for mankind" was taken by Neil Armstrong and later by his crewmate, Ed Aldrin, after Apollo 11's lunar lander touched down on the barren surface of the moon. It is interesting to realize that less than a century before we didn't even have automobiles, but here we were exploring the moon. It was a heady experience not only for the men who did it, but for all of us watching it on television here on Earth. I believe that it is safe to say that nothing since has captured the imagination of the American people to the same degree as placing an American on the moon. In the 1970s, we placed instruments on Mars and embarked on the Space Shuttle program.

We now have begun the search for a new initiative, for a way to grab the imagination and enhance the pride of the nation. We have witnessed the Space Shuttle make its maiden voyage and a subsequent nearly flawless operational test flight program. It has now been declared operational and has made three flights during which spacecraft have been launched. The President announced a new national space policy on July 4, 1982. Since then, a tremendous effort has been proceeding on a wide range of space policy issues, one of which is the question regarding the need for a national manned space station capability.

Before discussing the military implications of a manned space station, it is important that we understand the differences between the respective missions of NASA and the Department of Defense. As a predominately research and development organization, NASA is charged with the pursuit of advanced technologies and the scientific exploration of new horizons to further man's knowledge. The DOD is charged with a more fundamental mission--to preserve the freedom and security of the United States. In order to achieve this goal, it must deter conflict by maintaining a credible capability to fight across the entire spectrum of conflict from a small, local engagement to a nuclear war. Consequently, DOD must carefully channel its research and development efforts. The operational military commands state requirements that then drive the DOD R&D efforts to satisfy those requirements.

The military use of space has greatly contributed to satisfying those operational military requirements for more than 20 years. Space possesses certain characteristics, such as global coverage, acceptable overflight, vastness, and relatively free access, that make the use of space systems militarily attractive for a number of missions. As space technology has matured, the awareness of the advantage of performing some traditional military functions from space has resulted in a significant growth in the military reliance on space systems.

Currently, over two-thirds of the long-distance military communications are sent via communication satellites. The survivability of the U.S. strategic bomber force is greatly enhanced by the ballistic missile tactical warning provided by surveillance satellites.

All of these functions are currently being performed by unmanned systems. This is not to say that man does not have a role in space. However, having developed unmanned systems that work so well, coupled with our limited manned space flight experience, we still have many questions pertaining to the military man's role in space. The total U.S. manned flight time in space over the last 21 years is less than 3 man-years, and most of that time was on the three Skylab flights. We have witnessed man's inherent capabilities, such as the reasoning and dexterity abilities that proved useful in repairing Skylab. On the other hand, he is a frail creature in a hostile environment requiring complex life support systems.

DOD is very interested in determining man's role in space and how he might contribute in satisfying DOD requirements. We have noted with interest that the Soviet Union has an extensive manned space program. During recent weeks, Moscow television has broadcast several programs on their space station and the emerging space tug. In recent years, the Soviets have conducted a large number of civilian and military missions using their space stations. By the late 1980s or early 1990s, it is conceivable that the Soviet Union could have a much larger, permanently manned space station in low Earth orbit.

Determining man's role in space is not new to either DOD or NASA. The DOD has been interested in manned space flight since its beginning and has been looking for possible military uses of a space station since the 1950s. In the early 1960s large sums were spent on the Dyna-Soar program which was cancelled before it became operational because unmanned systems were determined to be more effective and less expensive.

DOD's next program pertaining to manned space flight was the manned orbiting laboratory (MOL) program approved by President Johnson in August 1965. MOL was to consist of two major parts in orbit: a modified Gemini capsule, called Gemini B, and the MOL itself. This configuration was to

*Deputy Assistant Secretary (Space Plans and Policy).

have been placed in low Earth orbit (LEO), several hundred miles above the Earth's surface, by a Titan III M booster.

As MOL was being developed, its purpose gradually changed from an experimental program to determine the value of military crews in space to a program with a definite operational mission. However, once this mission was further defined, it was again decided that unmanned space systems could do the same job at less cost. Thus, in June 1969, the MOL program was cancelled, but not before large DOD sums had been expended.

NASA was also interested in space station during the 1960s, particularly as the Apollo program began to wind down. This eventually led to the development of the Skylab space station. While DOD did not play a major role in Skylab's development, we did supply seven of the nine astronauts participating in the three Skylab missions during 1973 and 1974. Of course, since Skylab was an experimental manned program, we had a general interest in the Skylab experimental results.

These early efforts did establish one important fact -- manned space flight is expensive, but the value of man remains undetermined. This is true not only for military crews in space, but also for their scientific and commercial counterparts as well. Thus, the decision to regard the Space Shuttle as a national resource used to evaluate the role of man in space is a reasonable compromise between doing nothing and charging ahead to deploy a space station without a clear understanding of where we are going.

During the past decade, DOD has cooperated with NASA in developing the Shuttle and its associated expendable upper stages and launch facilities. In fact, DOD programs project some $15 billion in Shuttle-related costs by the end of FY '88. Much can be learned about the value of man in space from the Shuttle alone. During the 1980s, DOD will pursue this knowledge through four activities: 1) the manned spaceflight engineer program, 2) the space test program, 3) the space biotechnology program, and 4) the monitoring of NASA's manned space flight experiments.

The manned spaceflight engineer (MSE) program currently has 27 assigned military personnel. They are primarily engineers, assigned to specific national security payload program offices, rather than astronauts. The MSEs will fly aboard the Shuttle with their payload, participate in checkout, and then deploy the payload. They may also conduct experiments with equipment that remains on the Shuttle, either mounted in the Shuttle's cargo bay or in its cabin. The on-orbit experience gained by these engineers will have a significant impact on improvements in the design of future military satellites.

The space test program (STP), which has been in existence since 1966, uses the Space Shuttle and expendable launch vehicles to perform a wide variety of military experiments in space. These range from free-flying experimental spacecraft to Shuttle cargo bay experiments and using the Shuttle cabin as a manned laboratory. The engineers in the MSE program will play an active role in deploying and conducting the Shuttle STP experiments.

The space biotechnology program will conduct experiments to determine the capabilities of crews in space and how to overcome human limitations due to physical and psychological factors. The manned spaceflight engineers will also participate in this program.

DOD will also closely monitor and in some cases participate in NASA's manned space flight experiments. NASA will conduct a wide variety of experiments on-orbit to determine man's value in space, including the solar maximum repair mission in 1984 that DOD is helping to fund and the servicing of the Space Telescope in 1989. We have also started working with NASA to identify techniques, tools, and operational concept changes necessary for future spacecraft maintenance and repair missions.

In addition to these existing and planned programs, the DOD supports continuing the production capability of Orbiters to guarantee an adequate national launch capability. The Space Shuttle is rapidly becoming the nation's primary launch vehicle and, according to present plans, in most cases will be the sole means for getting into space. Consequently, the utmost attention must be directed toward the Shuttle program, which includes maintaining a continued production capability and providing a strong logistics support program.

Thus far, the Shuttle has been used primarily as a booster to place satellites on-orbit. While recognizing that the Shuttle is in the infancy of its operational lifespan, we should not lose sight of the fact that the Shuttle has been sold as much more than a manned booster -- its purpose is, namely, to enable on-orbit repair, and retrieval and to be a platform for an extended manned presence in space. The economic aspects, i.e., rapid orbit turnaround to reduce the cost per flight, tend to keep the time on orbit to an absolute minimum. In fact, many of the requirements one hears to justify the space station appear similar to those used in the early days of the Space Shuttle to promote that program.As we discuss the value of a space station, let us not lose sight of the existing value of the Shuttle and the nation's reliance on this vital capability nor neglect to adequately fund the Shuttle operations and improvements.

Given the above discussion as a background, how does a space station fit into an evolutionary scheme to gradually expand the capability of military crews to operate in space? I will first discuss how DOD has examined this question, describe how a space station could become part of a national space infrastructure, and, finally, indicate whether military missions seem appropriate for an initial, joint-use space station.

During the past year, DOD has undertaken an extensive review of military requirements for the space station concept. The Air Force participated with NASA in funding eight industrial contractors to expand their studies to include identifying potential military requirements for a space station. We reviewed the results of these study contracts and other analyses through the formation of a DOD Space Station Working Group and through the formation of an ad hoc committee of the Air Force Scientific Advisory Board. Representatives from the Army and Navy were also invited to participate. DOD is also participating in discussions of a space station with representatives of other government agencies through a working group of the Senior Interagency Group (SIG) on Space.

While many believe inherently there should be military utility for a manned space station, DOD's concerted efforts have been unable to identify any military requirements that uniquely justify such a capability. These results should not be surprising, since most of this country's space systems have been unmanned and a great deal of knowledge is still needed to determine exactly what man's role should be.

Lacking any specific military requirements, DOD currently believes that an evolutionary approach is preferred in determining the operational value of military crews in space. The first step in this evolutionary approach would be to increase the Shuttle's on-orbit time for experimentation from 7 to 20 days. A power extension package such as a large solar array could increase the Shuttle's on-orbit time up to 20 days.

Another evolutionary step might be to develop a teleoperator maneuvering system (TMS) that would increase the range of on-orbit servicing and repair to include satellites several hundred miles from the Orbiter. These steps would give us the on-orbit experience needed to design our systems to include the presence of man, if warranted. Prolonged time on-orbit could answer questions pertaining to space physiology. Thus far, 7 of the first 16 Space Shuttle crew members have experienced space motion sickness.

Should this evolutionary approach lead to a future decision to develop a space station, TMS and orbital transfer vehicles (OTV) could increase the range of servicing and extend our manned space flight capability from low Earth orbit to geosynchronous orbit. The OTV could also extend the range of our satellite servicing and repair capability to high Earth orbit (HEO) if the TMS could be mounted on and used in conjunction with the OTV.

The Space Shuttle will provide the first capability to assemble structures in space. In fact, structures are not limited by the size and weight limitations of the Orbiter's cargo bay, since several Shuttle flights could be used to transport the necessary structure components into space. The space station would increase this capability by lengthening the time on-orbit to accommodate the assembly of large structures. A space station could be used to determine the operational feasibility and cost of using the station as a transportation node -- assembling large structures in space, such as large antennas for communications or radar systems, and satellite servicing and repair.

While many of these experiments can initially be conducted with the Space Shuttle, a space station would allow for longer duration missions and experiments on a larger scale. As a transportation node the station might someday be useful to berth, refurbish, and maintain reuseable spacebased OTV spacecraft; store and possibly manufacture fuel; and transfer payloads from the shuttle to the OTV.

The station, as a test facility, would be a shirt-sleeve laboratory permitting long-term experiments with continuous manned presence for monitoring the progress of the experiments. A space station could be used to determine the physical and psychological effects of long stay times in space on military crews, i.e., longer than three weeks, something that the Shuttle cannot accomplish.

Platforms could be attached to the station on which the station's crew members could place experimental equipment for any length of time desired and adjust the equipment as necessary. Tethered experimental equipment could be maintained by the crew members and, similarly, free-flying platforms containing experimental equipment that could be periodically serviced, repaired, or replaced. The platforms would not necessarily have to be in the immediate vicinity of the station.

The optimum place to service or repair a satellite might well be on-orbit, although this is yet to be proved. On-orbit maintenance capability might require a manned presence through a remote TMS or the OTV. Before this could occur, the spacecraft would have to be designed to be serviced and repaired. While military satellites are not generally designed to be serviced or repaired, as the value of satellites increase, it may prove to be desirable to provide periodic on-orbit servicing and repair.

From a military viewpoint, there may also be future advantages in keeping spare satellites on-orbit to insure against a loss of service to our forces by our space systems. Having the capability to service and repair satellites on-orbit might prove to be one method to guarantee the availability of space resources should an active satellite fail or be destroyed.

I have discussed where a space station might (and I stress the word MIGHT) eventually prove worthwhile militarily. The possible space station operations that might have military utility have yet to be proved to be economical or technically feasible and there are no validated military requirements. Contrary to what many seem to believe, DOD does not have unlimited funds and while it may have a relatively large budget, it has a large and difficult mission. DOD's priorities must insure that our nation's defense is adequate. A space station program seeking DOD support has to compete in that arena for those limited resources and space systems must be a part of the overall DOD force structure. Thus far, no requirements have been identified that warrant DOD expenditures to develop or help develop a space station. DOD believes that we should continue to support and expand the capabilities of the Space Shuttle until more knowledge relating to the exploitation of man in space can be obtained. As this knowledge is accumulated, future DOD requirements may emerge for which a space station would be a viable solution. In the interim, should other non-DOD requirements warrant development of a space station, then we will consider it as a potential solution as DOD requirements arise. The first such DOD space station requirements would likely be to use it in lieu of the Space Shuttle for some RDT&E areas to further our knowledge of how the military man in space can contribute to DOD's mission.

THE IMPORTANCE OF THE SPACE STATION AS AN ELEMENT OF THE TOTAL SPACE SYSTEM ARCHITECTURE

Ivan Bekey
Director, Advanced Programs and Plans Division
Office of Space Flight, NASA Headquarters
Washington, D.C.

Abstract

The paper describes the role of the Space Station in the ensemble of the space-based infrastructure necessary for achieving permanent manned presence in space.

The major elements of the architecture are identified, functionally interrelated and discussed in terms of their evolutionary development as an extension of current STS capabilities. It is shown that they require a logical succession of flight R&D experiments and demonstrations using the Space Shuttle and the Space Station as test beds.

The paper describes the interrelationships between the Space Station and the various major infrastructure elements and shows that the latter are indispensable for full operational utilization of the former. Only the total ensemble of the architecture provide all the necessary prerequisites for permanent manned presence in space.

I. Introduction

The National Space Policy of July 4, 1982 commits us to maintain world leadership in space transportation with a Space Transportation System (STS) capacity sufficient to meet appropriate national needs. To demonstrate this leadership, a strong, responsive, reliable and cost-effective STS operation must be maintained. NASA, through the Office of Space Flight, is currently placing top priority on this requirement as the Space Shuttle has entered operational service and additional Orbiters and other elements of the system are being delivered to complete the fleet.

But in order to assume and maintain leadership in space, we must also be able to operate routinely - i.e., safely, successfully, on schedule, and economically - in Earth orbits at low and geosynchronous altitudes and between orbits. This paper outlines the importance of the manned Space Station as an element of the total space-based infrastructure that is required to achieve this leadership goal.

II. Permanent Presence Infrastructure

While the technological infrastructure for establishing routine access to and return from low Earth orbit (LEO) is the focus of current efforts, there is a growing consensus that the achievement of a permanent manned presence in space as next step is a logical mandate for achieving that leadership goal. If full operational capability of the STS in its present form can be seen as Phase I, then the establishment of permanent presence would constitute Phase II in the nation's push for world leadership in space.

As a necessary consequence of this mandate, certain augmentations are required for the space transportation infrastructure which enhance our capability to (1) support the establishment of permanent facilities for science, R&D, commercial and operations support in LEO and geosynchronous orbit (GEO), (2) provide routine, economical and flexible access to all orbits, (3) institute routine checkout, refueling, repairing and upgrading of spacecraft in orbit, (4) capitalize on the STS as a space R&D test bed, and (5) devise innovative STS uses and missions (FIG. 1).

OSF ADVANCED PROGRAMS OBJECTIVES

- **SUPPORT THE ESTABLISHMENT OF PERMANENT FACILITIES FOR SCIENCE, R&D, COMMERCIAL, AND OPERATIONS SUPPORT**
  - MANNED AND UNMANNED
  - LEO AND GEO
- **PROVIDE ROUTINE, ECONOMICAL AND FLEXIBLE ACCESS TO ALL ORBITS**
  - MANNED AND CARGO PAYLOADS
- **INSTITUTE ROUTINE CHECKOUT, REFUELING, REPAIRING, UPGRADE OF SPACECRAFT IN ORBIT**
  - MANNED
  - REMOTELY CONTROLLED
- **CAPITALIZE ON THE STS AS A SPACE R&D TEST BED**
- **DEVISE INNOVATIVE STS USES/MISSIONS**

(Fig. 1)

To meet these emerging objectives, our operations in space will gradually expand from the simple orbiting, retrieving and repairing of small free-flying spacecraft of today, such as SPAS-01 or SMM, to the establishment of both manned and unmanned facilities in orbit.

Thus, as summarized on FIG. 2, the space infrastructure around the LEO Space Station in the early period of ca. 1985-1990 will include unmanned facilities that will be used to aggregate payloads in order to obtain inexpensive "housekeeping" support from standardized service buses as well as to be more accessible by fewer Shuttle flights with attendant lower transportation costs. These platforms, to be discussed later in more detail, will fly in both LEO and - in the later time frame of circa 1990-2000 - geostationary orbit (GEO), aggregating science and applications experiments and payloads in low orbit, and primarily communications payloads in geostationary orbit.

INFRASTRUCTURE ELEMENTS

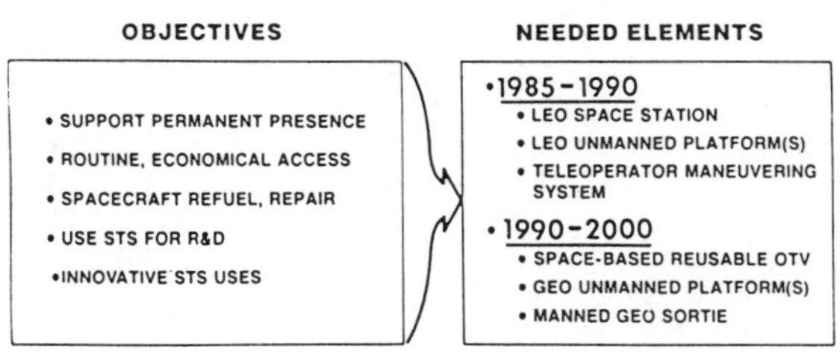

(Fig. 2)

Also a necessary element of the infrastructure for accomplishing the objectives of the early period of permanent presence in space is the Teleoperator Maneuvering System (TMS), a free-flying remote man-controlled vehicle for performing a wide range of payload service remote from the location of the human controller.

For placement of large unmanned platforms and other payloads at GEO, the infrastructure - in the later years - also requires the development of a reusable high performance upper stage for the STS for launch from and basing at the LEO Space Station. This system is referred to as an Orbital Transfer Vehicle (OTV). Addition of a crew capsule to the originally unmanned OTV will eventually - circa 2000 - provide manned sortie capability to GEO.

The crucial importance of the LEO Space Station to these elements of the space infrastructure is indicated schematically in FIG. 3. In essence, what the Space Shuttle is in importance to Earth-to-LEO transportation, the Space Station is to operations in space, becoming an indispensable transportation node to higher orbits and thus transcending its traditional scientific roles. As focal point and "mother ship" of the entire space infrastructure, it enables the efficient utilization of the very limited Shuttle fleet while ensuring economical and efficient access to all other elements of the infrastructure. Vastly superior to the Shuttle as an orbital operations base because it possesses more on-board resources and permanent duration, it serves the infrastructure also as a facility for free-flyer servicing, for storage and assembly, and as a communications and data processing node.

ELEMENTS OF SPACE INFRASTRUCTURE

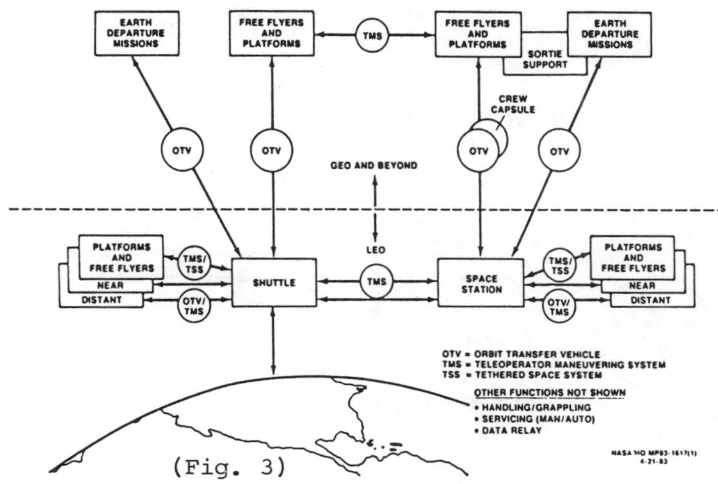

(Fig. 3)

In particular, as depicted in FIG. 3, access to space is provided by the Shuttle from which upper stages such as Centaur and early OTV's, expendable or reusable, may be launched for missions to GEO and other high orbits (with free-flyers or platforms) and to extraterrestrial targets in the solar system. While the Shuttle can directly launch these payloads as well as provide complete access to all elements in LEO and GEO by means of the Teleoperator Maneuvering System (TMS), its on-orbit staytime limitation will make these uses only temporary. To achieve and maintain leadership in space, operations more ambitious in orbital resource and staytime requirements must subsequently be relegated to a permanent Space Station.

As the right half of FIG. 3 indicates, the Space Station will be accessed by the Shuttle either through direct docking or indirectly via the TMS for transferring payloads. The Space Station, as the "hub" of the LEO/GEO infrastructure, not only supports attached (on external mounts) as well as intra-vehicular payloads but is also associated with a number of free-flyers and platforms both co-orbiting and in other inclinations for which it acts as the tending base either through direct docking or via the TMS and - over long distances - the OTV. For the latter, the Space Station assumes the functions of a "service station", basing, refueling and checking them out, attaching them to their checked-out payloads, and performing all necessary countdown activities for launching and sending them on their way.

As FIG. 3 illustrates, for successful operation of both the Space Shuttle and the Space Station permanent manned presence requires a number of components in addition to these systems which have to work in conjunction with them in order to enhance our space transportation capabilities in the time frame of ca. 1985-1995. Beyond its crucial importance to our utilization of low-Earth-orbital space, as indicated in FIG. 4, probably even more important in the long run is the Space Station infrastructure's role as both the key and the gateway for extending man's sphere of influence and presence in important new directions, encompassing

PERMANENT PRESENCE: KEY TO THE FUTURE

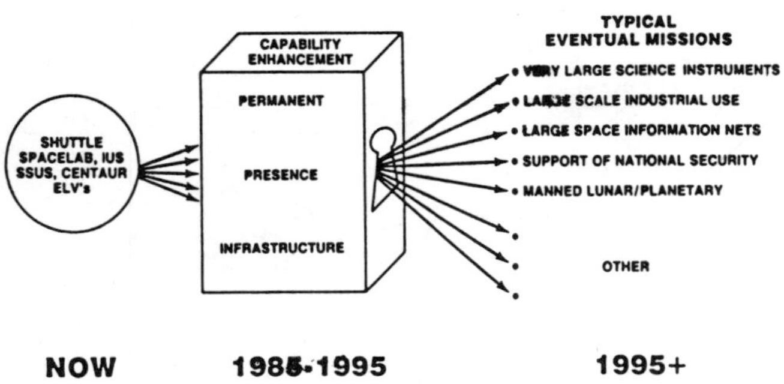

(Fig. 4)

such options as very large affordable space systems for science; large, new space information nets; new commercial products and large-scale industrial activities; support of enhanced national security; manned lunar and planetary initiatives; and others. Each of these exciting and rewarding missions would provide unprecedented scientific/technical benefits as well as a vivid example of the world leadership of the United States in space. And for each of them, the Space Station with its infrastructure would not only serve as the staging point but also as an essential national space test facility for developing the systems and operations technologies required for their eventual realization.

Thus, it can be seen that the objectives listed above are fundamental and ground-laying for achieving the goal of world leadership in space. In order to implement effective steps toward their accomplishment, we have identified four major thrusts as the cornerstones of permanent manned presence in space (FIG. 5):

- o Satellite Services
- o Advanced Transportation
- o Unmanned Platforms
- o Manned Facilities.

ADVANCED PROGRAMS THRUSTS

- **SATELLITE SERVICES**
- **ADVANCED TRANSPORTATION**
- **UNMANNED PLATFORMS**
- **MANNED FACILITIES**
  - **SPACE STATION (TASK FORCE)**
  - **CREW AND LIFE SUPPORT SYSTEMS**

(Fig. 5)

Satellite Services. Manned orbital activities for Satellite Services include such functions as placement, retrieval and in-orbit maintenance of free-flyers and platforms and for retrieval of non-stabilized satellites or space debris. These capabilities will be greater than those basic STS systems such as the Remote Manipulator System (RMS), the Extravehicular Mobility Unit (EMU) and the Manned Maneuvering Unit (MMU) possess, and they will require development/flight testing in orbit. They can be divided in Satellite Services at or near the Orbiter, and those remote from the Orbiter.

FIG. 6 depicts the evolutionary development of the first type of service systems as presently defined. Beyond the initial capability for satellite placement and limited retrieval provided by the Orbiter-mounted RMS, the MMU and the integrated spacesuit and backpack (EMU), improved and new services will be required for space systems such as the Long Duration Exposure Facility (LDEF), Multi-Mission Spacecraft (MMS), Space Telescope (ST), and for the Space Station and associated platforms. These include

SATELLITE SERVICES EVOLUTION (1)
— AT THE ORBITER —

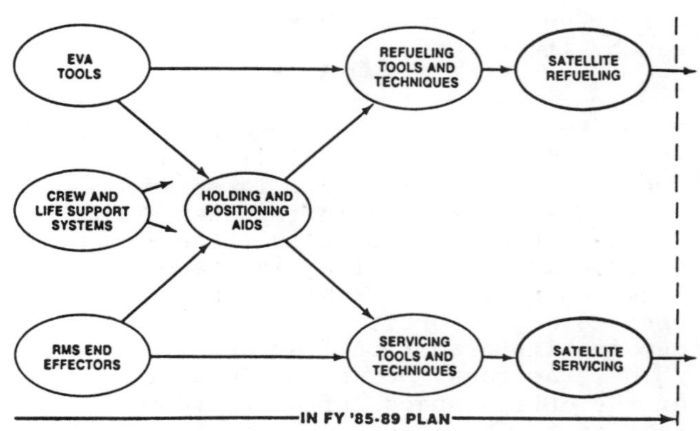

(Fig. 6)

provision of generic extravehicular activity (EVA) tools as well as refueling tools and techniques to permit full capability satellite refueling. Also needed are maintenance and repair tools, payload holding fixtures, berthing platforms, "smart" end effectors (mechanical hands) for the RMS, tools for crew members to use in EVA servicing operations, a remote work station called the Manipulator Foot Restraint (MFR) mounted to the free end of the RMS-arm, television systems, and equipment for large structures assembly and support (FIG. 7).

SATELLITE SERVICES AT THE ORBITER

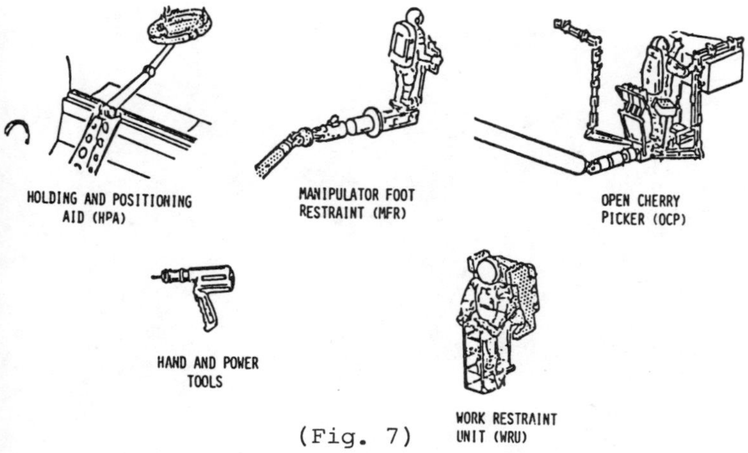

(Fig. 7)

The first major demonstration of manned orbital repair near the Orbiter will be undertaken during STS-13, planned for April 1984, when astronauts will attempt to repair the Solar Maximum Mission (SMM). Since this mission requires first operational use of the MMU, the latter will be flight-tested on an earlier flight, currently scheduled to be STS-11.

A second major demonstration is the Orbital Refueling Demonstration (FIG. 8) which will fly on STS-17, probably in late August 1984. By utilizing manned EVA for transferring hydrazine between two tanks in the Orbiter cargo bay, monitored and controlled from the aft flight deck, we

Orbital Refueling System

ORBITAL REFUELING DEMONSTRATION

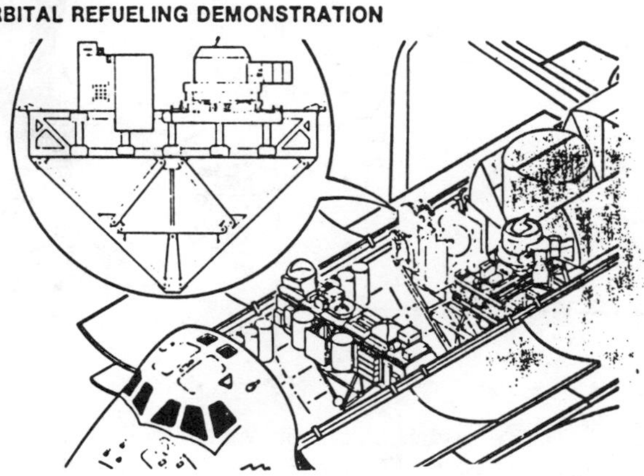

(Fig. 8)

will demonstrate the Orbiter's readiness for refueling satellites in space and in the process acquire unique data for formulation and standardization of a refueling quick disconnect for new generation satellites in order to facilitate on-orbit refueling operations.

The evolution of Satellite Services capabilities remote from the Orbiter is depicted on FIG. 9. Clearly, this aspect

SATELLITE SERVICES EVOLUTION (2)
— REMOTE FROM THE ORBITER —

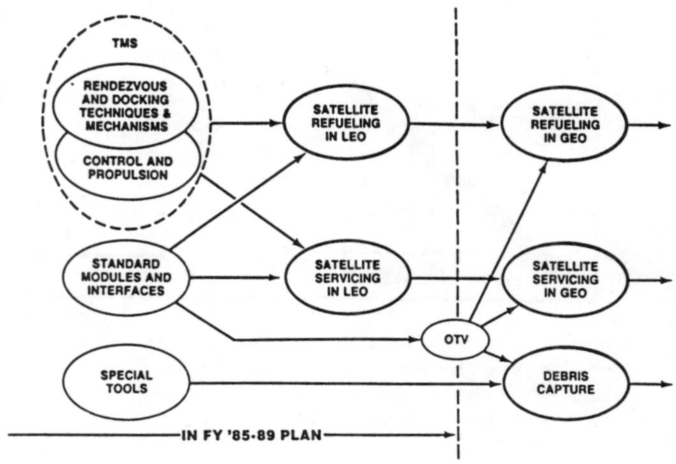

(Fig. 9)

of manned orbital operations will be dominated by the remotely controlled Teleoperator Maneuvering System (TMS) for servicing and refueling satellites in LEO and - when paired with a Centaur or, later, a space-based OTV - also for providing services in GEO as well as debris management. FIG. 10 indicates TMS options which may be pursued to acquire varying degrees of remotely manned servicing capability.

TMS OPTIONS/EVOLUTION

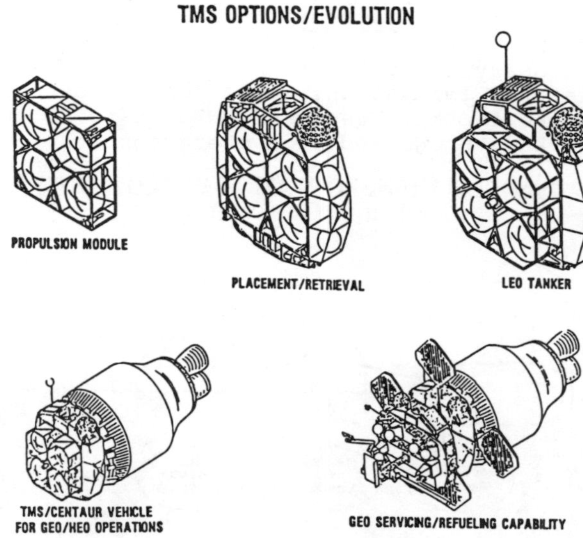

(Fig. 10)

As pointed out above, and as already introduced by the early development demonstrations on Shuttle flights mentioned, the achievement of the necessary new

Satellite Services capabilities will require the use of the Space Shuttle and - subsequently - the Space Station as a test bed for flight experiments and demonstrations. As an example, a preliminary sequence of successive flight experiments is offered on FIG. 11. For the regime of

FLIGHT EXPERIMENTS/DEMONSTRATIONS — SATELLITE SERVICES

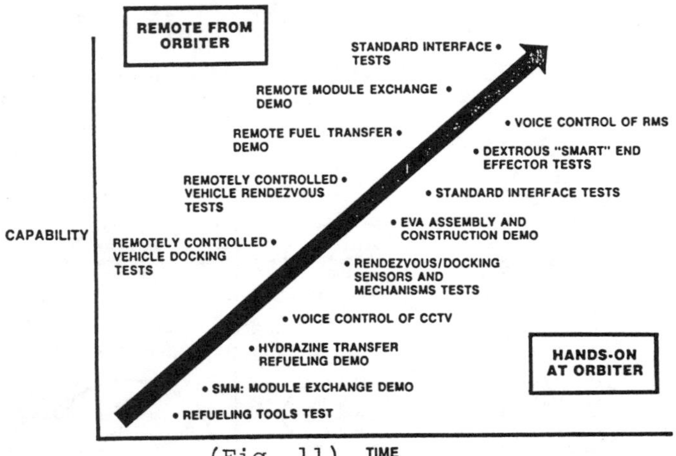

(Fig. 11)

Hands-on/at-Orbiter Services, the SMM repair and hydrazine refueling demonstrations have been discussed above as approved programs. Future activities would typically include Shuttle-borne experiments with voice-controlled closed-circuit television, rendezvous/docking sensors and mechanisms, EVA structures assembly and construction operations, dextrous "smart" end effectors, and voice control of the RMS. For Services far from the Orbiter, we see rendezvous and docking tests, as well as fuel transfer and module exchange, all under remote control, as typical flight experiments that will first use the Shuttle as test bed, then to be continued on the Space Station.

Advanced Transportation. Our thrust in Advanced Transportation provides the necessary elements of the permanent presence infrastructure for our leadership in the two main regimes -- Earth to LEO, and LEO to GEO and high energy orbits.

ADVANCED TRANSPORTATION EVOLUTION (1)
— EARTH TO LOW ORBIT —

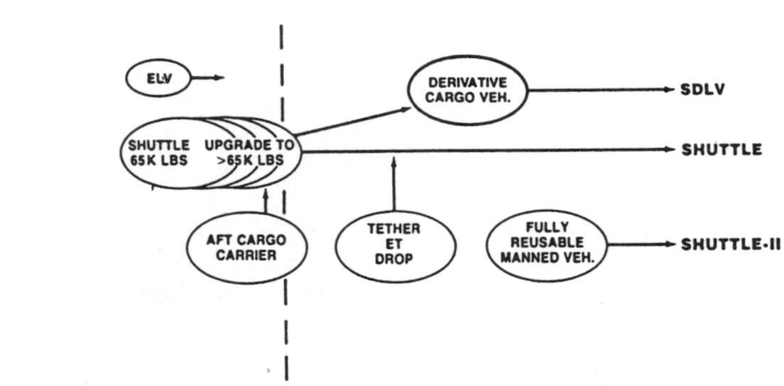

(Fig. 12)

FIG. 12 shows the potential evolution of our Earth-to-LEO transportation capability from the current Shuttle and ELV's to advanced and more economic LEO transportation. Our search for a practical means to carry large, heavy payloads to LEO has surfaced - for the near term - the Aft Cargo Carrier (ACC) modification of the Shuttle External Tank (ET), shown on FIG. 13, and - in the far term, if needed - a vehicle referred to as Shuttle-derived Launch Vehicle (SDLV) that borrows heavily from STS hardware and facilities but is unmanned and therefore need not incorpo-

(Fig. 13)

rate the Orbiter as such. The ACC concept appears of particular interest for the more frequent volume-limited Shuttle missions since it utilizes the base area of the ET for carrying a payload inside a protective container, while leaving the Orbiter cargo bay free for other payloads.

ADVANCED TRANSPORTATION EVOLUTION (2)
— LOW ORBIT TO GEOSTATIONARY AND HIGH ENERGY ORBITS —

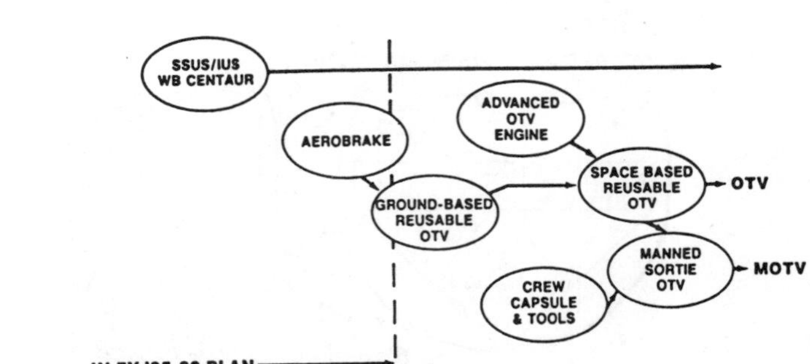

(Fig. 14)

The evolution of Advanced Transportation from LEO to GEO and high-energy orbits is illustrated schematically on FIG. 14. While the Centaur/STS as an upper stage will satisfy many GEO and Earth-escape requirements in the later 1980s, an OTV will become necessary to carry heavier payloads (or combinations of payloads) and/or reduce mission costs through recovery and reuse of the vehicle. Preliminary studies indicate that space-based OTV's may lead to about three times lower GEO transportation costs as compared to the Centaur/STS system. Large, reusable cryogenic OTV's will be needed to place heavy payloads (in the

(Fig. 15)

order of 7000 kg or more) in GEO, as required in the next 20 years for large GEO communications platforms.

One possible approach for a new type of OTV, shown in FIG. 15, could be designed to take advantage of the ACC modification to the Shuttle External Tank discussed above. This concept, designated the ACC/OTV, appears to be readily convertible to a space-based vehicle with the propellant capacity needed for manned trips to GEO (i.e., about 75,000 lbs of propellant).

(Fig. 16)

The use of an aerodynamic drag-type device, or "aerobrake", on a reusable OTV for employing aerodynamic forces during high-altitude braking maneuvers after return from GEO appears to merit further development efforts (FIG. 16). By dissipating the approximately 7800 ft/sec needed for circularization in Earth orbit, propellant requirements are halved for return compared to an all-propulsive version.

Such an OTV has a number of intriguing inherent characteristics which make it essential for a Space Station with full operational capability. One is the ability to access payloads in orbits very different from that of the Space Station, both in inclination and altitude. Whereas large plane changes at low altitude (and high velocity) are extremely demanding in propellants, a Space-Station-launched OTV could achieve very large plane changes with small propellant expenditures if it performs the plane change maneuver at an apogee altitude considerably beyond GEO and then returns to LEO with circularization there via aerobrake and small propulsive maneuvers. With such a vehicle, payloads may be placed in essentially any of the more commonly used orbits launched from the Space Station, with a total velocity change of less than 20,000 ft/sec. This compares with the 14,000 ft/sec required for reaching geostationary orbit from the Space Station. Thus, if such an OTV is available at the Space Station, the latter would have access to virtually any orbit -- with payloads that are about half the mass which the OTV could take to GEO in a straightforward transfer. It is clear, therefore, that the OTV is an indispensable element of the space-based infrastructure for permanent presence.

To develop our capabilities in the area of Advanced Transportation, discussed above, the use of the Shuttle, followed by the Space Station, as an R&D test bed for flight experiments and demonstrations would appear highly desirable for economy and efficiency. Some higher-priority technology milestones are tentatively

FLIGHT EXPERIMENTS/DEMONSTRATIONS — ADVANCED TRANSPORTATION

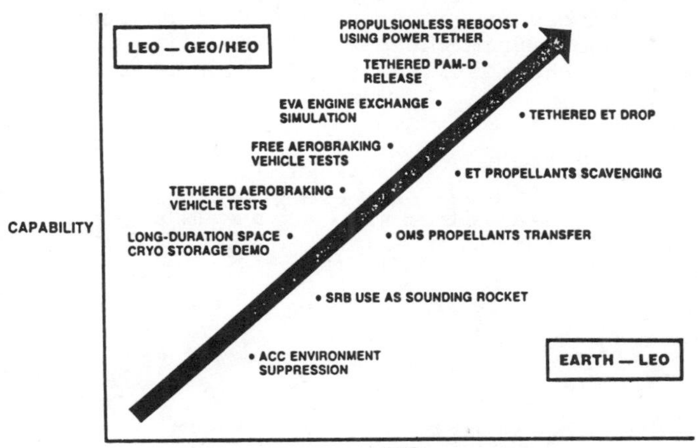

(Fig. 17)

listed in FIG. 17. In the Earth-to-LEO

regime, these experiments would concern itself with the ACC concept, the use of the Shuttle Solid-Rocket Boosters (SRBs) for carrying hitchhiker payloads (for high-altitude soundings during SRB reentry and return), and propellant transfer for eventual scavenging from the ET. In the orbit-to-orbit regime, demonstrations may focus on such suggested developments as long-duration cryogenic storage in space, aerobraking technology utilizing tethered "trolling" through the atmosphere, advanced EVA maintenance activities on propulsion systems, and advanced uses of space tethers.

Unmanned Platforms. Unmanned platform facilities operating in conjunction with the Space Station, because of their aggregation of payloads on common "housekeeping" carriers, offer the advantage of efficient logistics while minimizing adverse interaction among users and their experiments. Such facilities for observation or processing can be operated as free-flying subsatellites of the Space Station, co-orbiting with it or in different orbits so that each platform can be optimized for the function of its aggregate of payloads. It is this

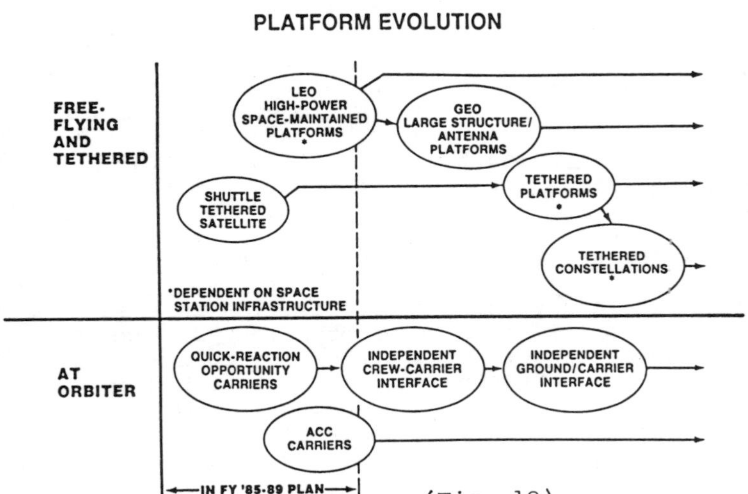

(Fig. 18)

combination of payload aggregation, standardized orbits and man-tending which enhances the economy of future payload support from the Space Station.

The most likely evolution of unmanned

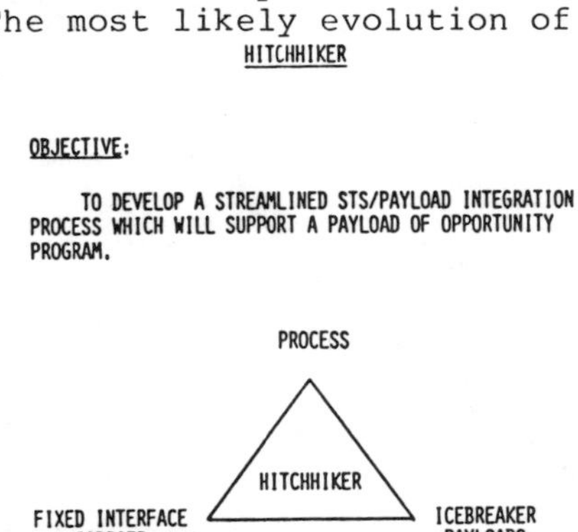

(Fig. 19)

platform is shown in FIG. 18 for the two types of Orbiter-connected carriers and free-flying and tethered platforms. In the former area, the ACC concept has been discussed before under Advanced Transportation. Quick-reaction opportunity carriers are a novel operative approach, currently under study. They are designed to allow late substitution into the Shuttle manifest of payloads and systems which might be selected only two to six months before a flight. This "Hitchhiker" concept (FIG. 19) is expected to reduce costs and lead time for experiments and make better use of space available in the Shuttle manifest.

As shown in the upper portion of FIG. 18, freeflying platforms will be evolving both in the LEO and GEO regimes. FIG. 20 illustrates two permanent or long-life space platforms defined by NASA study contractors which supply 10-12 kW (with growth up to 25 kW) of power, heat rejection, and housekeeping services to a number of Spacelab-type pallets of payloads. A geostationary

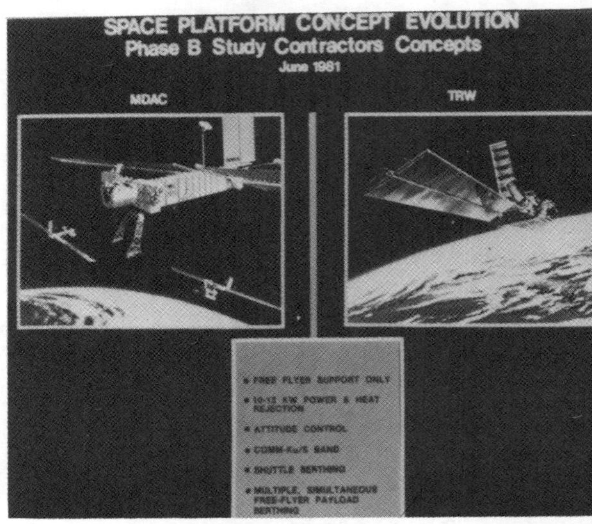

(Fig. 20)

platform designed primarily for the aggregation and support of communications payloads is shown in FIG. 21. Whereas the LEO platforms would be designed for manned access and servicing from the Space Station or the Shuttle via RMS or EVA, such GEO platforms - if not expendable - will be serviced by end effectors on a Servicer vehicle such as the TMS, remotely controlled from the Space Station. The latter also provides launching services for the OTV's carrying the GEO Servicer and the GEO Platform(s) to their destination.

A new type of space-based orbital carrier, the tether, is only now entering closer analysis. As FIG. 18 shows, such devices could evolve from the Shuttle-based Tethered Satellite System, currently under development by NASA and Italy (FIG. 22), to tethered platforms and entire tethered platform constellations. The advantage of such arrangements is that essentially no propulsion would be

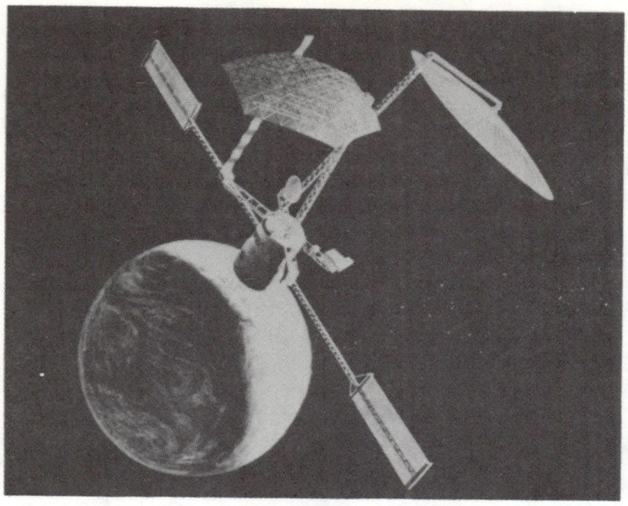

(Fig. 21)

(Fig. 22)

necessary for access to a particular platform from the Space Station since it could simply be reeled in. Also, two-wire tethers could carry power from a central source so that these platforms would not require their own power supply.

Other possible uses of tethers in space are suggested on FIG. 23. The possible evolution of space tethers over the next 20 years is shown in FIG. 24. Though tethering and its ramifications, such as the utilization of electrodynamic interactions, are still far less understood than other elements of the infrastructure, they appear extremely promising and should carefully be investigated.

As for the previously mentioned elements of the Space Station infrastructure, conducting flight experiments and demon-

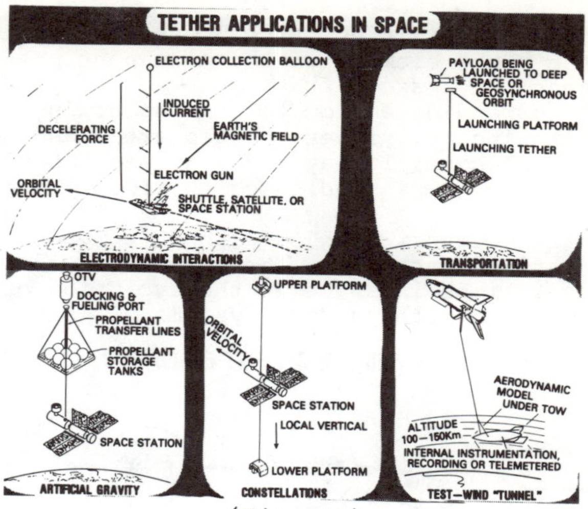

(Fig. 23)

EVOLUTION OF TETHERS IN SPACE

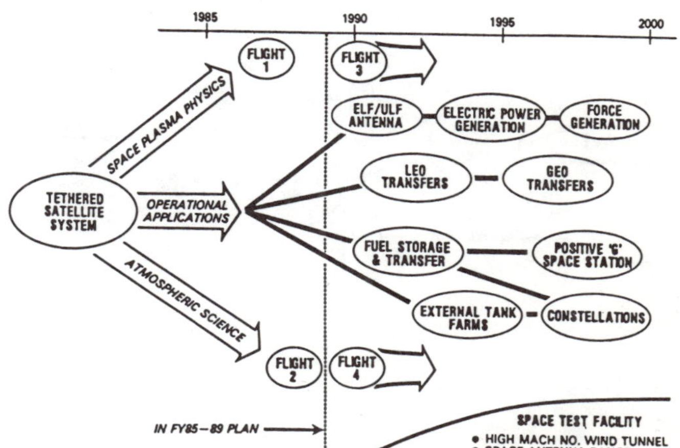

(Fig. 24)

strations on Shuttle flights and later on the Space Station appears extremely important for the development of unmanned platforms. Some suggested flight initiatives for at-Orbiter carriers and free-flying and tethered platforms are listed on FIG. 25. As mentioned above, an early implementation of the quick-reaction hitchhiker opportunity carrier concept is already being pursued.

FLIGHT EXPERIMENTS/DEMONSTRATIONS — PLATFORMS/CARRIERS

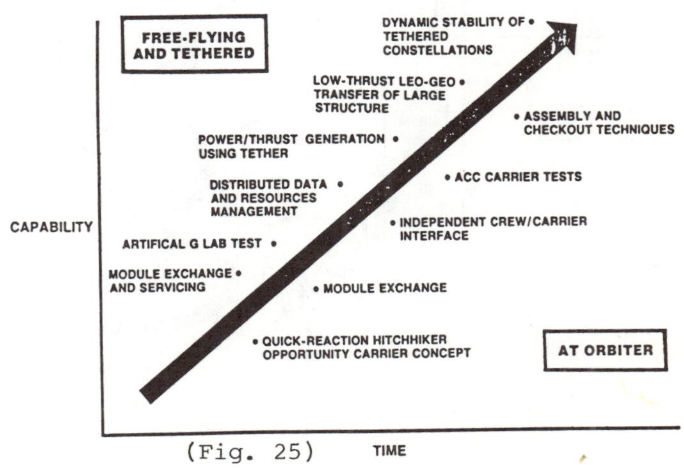

(Fig. 25)

Manned Facilities. The Space Station is the "hub" element of the permanent presence in space infrastructure. To provide focus and direction for the agency's planning activities in this area, NASA has established, in May 1982, a Space Station Task Force, as indicated on FIG. 5, above.

However, with or without a Space Station, central to the support of manned operations in space is the evolution of crew and life support systems. Their

CREW AND LIFE SUPPORT EVOLUTION

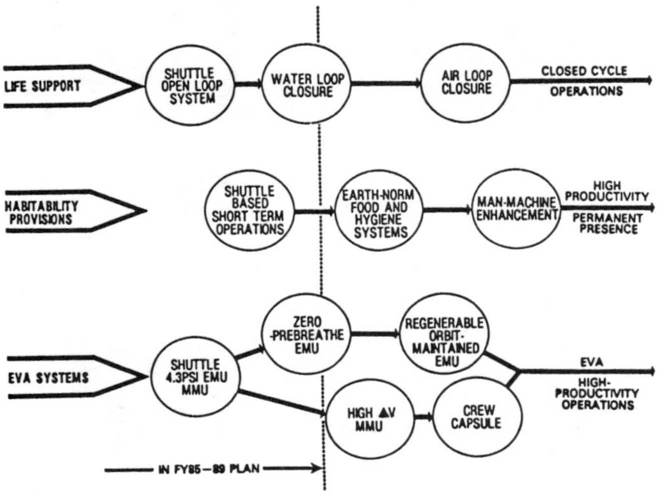

(Fig. 26)

anticipated development in the three major areas of Life Support, Habitability Provisions, and EVA Systems is shown on FIG. 26.

For on-board life support systems, successive closure of the water and air loops will be the target of efforts utilizing the Shuttle and - later - the Space Station as an R&D test bed, seen to result in the closed-cycle life support system shown schematically on FIG. 27. Permanent manned presence in space also requires habitability improvements which will enhance current Shuttle-based short-term

CLOSED CYCLE LIFE SUPPORT

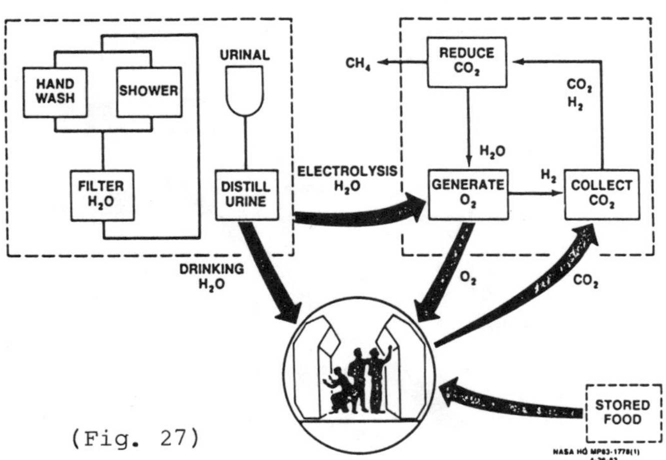

(Fig. 27)

operations towards eventually providing Earth norms in space food and hygiene systems as well as enhancements of man-machine interaction to achieve higher productivity.

As part of this aggressive drive toward higher human productivity in space, both Shuttle and Space Station will be important for the development of superior EVA systems such as the higher-pressure space suit to eliminate pre-breathing currently under development at NASA (see FIG. 28), a future regenerable orbit-

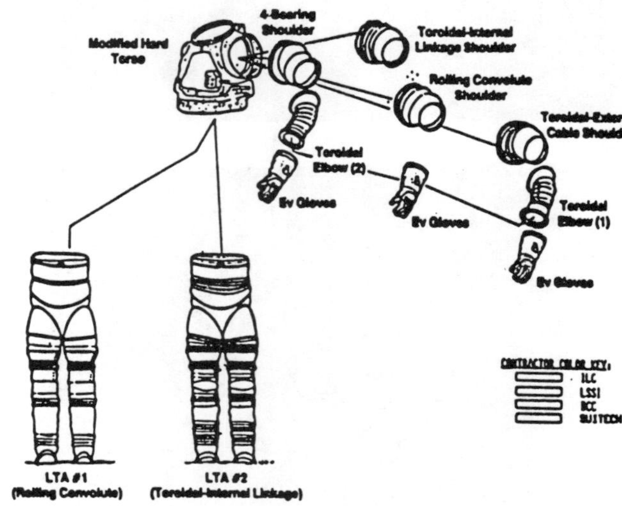

(Fig. 28)

maintained extravehicular mobility unit and backpack, an advanced manned maneuvering unit, and a crew capsule for manned access to higher orbits. Some tentative suggestions for flight experiments and demonstrations are listed on FIG. 29, including the high-mobility/tactility glove and possible development steps for subsystems that will provide closed-loop capability.

FLIGHT EXPERIMENTS/DEMONSTRATIONS — CREW AND LIFE SUPPORT

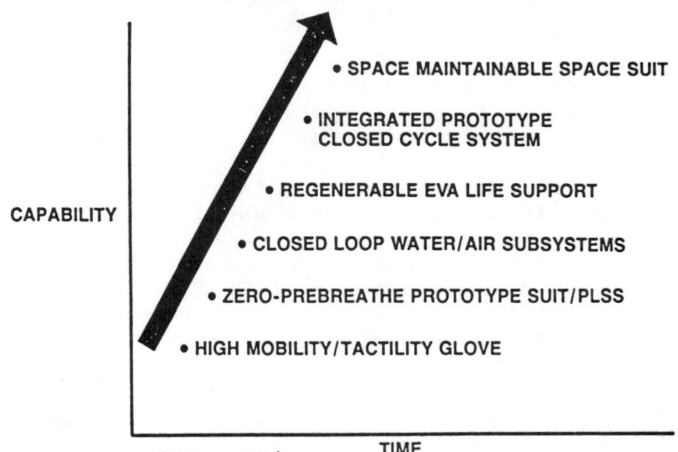

(Fig. 29)

III. Summary

A summary presentation of the evolutionary steps necessary for the development of a space-based infrastructure in support of permanent presence is shown on FIG. 30. These major elements are indispensable for a Space Station intended to make full use of its inherent space operations/services capabilities.

CAPABILITY EVOLUTION PLAN

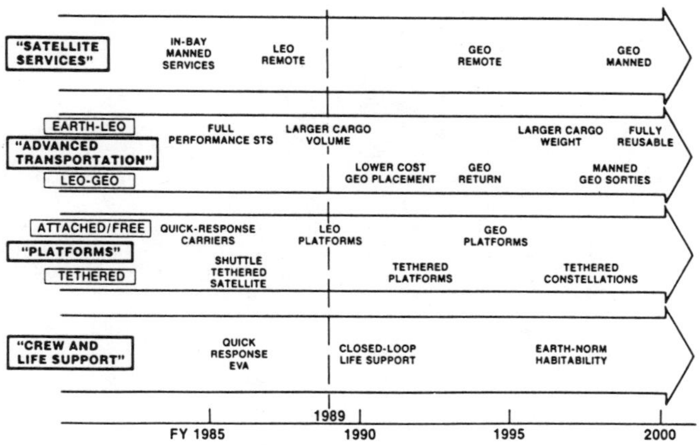

(Fig. 30)

It has been shown in this paper that these elements require a logical succession of flight R&D experiments and demonstrations using first the Space Shuttle as an orbital R&D test bed, to be then continued on the Space Station itself. It therefore becomes clear that the next 20 years beyond 1985 will undoubtedly represent a robust era for manned Earth-orbital activities in which both the Shuttle and the Space Station play key roles, the former as a universal "work horse" for carrying the Phase I capability of routine access-and-return into Phase II's permanent manned presence in LEO, the latter as indispensable node for supporting a wide variety of commercial, civil, international, and national security assets such as free-flyers and platforms in LEO and GEO.

This paper has shown that in order to attain the basic goal of achieving and maintaining world leadership in space, these support elements must include the Teleoperator Maneuvering System, the reusable space-based Orbital Transfer Vehicle, the Aft Cargo Carrier, advanced crew and life support systems, novel tether applications, and eventually manned sortie capability to geostationary orbit and beyond.

Conversely, a need has been identified for enhanced advanced development and intensified system definition activities focusing on these infrastructure systems.

ACKNOWLEDGEMENT:
The assistance by Jesco von Puttkamer in the preparation of this paper is acknowledged.

SPACE STATION TECHNOLOGY

Chairman: Raymond S. Colladay
Acting Associate Administrator for
Aeronautics and Space Technology
National Aeronautics and Space Administration
Washington, D.C.

Earlier sessions provided an overview of space station plans and progress as well as possible missions and utilization. We all recognized that Skylab was a space station and that we could build another station with current technology. However, its capability would be severely limited and even become obsolete over a 20-year lifespan. More advanced technology promises a significant opportunity for a more capable, productive, and cost-effective evolutionary space station. To that end, NASA has been pursuing technology development in a number of basic areas generically applicable to a space station.

We have time enough to expand the menu of technology options before committing to a specific station design. In examining options, some of the key issues are: the desired level of technology in the initial station; the risk, technical barriers, and "show stoppers"; and provisions for growth. The issue of evolutionary capability is particularly important. We must provide technology for an initial station configuration that permits evolutionary growth, fed by a pipeline of emerging mature technologies. The theme of this session is technology for productivity in a space station. The papers will address the systems technology considerations of productivity and specific technologies critical to productivity: automation and artificial intelligence, information systems, communications, and crew and life support.

PRODUCTIVITY IN AN EVOLUTIONARY SPACE STATION

John L. Anderson and Richard F. Carlisle *
NASA Headquarters
Washington, D.C.

Abstract

A space station will involve the formation and operation of a technological union of humans and machines with the intent of achieving sustained operation for a period of 10-20 years. A major goal of space station technology efforts is to enable both the initial and evolutionary productivity of the space station. While technologies improve productivity, space station productivity is to a significant degree a broad systems issue. Various functions and attributes of space station design, development, formation and operation and their potential effects on productivity are examined from a systems viewpoint.

Introduction

Until the fifth flight of the Space Shuttle, the 25 years of space operations of the United States had been experimental and exploratory. That quarter century of test flight led to the operational Space Transportation System now commuting to and from low-Earth orbit. It also has given us the foundation for our next major venture into space--the establishment of permanent human presence in space.

A space station will involve the formation and operation of a technological union of humans and machines. An indication of the potential complexity and scope of operations of such a union may be obtained from the alternative station configurations shown in Figure 1. While there is an element of experimentation in the forming of a space station, its purpose is to achieve sustained operation over a period of 10-20 years.

It is commonly accepted that the U.S. could build a space station with present technology; after all, Skylab was a space station. However, such a station's

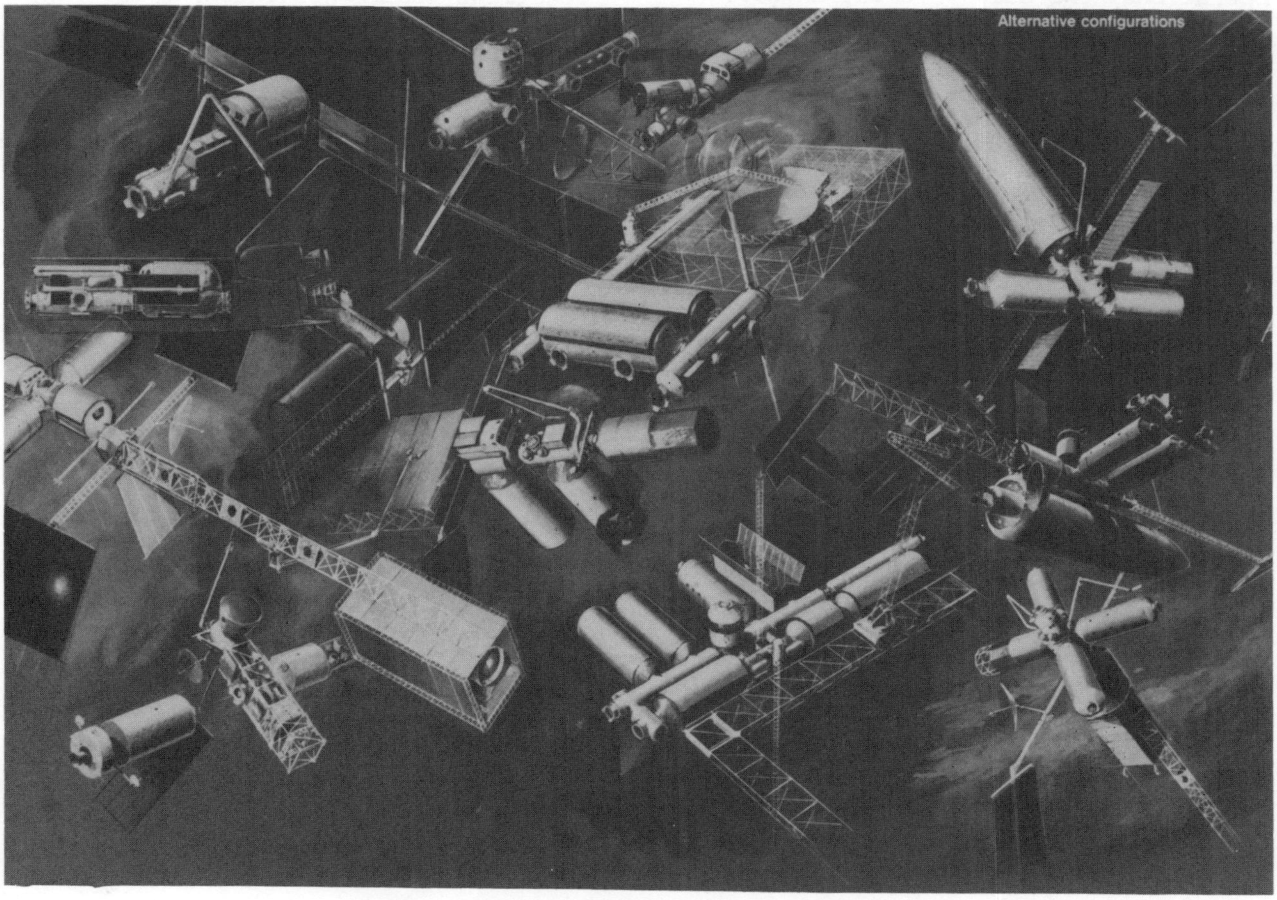

FIG 1. ALTERNATIVE SPACE STATION CONFIGURATIONS

* Manager, Space Station Systems Office
Member AIAA

capabilities would be somewhat more limited, less cost effective, and less long-lived than possible with selected advancements in technology. The technology issues addressed in this paper deal with the productivity of a space station, a subject of more importance for an operational space system than an experimental one.

What do we mean by productivity for a space station? As for precedents, home and office productivity are more relevant than factory productivity. In the home or office, technology has improved productivity through labor-saving devices and utilities, thus greatly reducing the parasitic demand of subsistence and upkeep. In the factory, technology has improved productivity primarily through high-multiple repetitiveness.

Three characteristics of the envisioned space station significantly affect the concept of its productivity, shown symbolically in Figure 2. It will be one-of-a-kind, fulfilling a specialized community role that really has no precedent. It will be evolutionary in that its configuration and capabilities will grow from some initial state. One pivotal technology, the data system, will be the valve through which all operational and user information will flow.

The productivity of a space station might be directed toward achieving the highest output to users for the least cost with the most humanistic use of crew and ground support personnel. A precise definition of each of these terms would neither be agreed upon by everyone nor be very helpful if it were; the systems considerations and tradeoffs are too complex. Therefore, this paper treats space station productivity as a broad systems issue. It explores relevant systems concepts, mapping some areas, or identifying critical features, opportunities, and pitfalls, and noting where added exploration is needed. The paper does not specifically address technologies which improve productivity; it covers the functions and attributes of space station development, formation and operation that could strongly affect productivity.

Systems Design

Extensive mission analysis, program planning, and technology studies are being conducted prior to system design. A major output of these studies is to determine an optimum way that subsequent space station system design can provide for mission flexibility, technology advancement, maintainability and evolution, that is, to enable the productivity of the system to improve during its useful life. The initial focus of Space Station technology activity is on the application of advanced technologies to the core functions (basic housekeeping and utilities) of the station. Subsequent attention will be given to technologies that will enable specialized functions and mission performance as they are defined. A major goal of space station technology efforts is to enable both the initial and evolutionary productivity of the space station.

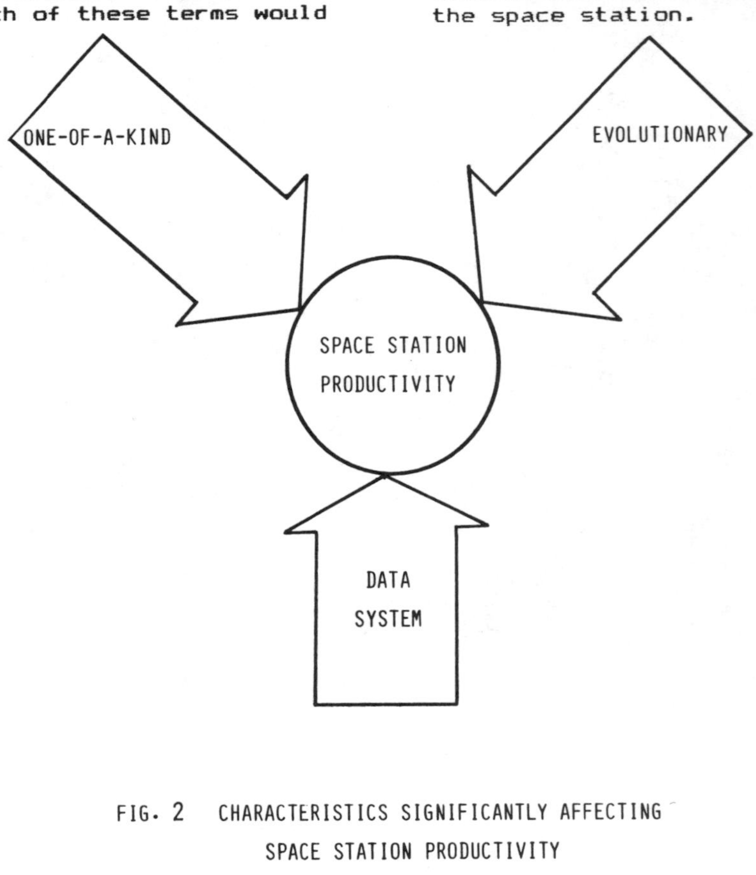

FIG. 2 CHARACTERISTICS SIGNIFICANTLY AFFECTING SPACE STATION PRODUCTIVITY

Productivity of humans is a function of many factors: experience, training, motivation, environment and operational aids which include standards, procedures, tools, fixtures and machines. While this paper treats the systems aspect of productivity, it is appropriate to at least identify the technology aspect. Maturing technologies applicable to improved productivity are shown in Figure 3. The functions they enable can be included in the initial station if the technology is available at acceptable risk. If the technologies are too high in risk for the initial station then the architecture and configuration of the initial station should provide for their economical incorporation at a later date. Later inclusion can be accomplished either by add-on or by replacement or modification. This anticipated introduction of new technology must first be anticipated in the system design.

The goal of enabling multi-generation productivity applies to the entire human-machine system, both space-borne and ground-based. The productivity of a station will be derived from the efficiency or interactive effectiveness of the ground support crews and data systems, the flight crew and systems, and the users and their systems. The incorporation of these elements into a highly productive system will involve tradeoffs and integration of many design alternatives that acknowledge the characteristics of both humans and machines which promote productivity (Figure 4).

Systems Risk

Boundaries on the productivity of a space station are drawn, some crisply, some more vaguely, by the philosophy of the system risk--even before the design itself. For example, design engineers typically will incorporate familiar components or design approaches which may already be behind the SOA that is familiar to other industries. This conservatism of familiarity, while minimizing risk, may be unnecessarily sacrificing more advanced productive technologies. The acceptable measure of risk for a long-lived space station may have to be defined differently than for an experimental space vehicle. The concept of an evolutionary system might incorporate a design that would permit higher risk components to be chosen at the outset or inserted at a later time into the operating system. That is the isolated introduction of a high risk, high productivity technology into a buffered operating system protected from cascading failures would be permitted.

PRODUCTIVITY FACTOR	ENABLING TECHNOLOGY	FUNCTIONS PROVIDED
EXPERIENCE	INFORMATION SYSTEM HEADS-UP DISPLAY (EVA) EXPERT SYSTEMS	OPERATIONAL HISTORY PROCEDURAL REVIEW & TUTELAGE CONSULTATION; EXPERT ADVICE
TRAINING	LIFE SCIENCES COMPUTERS COMPUTERS & DISPLAYS	HUMAN PERFORMANCE PLANNING & MEASUREMENT; TEACHING TECHNIQUES OPERATIONAL SIMULATION INSTRUCTION; PROCEDURAL MANUALS
MOTIVATION	FOOD; HYGIENE; RECREATION AUTOMATED WORK STATIONS	PSYCHOLOGICAL WELL-BEING USER FRIENDLY INTERFACING
ENVIRONMENT	LIFE SUPPORT SYSTEM EXERCISE EQUIPMENT	PHYSIOLOGICAL WELL-BEING PHYSIOLOGICAL WELL-BEING
STANDARDS & PROCEDURES	DATA SYSTEM STANDARDS COMPUTERIZED DIAGNOSTIC PROCEDURES	SYSTEM CONSISTENCY; MULTIPLE USERS TROUBLESHOOTING
TOOLS & FIXTURES	EVA-GLOVE-HAND-TOOL TELEPRESENCE AUTOMATED WORK STATIONS	MANIPULATION REMOTE VISION & MANIPULATION CONTROL
MACHINES	TELEOPERATOR MANEUVERING SYSTEM TELEPRESENCE	SPACE-MOBILE REMOTE PRESENCE REMOTE PRESENCE

FIG. 3 APPLICATIONS OF MATURING TECHNOLOGIES TO HUMAN PRODUCTIVITY IN SPACE STATION

Furthermore, the inclusion of even a mature technology component into a relatively complex system carries with it a systems risk. Its systems interactions and performance cannot be completely anticipated. If one considers multiple interacting complex systems, the uncertainty of component interactions is greatly compounded.

Systems Management

Throughout the design, development, and long-term operation of a space station an engineering data base will be needed as a technical reference. The use of computer aided design and engineering (CAD/CAE) will offer an integrated common data base. By combining that integrated base with the operation manuals, the complete system and technical specifications for the station can be computer accessible. This ready access to the system description will be important because of the increasing degree of autonomy of some subsystems, the sheer amount of information (Space Shuttle requires 80 printed volumes stored on board!), and because of the need for quick access and assessment of unexpected events or anomalies. Furthermore, an evolving system should have some "corporate" record of its specifications, health, modifications and past history so that as new modules are added, 1 or 2 decades later, the information systems are readily integrable.

Systems Operations

Automation may be defined as the self-operation (without outside control) of a process or system under normal, usually repetitive or routine conditions. Autonomy is automation with the added capability of self-government; that is, with problem-solving and decision making under abnormal conditions.

An area to consider for improving productivity is the automation of tasks and subsystems in order to reduce ground support personnel and to free the crew for activities that make better use of human abilities. However, limits and guidelines for doing this must be determined.

Some functions and subsystems will be easier to automate than others. Some must be automated because of response time and critical nature. Some should be automated because they are boring, repetitive and non-critical. It may be less productive to automate tasks that require substantial human judgement, tolerance for ambiguity, or infrequent complex procedures.

The components, subsystems and systems of a space station will have to be maintained--monitored, serviced, repaired and replaced on orbit. A degree of modularization appears to be a productive approach to equipment design. The use of

HUMAN	MACHINE
SELECTIVE RECALL	LARGE NUMERICAL DATA STORAGE
JUDGEMENT	PRECISE REPETITIVE OPERATIONS
IMPROVISATION	QUICK RESPONSE
SITUATION ADAPTATION/FLEXIBILITY	DIFFERENT SENSOR RANGE
GENERALIZATION	CONTROL OF LARGE FORCES
PATTERN RECOGNITION	NO FATIGUE OR DISTRACTION
AMBIGUITY TOLERANCE	RAPID COMPLEX ANALYSIS (PREPROGRAMMED)
GLOBAL AWARENESS	NO PHYSIOLOGICAL BOUNDS

FIG 4. CHARACTERISTICS PROMOTING PRODUCTIVITY

fault tolerant computer systems and automated monitoring and diagnostic instruments for detection and warning of impending failures should reduce downtime and inopportune disruptions, thus making the system more productive. The maintenance data base must be a part of the station engineering data base. Computer aided learning of troubleshooting and maintenance procedures can significantly increase productivity. The number of subsystems, failure modes, subsystem interactions and corrective procedures seem to favor a computer tutor, perhaps using expert systems. The need for this type of system is accentuated by two other factors: 1) the large number of crew members who will need to learn and use the maintenance procedures over a 1-2 decade life and 2) a record of the maintenance history over the life of the station will not only be valuable but it will be used to adapt and update the maintenance procedures as operational experience is gained and as the system ages and evolves.

The users, scientific and engineering investigators, which the station will be designed to serve must find the station productive enough as a laboratory for industrial, commercial, and university use for 1-2 decades. Thus, the station must be able to grow in capability, improve its operation as any Earth laboratory, and provide support services.

On-board utilities may be classified as part of systems operations. These utilities will provide basic space station functions and will substantially influence productivity. The most critical utility is life and crew support. Taken in the larger context of habitability, this utility must not only support physiological and work needs but also the crew psychological needs. A second critical utility is communications, which must include intra-station module, inter-vehicle and space-ground links.

Most functions of systems operation must be either monitored or directed through a single technology--the human/machine interface. Thus, human input and output must be matched carefully with the machine output and input. Beyond just being user friendly the machine input/output should exhibit a high degree of intelligence. The space station will be an information-rich system; to manage and maintain it, to keep it productive, and to serve its user functions effectively will require an advanced human-machine interface.

Progressive Autonomy

When first launched, subsystems and modules will require checkout and human control until the performance is validated. For critical functions, a period of close human control may be necessary during which time the crew establishes enough trust in the machine's performance to turn over portions of a function to be performed autonomously by the machine. Ultimately, however, many functions must be done autonomously to free the crew for more productive work. In other words, the systems and the technology must be planned to progressively offload tasks from the humans to the machines. This off loading should be guided by the characteristics of both humans and machines which promote productivity (examples of which are shown in Figure 4) as well as the constraints and limiting characteristics which inhibit productivity. Humans will remain in the control loops of the various systems, if only to respond to emergencies.

There is an analog to learning which indicates we should make some provision for teaching the automated system to handle operational conditions of increasing complexity. Enough flexibility and expandability should be built into the system so that the data management system can be configured and programmed for new control capabilities as needed. The lowering cost of microprocessors, expanding capability provided by gate arrays and microprogramming, and the emerging technology of artificial intelligence can permit the system to learn and therefore enhance the systems performance and productivity.

Technology Upgradability

Any technological system, particularly a computer-oriented one dealing in research, cannot expect to stay current, productive, or effectively utilized if it cannot upgrade its capability as the state-of-the-art of critical technologies advances. The pivotal technology affecting the utilization and productivity is probably the data management system with its associated networks of computer hardware and software. That technology also happens to be advancing at an explosive rate which means that even at launch a space station's computing capability will be somewhat behind the state-of-the-art and within a decade it might be considered archaic.

A major deterrant to technology upgrading is the tight integration of multiple complex systems for a spacecraft, especially one that is life supporting. Any change, such as increased capability through a new or improved technology, will cascade through the entire system. Designs and modular approaches are needed which permit the "impedance matching" of new technology systems with old.

An assessment must be made about the degree of technological upgradability to be allowed. This is particularly true for data systems because of the continuing rapid advancement of microelectronics and computer technology. However, because of cost, operational difficulties, and systems risk associated with technological changes (hardware) the payoff of such changes and upgrading should be high. In particular, substantially increased productivity, data throughput, number of users accommodated, sophistication of payloads and experiments of a national laboratory asset should warrant technology upgradability.

Evolution

One concept of a space station treats the station as evolving in modules over several years. The sequencing of modules and their functions will be orchestrated from early in the design phase of the first module. Modular growth raises a number of systems and technology issues. Should the utilities be reconfigurable? The data, communication, fluid, thermal and power systems will have to accept an expansion or linkage to the additional modules as they are introduced into the station configuration.

As the configuration changes the attitude control and stabilization system will have to account for different system dynamics. Automated control will necessarily have to accommodate or be reprogrammed with revised control laws. As the system grows to include space vehicle traffic, intra- and extra-vehicular activity, stability becomes a more complex system operation and perhaps a more critical one because of onboard experiments.

The system problem of creating the proper modular interfaces is one that must be addressed from the beginning. From the data system to the mechanical and electrical systems interfaces, successive modules should merge. One solution is to establish a set of standard, tight interface specifications from the outset. Another solution might be to have looser interface specs but develop a separate buffer or adaptor technology that allows on-orbit fine adjustment during the mating of modules.

In Flight Test & Integration

The Space Shuttle has an exact duplicate, non-flying version of itself on the ground at JSC/MSFC as a simulation/integration test bed. There, all the planned additions or changes, software or hardware, to the Shuttle avionics systems may be completely tested before implementation. It is likely that such a duplicate system for a space station will be too costly, both initially and over twenty years. A complete software simulation might be feasible, but a careful assessment is needed because the software cost might be comparable to that of a hardware simulation.

Technologies and systems approaches are needed that will permit to some degree the test and integration of a new subsystem to take place in orbit--an in-flight test. Some of the integration problems will simply not be appropriate for ground test--such as attitude control and stabilization.

Conclusions

We may expect one of the purposes of a space station will be to provide an effective, productive space laboratory for research and experimentation over 1-2 decades. The technologies and systems concepts identified here are early explorations into the concept of productivity in a one-of-a-kind, operational, evolutionary system.

THE ROLE OF AUTOMATION AND ARTIFICIAL INTELLIGENCE

Roger T. Schappell*
Martin Marietta Aerospace
Denver, Colorado

Abstract

The projected costs for designing, fabricating, launching, operating, and maintaining a future space station, based on Shuttle technology, will be prohibitive unless we capitalize on current and near-term automation and artificial intelligence technology. Although we have much of the technology in hand, we have not yet arrived at a plan which will optimize use of our limited resources, which, in turn, will enable application of the required technologies. This paper was prepared with the intent of illustrating how we might strengthen our automation technology base and apply it in order to meet the challenges that face us.

Introduction

Our traditional role as an aerospace contractor is to "make it happen," i.e., upon receipt of an RFP, we will design and build the launch vehicle or spacecraft using "current" technology. At the same time, we are required to build it in a manner which is as cost-effective as possible. In order to maximize this objective, realistically only those advanced technologies sufficiently mature at that point in time should be seriously considered for utilization; otherwise, probable program cost and schedule impacts will result.

With that in mind, how do we prepare for, fund, and take timely advantage of the emerging technologies? How do we implement them in a space station that is to be operational in the early 1990s? What are the high payoff technologies to be considered? I shall begin my presentation with a synopsis of the technologies requiring immediate attention and then attempt to address the questions I have posed.

It is my contention that had the nuclear power industry performed more extensive materials testing; developed more forward-looking remote maintenance, inspection, repair, and sensing techniques; applied relatively basic decision aids and computers; and educated the public; their contribution to the world would be better understood, appreciated, and perhaps, more universally accepted. The reasons generally given for their current dilemma are competitive pricing, training, limited resources, the requirement for minimal technological risk, and government regulations. Although NASA is also constrained by most of these parameters, NASA does have the opportunity to move out in a timely fashion and reap the benefits of specific emerging technologies which will enable lower design, development, fabrication, operational, and maintenance costs.

*Manager, Advanced Automation Technology

Critical Technologies

The technologies presented here are emerging technologies which have not been part of our mainstream design processes but promise to provide unusual potential for a more cost-effective space station. I am convinced that, in the time frame of interest, these technologies will be in common usage and will not require another decade to mature from technological elegance to operational utility. These technologies offer multiple usage and therefore, must be exploited in a plurality of applications. Furthermore, these technologies were selected because of their potential payoff as contrasted with subcritical technologies which would merely result in a dissipation of resources. It is also important to note that this is not intended to minimize the importance of other technologies such as propulsion or power but has been kept within the realm of the author's primary knowledge and the confines of the requested paper. The specific technologies which I consider critical are as follows:

(1) Artificial Intelligence
(2) Electromechanical Systems - Robotics and Teleoperators
(3) Sensory Perception
(4) Integrated Design, Simulation, and Analysis Tools
(5) Advanced Control System Architectures
(6) Sub-Micron Microelectronics (VHSIC)

Although the latter two are essential automation-related technologies, they are the subject of another paper and will not be discussed further herein.

Artificial Intelligence (AI)

This burgeoning area has potential for multiple real-time applications such as autonomous satellite control, resource planning, mission timeline replanning, automatic decision making for fault isolation, failure mode workaround, etc.

> "Artificial intelligence (AI) is that branch of computer science concerned with the design and implementation of programs which make complicated decisions, learn or become more adept at making decision, interact with humans in a way natural to humans, and in general, behave in a manner typically considered the mark of intelligence.
>
> Intelligence, here, is to be understood not as a property that, for example, gifted mathematicians posses; but rather as a property all humans and some animals possess. Intelligence, in this sense, is the ability to understand and process large and varied amounts of information. It is the ability to meet and cope with novel situations by adjusting behavior, the ability to comprehend the interrelationships between facts and concepts, and

the ability to generate new concepts and relationships from those already known (e.g., already in the data base). The artificiality of the intelligence means merely that the intelligence is achieved by means of technology.

The scientific research done in AI covers a large area of theoretical topics such as knowledge representation, knowledge acquisition, problem solving and search, vision, theorem proving, and natural language. Though each one of these topics can be researched from the human ability perspective, i.e., by asking how humans represent knowledge, acquire knowledge, solve problems, see objects, communicate, etc., researchers in AI are concerned with implementing the given ability in computers. AI is not only a theoretical enterprise--it has definite and robust applications. The primary concern in the applications arena is the design and implementation of expert systems and natural language interfaces.

Aside from the general scientific curiosity of wondering how to design and implement a computer program that "learns," what advantages might one obtain from the application of AI? Here are some rather broad, obvious ones.

(1) Augmenting our ability as humans to come to grips with the enormous and increasing amounts of information that we are generating.

(2) Increasing the efficiency in man/machine interfaces (the ability to communicate with a computer in English) enables humans to get more work done and obviates the need for specialists for those hard-to-use formalisms known as modern computer languages and data base query languages.

(3) Creating systems (such as space vehicles) that can make crucial decisions on their own when they have to.

(4) Decreasing the effect of such human problems as forgetfulness, fatigue, and emotional turmoil.

(5) More rapid problem solving and strategic and tactical planning in a wide variety of domains."**

The Artificial Intelligence Unit at Martin Marietta Denver Aerospace is currently concentrating on the application of expert systems to a variety of DOD and NASA interests. We actively maintain communications with many members of the AI research community and continually stay abreast of state-of-the-art advances in expert systems as well as other areas of AI. Our philosophy is to maintain a core organization of AI research scientists. These scientists actively maintain their knowledge of various areas of AI while, at the same time, studying different applications of expert systems. We conduct research to extend AI technology in consideration of applications of interest. Since Martin Marietta has expertise in a variety of areas, it provides a fertile base for the development of applied expert systems. To this end, the scientists in the AI group have constructed an expert system laboratory which includes software and hardware (including VAX 11/750s and LISP machines) that supports the development of expert systems.

To further illustrate our intended use of AI technology, Table 1 summarizes our current objectives.

AI can be usefully applied to a wide variety of problems. Because of the field's newness, however, the applications must be cultivated with directed research.

Table 1 - Artificial Intelligence Technology Objectives

EXPERT SYSTEMS
- o Demonstrate a fault isolation expert system for a space station power module; subsequently, develop expert systems for other modules (thermal management, etc.) and for other C^3 applications (ground support systems, etc.)
- o Develop tools to facilitate rapid expert system development.

NATURAL LANGUAGE
- o Develop a robust natural language data base query system for large distributed C^3 data base applications.
- o Develop a text synopsis capability.
- o Incorporate voice recognition/synthesis capability.

PLANNING SYSTEMS
- o Develop an AI planning system for real-time applications such as robotic control and subsystem resource scheduling.
- o Apply AI planning techniques to the C^3 strategic tactical environments.

IMAGE UNDERSTANDING
- o Develop a system for acquiring and reasoning about three-dimensional data for robotic control, autonomous navigation, etc.
- o Develop a capability to perform adaptive reasoning utilizing multisensor data.

**This brief dissertation on artificial intelligence was prepared by Jeffrey Van Baalen from Martin Marietta Aerospace, Denver, Colorado.

Those of us involved in the AI effort at Martin Marietta are enthusiastic about the applications of expert systems and natural language parsing and see a future where artificial intelligence and automation in general play significant roles in everything from manufacturing to space exploration.

Electromechanical Systems, Including Robotics and Teleoperators

This represents a potentially high payoff area because it can provide for the optimized use of man and operation in hazardous and remote environments. It can also enable direct interaction with space station experiments by the experimenter at a remote control station.

Intelligent Robots. The state of the technology is best illustrated by a current AFWAL/DARPA activity wherein a dual-arm manipulator system is under development which will demonstrate the feasibility of intelligent robotics. Intelligent Task Automation (ITA) represents a challenging merge of optimized sensory perception, artificial intelligence, adaptive control, high-speed processing, and precision mechanical components technology to demonstrate the feasibility and potential of intelligent robotics. It will lead the way to near-term applications of intelligent robotics in the areas of aerospace batch manufacturing, battlefield robotics-front line and rear echelon support, and future military and NASA space applications. It is also significant to note that ITA represents a coordinated government, industry, university, and research institute activity. This project (ITA), shown in Figure 1, is characterized by the following technological advancements.

- Tactical and strategic AI planners
- Real-time sensory perception via three-dimensional laser imaging, stereo vision, tactile sensors, proximity sensors, and high-speed processors.
- Adaptive control system for manipulator control.
- Multiple arm coordination control.
- Peripheral processors for high-speed coordinate transforms.
- High-speed preprocessors for real-time operation.

Furthermore, a very important consideration of ITA is safety with respect to man and the work station. Various techniques are under investigation which will insure that the robot cannot destroy the piece part or work station. This is of paramount interest for space applications. It is one thing to have a "dumb" preprogrammed robot destroy a relatively inexpensive piece part in the factory, but is could be catastrophic to have a robot damage a satellite, space station, or human. Therefore, as with current DOD robotics projects, NASA has very similar technological requirements for space.

Teleoperator. Many of the technologies under development for intelligent robotics will also be of value to telepresence systems of the future. However, even with direct control, real-time sensory perception for safety is a must. The comparison of performance of space operations by man at the worksite versus man at a remote control station is essential. Although the benefits of having man at the worksite are known, recent initiatives indicate that intelligent operational aids can enhance operator productivity, either at the worksite or at a remote control station.

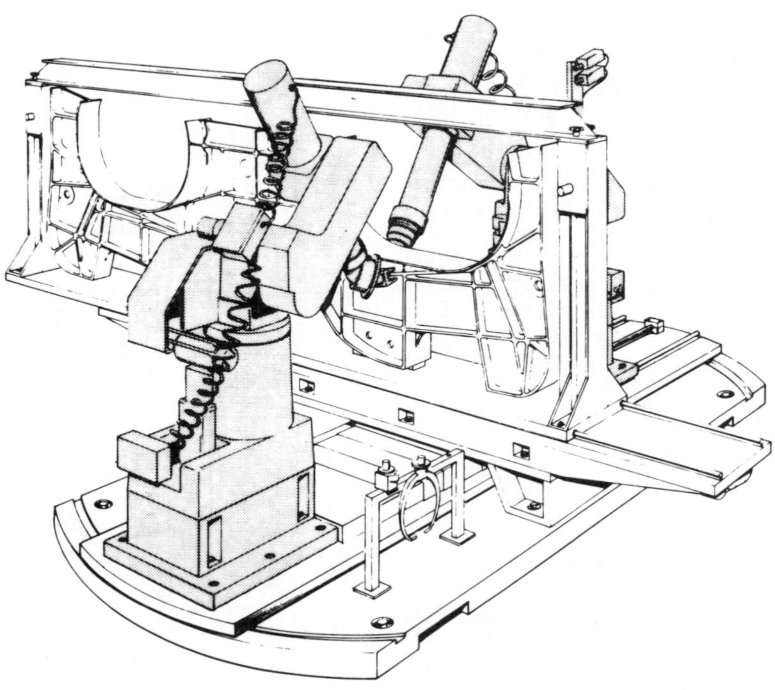

Figure 1 - Intelligent Task Automation (ITA)

A recent investigation has resulted in the preliminary design of a three-arm Remote Orbital Servicing System (ROSS), consisting of two major elements: a servicer kit and a ground control station. An artist's concept of the servicer/TMS is shown in Figure 2. The human's ability to support remote operations could significantly impact the design techniques and support equipment required for the development of future space projects. Having the ability to make decisions and perform specific tasks from a remote location, i.e., Space Station, Space Shuttle, or earth, provides considerable advantages in the areas of reduced training and life support costs, and crew safety and availability.[1] The man-in-the-loop control mode is a key element in this concept. With adequate sensory feedback and the ability to continually issue new commands, systematic errors in actual task performance can be reduced.

Figure 2 - Remote Orbital Servicing System

Hybrid Systems. It is probable that the automated space robotic systems of the future will embody a combination of the technologies represented by telepresence systems, preprogrammed robots, and intelligent robotics. As an example, a preprogrammed robot, Integrated Orbital Servicer System (IOSS), as shown in Figure 3, has been built and tested at NASA's Marshall Space Flight Center. It can be programmed to perform specific functions. It is anticipated that future versions of the IOSS will require some sensory perception such as proximity and tactile. It is also anticipated that future servicers will also enable real-time manned interaction from a remote control station. This would be necessitated by unanticipated repairs or unknown state of the work station due to a fuel cell rupture or other failure. Both preprogrammed and telepresence systems are feasible today; however, further laboratory and flight experiments are required in order to investigate specific mission and hardware related concerns.

Sensory Perception

Automation, whether it is applied to robotics, vehicle control, or spacecraft housekeeping, cannot be accomplished in an "intelligent" and/or safe fashion unless sufficient sensory perception is available. There are many sensor systems to be considered including vision, tactile, proximity, etc. Many matrices have been compiled and much research is ongoing relating to sensors and sensor data processing and fusion, but this is beyond the scope of this paper. Suffice to say that workshops such as the recent AF/DARPA Robotics Workshop held in Denver in March of 1983 have resulted in summations of the state of the technology, and the resulting data is available to the potential user.

Integrated Design, Simulation, and Analysis Tools

The need for advancement in the field of automation technology is apparent. Man's inherent limitations for decision making and assimilation of diverse and vast quantities of information, especially with regard to spacecraft operations and data processing, has led to the conception of highly integrated automated systems and automated decision-making tools capable of precise and quick responses in the space environment. The evolution of these advanced systems has also been a result of economic factors, the need to surpass man's limited strength and sensory perception, and the necessity for devices capable of existing and working in hazardous environments.

At present, one of the most pertinent activities is the NASA-sponsored project for the evaluation of automated decision-making methodology and the development of an integrated robotic system simulation (ROBSIM) tool for analysis of a broad spectrum of advanced robotic systems. This simulation will facilitate the design, construction, and implementation of automated decision makers and multipurpose manipulators. It is anticipated that all future NASA missions will benefit from these advancements in automation technology. They are urgently required to provide a cost-effective and timely resolution to the problem of constraints placed on space missions and in ground-related activities. The ground support systems, tracking, data retrieval and analysis, testing, and system design are examples of gross areas appropriate to the application of this task in NASA missions. As an example, the long-range ROBSIM goals encompass the following:

(1) ROBSIM will act as a design tool having the ability to evaluate varying levels of automation and various sets of physical parameters as a function of mission requirements.

(2) ROBSIM will be an invaluable tool for parametric studies and eventual error analysis.

(3) ROBSIM will also perform as an optimization tool for trajectory and control systems similar to that implemented in the NASA LaRC/Martin Marietta MOD6MV Viking dynamic analysis simulator program.

(4) ROBSIM has real potential as a research tool for the evaluation of artificial intelligence technology in automated decision making and as a research tool in the application of "smart" and sometimes adaptive sensor system utilization.

(5) ROBSIM will eventually accommodate the simulation of manipulators, mobility systems, sensor systems, information extraction techniques, decision algorithms, man/machine interface, end effectors, physical environment, and constraints.

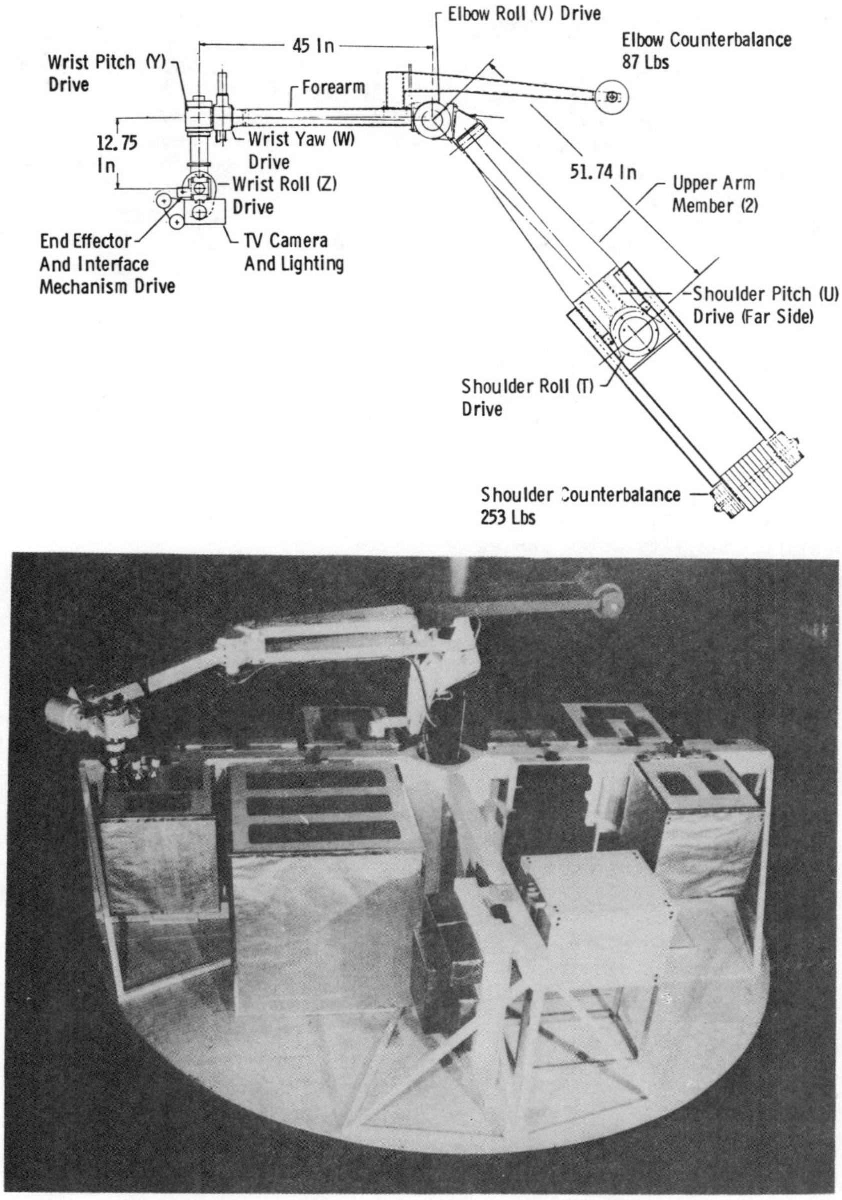

Figure 3 - Integrated Orbital Servicer System

Acquisition of the Critical Technologies

It has already been stated that "We have only a few years to get technology to the necessary readiness (for space station) without incurring excessive technical risk or residual development cost. We must array the technology options and set priorities to use available resources effectively."[2] The question which I posed earlier relating to taking timely advantage of the emerging technologies can be answered in the following manner. Within NASA and other government agencies there are a number of relevant activities which must be brought to bear on the space station technology acquisition and implementation problem. Several methodologies which should be intensified are as follows:

(1) Coordinate with industry on their IR&D activities. Become "active" reviewers and plan for annual "Industry Looks at NASA" workshops similar to the RADC format. This could be accomplished by the respective centers.

(2) Become active participants in the relevant DOD workshops and if possible, contractor reviews.

(3) Implement more jointly-funded NASA/DOD technology contracts in order to achieve a "critical mass" investment similar to the joint DARPA/Air Force/Army/Navy projects.

(4) Emphasize joint industrial/university programs. It not only enhances industry's and the government's return on investment, but the faculty and students also benefit via research grants, contracts, etc.

The Role of Automation and Artificial Intelligence

My third question relating to technology implementation into space station can be answered in a straightforward manner. To begin with, centers of expertise for each of the critical technologies are essential within NASA. With that as a given, plans can be, and in most cases have been, created to investigate the application of the requisite technologies for the time frame of interest.

It is important to understand that the perceived application of automation technology for space station is not to replace man, but to augment or enhance his capability in a technologically complex and sometimes hostile environment. It is analogous to the development of diagnostic tools (expert systems) for the medical profession. In this case, there is no intent to replace the physician but rather to increase his data base and to enhance his diagnostic capability using the experiences of others with the user-friendly tool. Consider that we were able to design, build, and operate flying machines long before the advent of the computer; but with the computer, we have been able to extend our flight regime well beyond the solar system.

With respect to the development and application of automation technology, consider the following concerns:

(1) Although we have built and tested expert systems and natural language front ends for various applications, only recently have we become concerned about the real-time operation aspects of expert systems. Therefore, it must be understood that much work is required if we are to be able to successfully implement automated real-time decision making processes such as in the current Automated Management of Power Systems expert systems development project[3].

(2) With regard to robotics and teleoperators, once again the emphasis is not on replacing man in space but rather on enabling more dextrous, remote, and cost effective activities. To accomplish this, significant laboratory work is required in order to evaluate potential control modes, human interaction optimization, etc. Furthermore, unique laboratory and space facilities are required to investigate these potential capabilities and to define the system requirements.

In summary, with specific and timely automation-related initiatives, we will be able to take a significant step beyond, wherein we will extend man's reach and utility through the use of onboard intelligence, robotics, sensory perception, and unique processing and packaging concepts (VHSIC). We will continue to pursue optimized use of man with the goal of increased productivity in space.

References

1. Spencer, R. A. and Depkovich, T. M., "Satellite Servicing Through Space Telepresence."

2. Olstad, W. B., "Targeting Space Station Technologies," Journal of Astronautics and Aeronautics, American Institute of Aeronautics and Astronautics, Inc., March 1983.

3. Imamura, M. S. and Bridgeforth, A., "Automated Management of Power Systems," presented at the AIAA/NASA Future Space Systems Technology Workshop at Langley Research Center, 1980.

SPACE STATION INFORMATION SYSTEMS

William L. Swingle, Assistant Chief
Avionics Systems Division
NASA Lyndon B. Johnson Space Center
Houston, Texas

Dr. Charles W. McKay
Professor & Director of High Technology Lab.
University of Houston at Clear Lake
Houston, Texas

Abstract

The space operations information system is defined and characterized in a wide perspective. Interactive subsets of the total system are defined and discussed. Particular attention is paid to the concept of end-to-end systems and their repetitive population within the total system. High level program goals are reviewed and related to more explicit system requirements and user needs. Emphasis is placed on the utility and cost effectiveness of data system services from a user standpoint. Productivity, as a quantitative goal, in both development and operational phases is also addressed. Critical aspects of the approach to successful development of the data management system are discussed along with recommendations important to advanced development activities. Current and planned activity in both technology and advanced development areas are reviewed with emphasis on their importance to program success.

Introduction

Though the space station is commonly viewed as a potentially "new" program, it is in fact an old one in terms of planning with its origins somewhere in the middle to late 1960's.

The NASA has consistently contemplated the utilization of space from the perspective of integrated space operations supported by an integrated space transportation system (STS). The space shuttle represents the first developed flight element, providing economical transport to and from earth orbit. While the space station, by itself, will be a complex hierarchy of information systems, it forms but one more element of a potentially large and complex network of distributed elements in a space operations system. Fig. 1 gives a graphic description of the expected scope of the described distribution network. Besides the space shuttle, the STS will include:

- o Manned space stations or space based operation centers for transport staging, production, payload servicing, and the support of science and technology projects.
- o Orbital transfer and proximity vehicles (OTV's) for transport operations throughout the earth orbital domain.
- o Dispersed unmanned platforms and tended free flyers for the support of projects best served with space station unattached environments.
- o Ground centers explicitly related to the support of STS functions and services.

In addition, there are other operations elements that are interrelated to the STS from an information standpoint:

- o Attached and unattached payloads
- o Supporting service satellites (TDRS, GPS, etc.)
- o Ground data receiving and distribution networks
- o Payload operations centers
- o "User" facilities

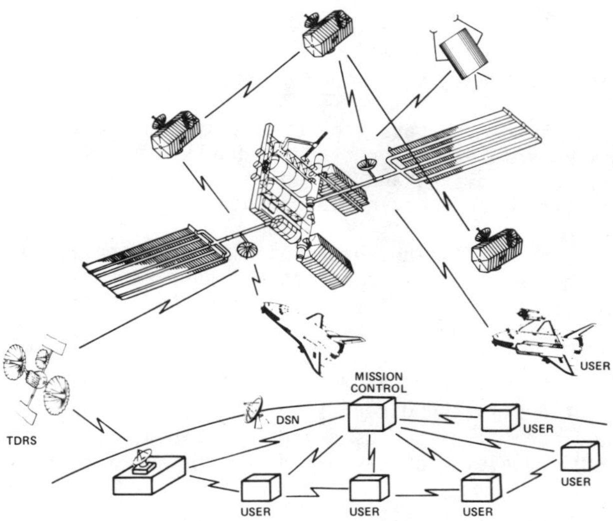

Fig. 1 Space Operations Distribution Network

The Space Operations Information System

Description

The space operations information system is viewed as the total integrated network of information paths and interfacing elements and nodes associated with space operations, including all flight and ground elements. Fig. 2 presents a functional representative of flight and ground elements of the system. Note particularly the development support activities which have large potential for both direct and indirect integration benefits.

Many of the elements are already designed and in operation in some current design configurations. New elements to the system, such as space station are nearing development start. The continuing task of sustaining while evolving the system to improved and more productive configurations is both ominous and challenging. It has been both popular and proper to approach the STS and space station information systems as integral elements of a space operation system to insure adequacy of functions and performance as well as a base to support both design and physical integration.

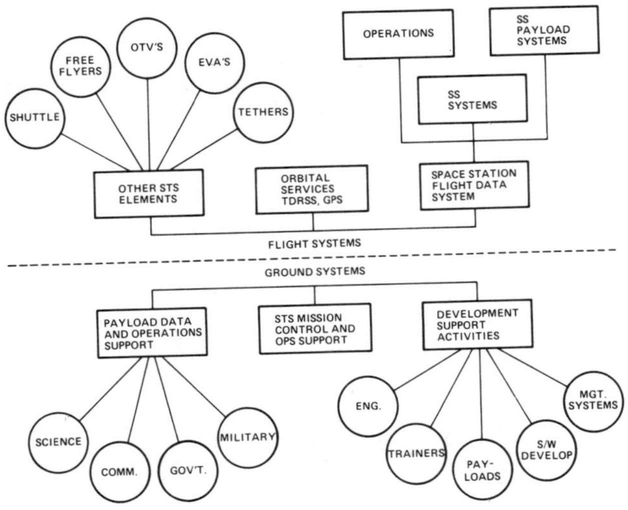

Fig. 2 Space Information System Elements

Major Information System Subsets

Below the total operations system level, there are some major system subsets that are frequently isolated and for which it is convenient to assume some definition or boundaries.

a. STS - The system of elements associated with the physical transport of or between elements in space or the ground, and the support operations thereof.

b. Payload Systems - Those elements associated with the direct servicing, interfacing, and control of payloads and the communications and processing of payload data and information.

c. Space Station System - Boundaries defined by the space station, physically. Associated with the communications and processing of information within those boundaries containing both STS and payload functions.

1) Core Systems - Onboard systems associated with the operation of the habitat and the crew work functions

2) Payload Accommodation - Onboard systems associated with the direct support of payloads.

d. STS Element Systems - Same as for space station systems for the applicable elements.

e. Ground Systems - Those systems located on and associated with ground-based activities.

Different boundary assumptions (geographical, functional, etc.) are appropriate depending on the type of system problem being worked. End-to-end system concepts apply everywhere and are essential to requirements and design analysis and development.

End-to-End Systems

It is not difficult to view the total function of the space operations information system as being so boundless and complicated as to limit consideration as a whole.

Important to the analysis of the system is the functional end-to-end system concept, a generic term defining some total information system between a user on one end and some user domain of interest on the other. An end-to-end approach is important to the analysis of requirements and performance and for the functional modeling and structuring of the information system process. A confederation of many functional end-to-end systems will exist throughout any of the defined system boundaries, with a multitude of overlays with the data management system. An elementary model depicting the principles involved is shown in Fig. 3. That portion of the system comprising the application processing service and data communications and handling service are commonly referred to as the data management system or utility. It is associated with the communication and handling of data without changing information content and with the providing of processing services which alter content in accordance with user or application specified requirements.

A few illustrative examples of end-to-end functions are shown in Fig. 4. Note that the attitude control end-to-end system (EES #1) is totally within the GN&C local system. The domain of interest is spacecraft attitude and the user is an automatic control algorithm in the GN&C applications processor. A payload end-to-end system (EES #2) may typically extend from some scientific experiment in space to a remote controller and observer of results on the ground. Likewise, a ground controller monitoring the status and health of space station subsystems is another end-to-end path (EES #3). User access may be by simple direct circuit or through multi-internetwork gateways including the RF communication path to the ground.

Note that the data management system can also be viewed as an end-to-end function between any two points of application processing where any one processor's information is the domain of interest of another processor's user function. Reference (1) provides additional detail of current activity and information related to the development of high level reference models and standards for data systems in an end-to-end context.

"End-to-end" should not be used as a unique descriptor to connotate the whole system but as an approach to properly address system functions between logical end points.

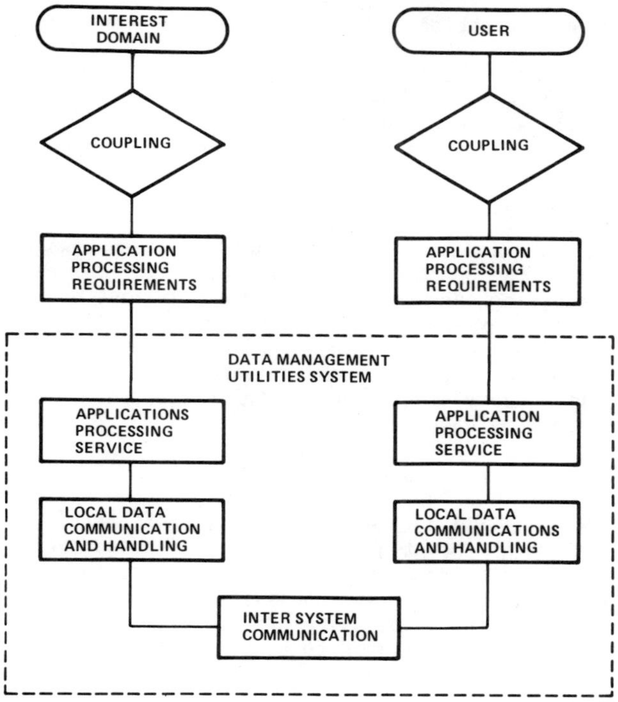

Fig. 3 End-to-End Reference Model

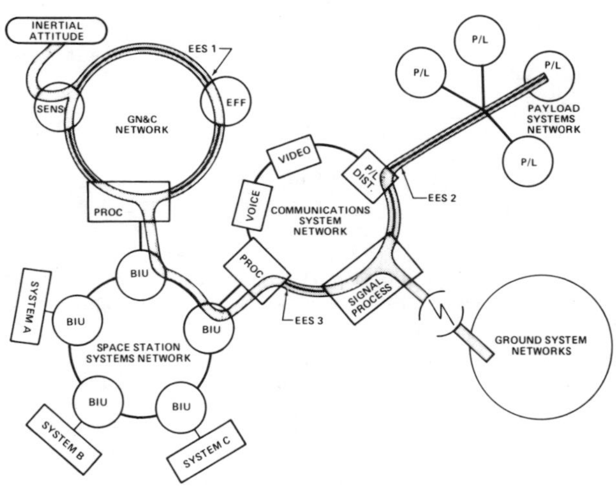

Fig. 4 Interconnected Networks Sample

High Level Program Goals and Challenges

Program goals for information and data systems are common to all topical areas and are simply and oft repeated as:

- o Satisfy mission demands.
- o Low life cycle costs
- o Evolvability

The first and second are the dreams of all program managers; meet the requirements at minimum cost. Evolvability is somewhat of a paradox in that it implies both change and uncertainty over real requirements. Both situations normally conflict with low-life cycle costs. It has become increasingly apparent, however, that sequential development, over a long operational period (25 years) along with constantly evolving and challenging requirements are likely. System and program approaches need to deal with this reality in the best way possible.

It is acknowledged that the space station is primarily a user's tool for the utilization of space. If it is to be productively utilized, it needs to be a marketable commodity in terms of its accessibility, utility, and value to the user over its operational life.

The major problem in achieving these goals are the usual ones of resources; the state-of-the-technology base, time, and money.

It is expected that maximum performance in these "goal" areas within currently expected cost and schedule constraints will be difficult. We can, however, use all of these resources to select the right system and program approaches to make the best use of the expected technology base and to "grow" into our goals as time, money, and the evolving technology base permit. The key word in the system and program design now becomes flexibility.

Productivity Considerations

Life cycle cost effectiveness and productivity have been space station watchwords ever since its re-emergence as an active candidate program. These two goals are not necessarily synonomous.

Productivity, from a space station and data system viewpoint, implies a utility to users that has a high ratio of output over input. Output must be measured as user product value and input must include both user access cost and utilization costs. Life cycle cost effectiveness is the real program target and must include effects of the initial investment. Huge initial investments to develop a productive or efficient utility for a very limited user market and service life is not cost effective. The opportunity for application must be consistent with the investment risk. Just how businesslike we are liable to be in pursuit of these goals is unknown. Positive impact to productivity will probably be pursued in tolerance to funding rate constraints and schedule. This is one reason that evolvability is so important; it allows the reasonable opportunity to upgrade if it too does not become victim of near-term constraints. A flexible and evolvable system needs to be accepted as a near-term requirement.

Data Management System User Needs

There are a number of achievable attributes of the data management system (DMS) that are important to users from the standpoint of their goals and the productivity of their operations.

Accessibility - In the short term, user access to his "ends" through the DMS should be rapid and responsive. He should have access to distributed archival data (including library listings) through active network links. Long term access to and response of the "program" should be a reasonable procedure not requiring years of paperwork and supplication.

User Transparency - The data utilities should be user transparent. He should not have to understand its complexity or how it works, only what it does and how to use it.

Utility - It should provide a range of useful services and the user should be able to use it productively with a minimum of special education and training.

Non-Interactive - The data system utility should have sufficient resources and fault tolerance that user systems will not suffer disruptive delay or limitations within specified requirements. The traffic in the retiring or addition of user systems should not significantly disrupt or imperil the operation of active systems.

Standard Interfaces - Simply structured, layered, and standardized data interfaces will enable a common stable basis for interfacing the utility to users and for possible direct interfacing between users.

Processing Service - Where supplied or shared processing resources are used for application processing, the user should have access through user-generated but DMS specified higher order language code with all software environmental and verification services provided. Tools, procedures, and related data bases will be available through remote terminal access.

These attributes are important not only to payload users but to space station operations and subsystem users towards timely and cost-effective development and utilization of their systems. Taken together, these attributes define the automation requirements for the DMS.

Areas of Opportunity Toward Program Goals

A number of important areas of emphasis relevant to cost effectiveness and evolvability goals for data systems have been identified and discussed in activities to date. the right mix against program goals is certainly not clear at this time though in some areas the direction seems apparent.

Requirements

A key driver for all cost areas. Incomplete or poorly interpreted requirements as well as over-specification can all adversely affect program costs and be devastatingly disruptive to program schedules.

Once the general range of mission applications have been determined, a compilation of functional requirements (detailed as possible) of all information system related requirements should be undertaken. For space station, the habitat system type requirements will be more easily determined than for the payload systems. All functions for the space operations system should be considered so that an accountable distribution among the system elements can be attempted. Successive partitioning of functional requirements into subsets in consonance with some logical criteria will eventually enable a characterization of performance requirements both within and between network elements.

Early estimates indicate that performance requirements for the habitat type systems are easily within the bounds of today's technology. Some early estimates of payload needs are more extreme and will be somewhat dependent on technology upgrade.

Strong consideration should be given toward making the onboard system resource rich (memory, processing throughput, data rate) in light of expected expansion and growth. Resource limitations have an explosive effect on software costs. The price, physically and dollarwise is not particularly significant in the projected technology base.

System Architecture

McKay (1981) defines computer systems' architecture as the structural organization and the interrelationship of the hardware and software elements that comprise the system. The selected architecture of the system is vital to and must be properly determined to enable certain key objectives of the DMS such as:

- Modular expandability and extensibility in space
- Efficient system development and verification
- Fault tolerance through the semiautonomy of nodes
- In space functional processing growth.

Particularly important in the architectural structure is the basis for the separation and distribution of the total functional requirement set. The prime criteria in this partitioning process is the functional independence of subsets that not only enables separation of the whole development task into relatively autonomous subsets but also insulates them from each other during the evolutionary process and failure situations. We must search out the clean functional boundary layers.

There is no known argument to the general need for a distributed architecture. It is geographically dictated at the space operations or STS level and unopposed at the space station level for many reasons. According to Enslow (1974): "A distributed network consists of a collection of nodes and their associated data links and physically distributed components, configured such that five key criteria are all present in some degree."

- A well-defined set of policies are used to govern the integrated operation of the total system.
- Nodes operate and interact in a networking environment of cooperative autonomy.
- Cooperative protocols are used to control the transfer of information.

o There are multiple shared resources that can be dynamically reconfigured to meet the needs of specific tasks.

o There is system transparency which permits the user to request services by name without specifying which physical or logical components are to provide the service.

Pursuit of these architectural principles should give us the best chance to realize objectives. Some compromises at isolated hierarchal levels may be required due to issues peculiar to real-time, flight-critical segments of the system.

At least two types of overall structure are at issue, global vs. hierarchal. The latter is almost dictated at the space operations level due to the spatial separation of regions (and architectures). Fig. 5 displays the general difference in the two approaches. Hierarchal provides more explicit separation of functions and operational interaction, on local levels, while global provides greater flexibility and more efficiency in resource sharing and utilization, respectively. Both forms will occur over the scope of the space operations systems. Choices must be made at local levels.

A particularly important issue to the success of the system will be the implementation of data base distribution and management.

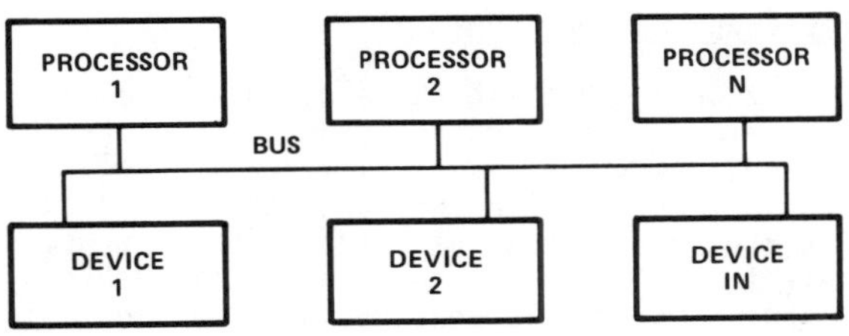

GLOBAL SYSTEM

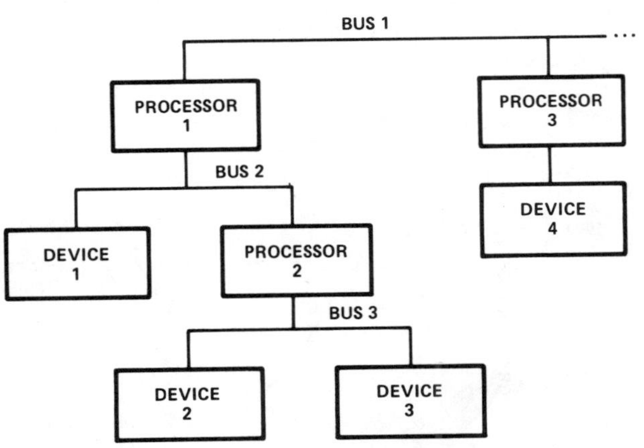

HIERARCHIAL SYSTEM

Fig. 5 System Architectures

Distributed Network Operations (NOS)

Given the premise of a complex distributed system architecture comprised of both local area and remote area networks, the management of such a system in the context of the evolutionary life cycle expected for space station becomes a serious challenge. Even with the presumed autonomy of network nodes or objects the effectiveness of the network operating system will be a key to the achievement of productivity and evolvability goals.

According to McKay (1979):

"An operating system consists of

o A set of policies to govern the integrated operation of all categories of computing resources.

o A set of management modules (hardware, software, firmware, and documentation) which administers the policies."

In any computing system, the operating system must manage processors, main memory, devices, and information. In distributed systems configuration (hardware and software) management and communications management are added to the list.

Objectives for the NOS that support the achievement of program goals should include:

o Standard hardware and software interface to the network.

o Standard data formats and message protocols.

o Support high level, object oriented directories for distributed data storage, routing, and accessing.

o Data network operations transparent to the applications hosts.

o Effective management of system processing.

Data Communications. The ability to expand or extend the system in any dimension (hardware, software, function, performance, etc.) in a cost-effective manner will require a structured and modular implementation of the data communication and handling function. This can be achieved by the development of appropriately structured and layered interface models and of standards for the layer interfaces. The Open System Interconnection (OSI) model sponsored by the International Standards Organization appears to be evolving as the generally accepted standard and is a strong candidate for space station application. The OSI refers to a class of systems with information transfer relationship defined in accordance with the ISO model, shown in Fig. 6.

Such models are usually composed of computers, devices, operators, and computing processes which may be physically separated. Models in general and this one in particular o Provide a common basis for the coordination of standards development

o Allow existing standards to be placed in perspective to the model

o Identify areas for developing or improving standards

o Provide a common basis for correlation of related protocols (ISO, 1978).

The ISO model has seven layers. (For more details, see: ISO, 1978; Tannanbaum, 1981.)

The applications layer is the interface to both user and system application processes. Lower layers exist only to support this layer. Application processes are the ultimate source and sink for information exchanged through the lower layers.

The presentation layer provides services to applications related to the representation and manipulation of structured data. File format conversion between remotely associated nodes is an example.

The session layer binds and unbinds two or more application processes into a relationship while controlling the information exchange during the session.

Whereas the session layer could care less as to how the data is transported, the transport layer optimizes the use of available resources to connect the session entities.

The network layer keeps the transport layer free of routing and switching considerations. As such, some nodes may serve as either terminal or intermediate sources of data.

The data link layer protects the network layer from the error prone physical layer. Data exchanged with the physical layer is processed to facilitate error-free transmission and reception.

The physical layer is responsible for transmitting and receiving bits over some physical media. This layer defines the mechanical, electrical, functional, and procedural characteristics to make, maintain, and release data circuits (ISO, 1978). Further information as to the standard nomenclature applied to model and system elements and processes can be found in the ISO literature. Information relative to proposed standards can be found in the IEEE 802 Standards for LANS and the CCITT X.21 standards for remote network/node communications.

Notice that the model shows effective communication directly among corresponding layers between hosts. Each layer has its own protocol for conversing with a corresponding layer. In reality, data must be passed vertically and between hosts via the physical layer; however, each layer is unaware of the protocols or information headers related to other layers. Functional transparency exists between layers. The layers, interfaces, and protocols comprise the network architecture.

While the OSI model has been specifically elaborated, the important message is that the formal reference models and standards are mandatory for system order and evolvability. Other viable candidates, particularly those with pedigree, should be considered.

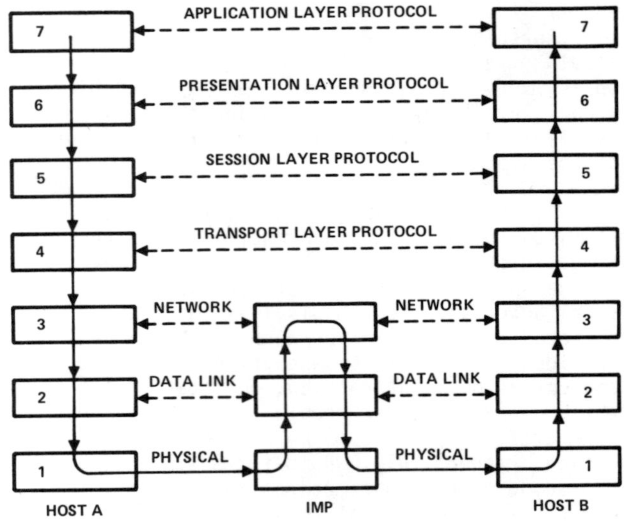

Fig. 6 The Seven Layer ISO Reference Model

Software Engineering

During the last three decades, the revolution in digital processing devices related to speed, functional density, and power requirement has resulted in the infiltration of such devices to almost all areas of information systems. Within NASA, Fig. 7 shows the relationship between the development costs for recent major programs and that portion devoted to the development of onboard flight software. The main point is that software is becoming an increasingly larger portion of total development costs. It is probably an even larger portion of the maintenance and operations budgets. Due to the proportional weight of software effects in both the development and operations phases, it warrants major attention.

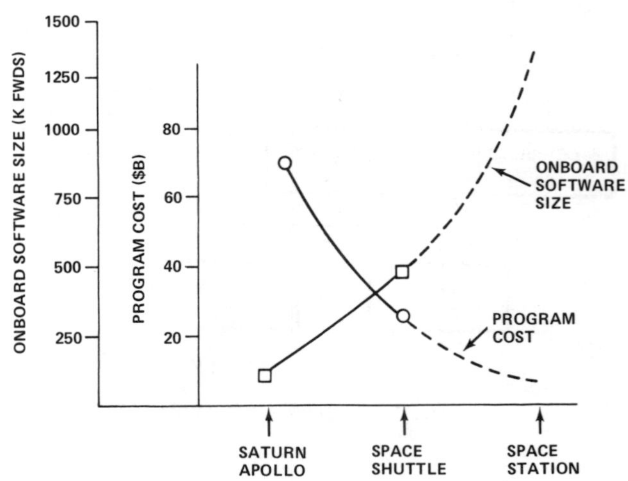

Fig. 7 Flight Software Cost Histories

In spite of the many complaints, software is the most convenient and effective medium to implement new functions and to accommodate and provide flexibility for evolution and change. It needs to be structured and developed to do so gracefully.

Early in the program, software tools, methods, and experience libraries, both government and commercial, need to be searched and examined for resources that can be extracted and adapted for use on space station.

Areas of specific potential that should be investigated include:

o Choice of higher order languages. The fast developing ADA programming support environment is a product of modern software engineering and a process bound to re-enforce the use of good software engineering methods. Its capabilities are tailor-made made for space station. Maturity may be an issue.

o The concepts of reusable software modules or components, and layering.

o Methods for establishing and maintaining a requirements, design, and configuration data base with requirements traceability and relational query capability.

o The use of object-oriented data interface communication with users.

o The development of standards for data systems design application to all new developments in the STS and maybe the whole operations system.

o Functional partitioning and allocation of resources to criteria keyed to program goals. Criteria may include modular autonomy and function criticality.

o Methods, models, standards, and plans to cope with both planned and unplanned change traffic.

o The application of automation to software development and verification processes. Efforts should include both build, analysis, and verification tools.

o How to exploit commonality as a productivity and a reliability weapon. All elements of the system should be considered. Areas to be addressed should include both software products and development facilities. Fig. 8 illustrates a commonality approach where both resources may be shared via user terminal access. Users could develop and functionally verify their own code via a HOL interface.

o Alternate verification criteria including graduated requirements based on criticality.

o Management policies.

The preparation and planning we devote to the area of systems, software, and policies will produce the biggest rewards.

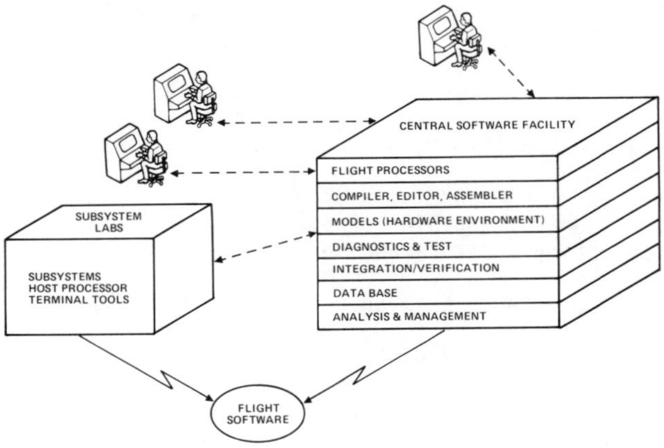

Fig. 8 Shared Software Facilities

Commonality

The advantageous use of commonality should be pursued in both hardware and software areas. "Popular" thinking tends to stress the principal of node, subsystem, or payload development autonomy. However, the potential value of commonality is ominous as relates to:

o Development and acquisition costs

o Integration

o Ground logistics support

o Training and technical support

o Onboard maintenance and sparing requirements.

Maintainability - A key area where in past programs we have usually given insufficient attention during the development phase. The spaceborne portion of the system including space station will not recycle to the ground at high frequencies but must be maintained in space. The potential effects on productivity through direct costs and operations down time is evident. Fault tolerance and the capabilities of the system management implementation will be important not only from a safety standpoint but will also be a "productivity" driver from a maintainability aspect. This is a productive area for the consideration of options.

The Technology Box - Data systems technology activity has been widely documented in other literature. Current programs are aimed at a demonstrated technology base for the 1987 time frame that features:

o Optical data transmission to 200 m bits/sec.

o Fault tolerant bus operation with autonomous node and automated fault detection and network reconfiguration.

o Space qualification of components

o High density, electronic auxiliary data storage (bubbles)

o Fault tolerant computers and computer modules

o Various software engineering activities including participative evaluation and application prototyping of ADA software development environments

o Prototyping of distributed network operating software and data base management

o Fault tolerant validation methods

o Optical disk data storage.

The technology program is typically but not uniquely formulated to build on the general commercial and military technology bow-wave with adaptation and proofof-concept projects to enable their use in the initial development phase. There are many options for more technology mature but adequate approaches. Meaningful risk vs. rewards assessments will be needed.

Advanced Development

Advanced planning is currently based on the broad assumption of initiating phase C procurement by FY-1987 and deployment of the first manned space station in FY-1991. The associated lead time is available to prepare the technology base and to carryout the necessary advanced development activities to enhance a technical succcess position by the phase C start milestone. Key parts of the general approach to this task within NASA are:

o The formulation and implementation of a technology plan (including flight test) to achieve a targeted technology readiness base no later than the phase C PDR milestone.

o The development and utilization of advance development test beds, by major discipline areas to serve as a basis to:

o Develop and demonstrate system concepts, configurations, and methodologies upon which to base system specifications.

o Receive technology products for demonstration in an integrated system environment to enhance their risk acceptability to the program.

At the risk of being repetitious, it seems clear that the road to success in the information and data management systems area is a well-ordered, structured, modeled, standardized, understood, and documented system design, with supportable decisions on relevant program policies and a compatible development and operations plan. Given the probable interval between the phase C and first deployment milestones on the program, extensive preparation during the advance development phase in terms of planning, systems analysis, documentation, and product and facility development is required. Planning is the unavoidably necessary underlayment to the success of this activity.

A formalized systems analysis and planning activity has been formulated, involving separate but coordinated NASA/industrial teams to work this advanced development area. The general tasking to these teams is to consider valid current and generated options in the areas of system definition, requirements, programmatics, policies, and the technology base, against the goals of the program. Based on this analysis, they will converge towards some initial requirements, technical systems definition, and planning basis for the program. The planning basis will include a program schedule template identifying all key decisions and product milestones necessary to support phase C readiness in consonance with the requirements and technical base. The results will be iterated periodically based on review inputs and the results of other related activities. The scope of the object system includes both onboard and ground portions of the data system associated with the "core" space station and "user" accommodation functions. The selection of industrial participants for these analysis teams is currently in process via competitive procurement.

Besides their inherent necessity to the program flow, the systems analysis activities will help converge needs and directions for the technology and test bed programs. However, the systems analysis and initial test bed activities do not have that much of a front end overlap. The initial test bed activities must include detail design and test of the onboard data communications functions. These are not particularly constrained by system architecture other than the presumption of distributed systems. These leading test bed activities need to progress in order to merge with study and technology outputs at the appropriate time.

Conclusions

1. The large scope and complexity of the information and data management systems associated with space operations and the space station are visible and understood on a high level.

2. The importance of understanding and satisfying user needs as a requirement to program goals is accepted and committed.

3. The projected technology base coupled with the right systems and management approach will support program goals through an evolutionary process.

4. Time is of the essence to plan, organize, and implement advanced development activity to support the needed system design and development state at phase C start.

Acknowledgment

My sincere appreciation to Jack Garman, Ed Dalke, and Bill Mallary of the Johnson Space Center for their frequent willingness to consult and advise and to Donna Pivirotto and her team at the Jet Propulsion Laboratory, whose material I use frequently.

References

1. A. J. Hooke (Jet Propulsion Laboratory): Reference Model of Space Data Systems (Draft). May 19, 1983.

2. Space Station Systems Definition (Book 5). November 1982.

3. C. W. McKay, MSI, LSI, IO and Modular Memory Systems, Prentice Hall, 1981.

4. ISO Reference Model of Open Systems Architecture: Version 3, International Organization for Standardization, 1978.

5. P. H. Enslow, Jr., "What is a 'Distributed' Data Processing System?" Computer, Vol. II, #7, January 1978.

6. C. W. McKay, "A Design for Loosely Coupled, Locally Distributed Networks of Reconfigurable Microcomputer Resources, Proceedings of the First Annual Conference on Computer Developments, University of Houston at Clear Lake, 1979.

7. A. S. Tanenbaum, Computer Networks, Prentice Hall, 1981.

SPACE STATION COMMUNICATIONS

C. Louis Cuccia
Ford Aerospace & Communications Corporation
Palo Alto, CA 94303

Abstract

This paper presents a concise history of the various types of communications that have been used in low earth orbit vehicles that form the basis of the various types of communications and communication requirements that can be realized in space station developments over the next decade. The space shuttle can be assumed to be a prototype space station in the tradition of Apollo and SPACELAB. The shuttle communication operations require earth-to-ground support communications, EVA communications, internal space station communications, and communications to and from other spacecraft (TDRS) and free flying vehicles for experiments (SPAS-01). These basic communication requirements will be shown to expand to the point where the man-computer alliance in the space station will transform the station into a space communications and computer center capable of providing data processing and storage in association with ground based distributed processing along the growing terrestrial 1SDN global digital highway. The space station will also provide unique means to obtain data and information from one part of the earth or space and transport them to another point on earth.

I. Introduction

The space station must not be confused with a geostationary communication satellite relative to design, purpose, and communication requirements. As set forth in Table 1, the geostationary communication satellite is designed to remain over a particular area of earth surface and to provide continuous interconnectivity between users based on traffic and economics in the case of a commercial Satcom, or reliability or survivability in the case of a military system. A moving low orbit space platform, such as a space station, will circle the earth around eight times per day. Because it will move from horizon to horizon at a point on earth in a matter of minutes it cannot provide the continuous interconnectivity of a geostationary Satcom. It is this lack of continuous interconnectivity to earth (except via TDRS) to support conventional data, voice, and video communications that has created the concept that the space station is a communication service only in support of itself and of associated experiments and free flyers and not as a viable space node between two or more points on earth.

The space station can indeed include a communications service to interact with earth-based command and control, to interact and process data related to free flyers and experiments, to support EVA, and to support assembly, repair, and sortie operations in space. However, this paper will show that the space station, as a *moving node* in space can function as a key local-area-network either self-contained in the space station or as a part of a distributed computer network on the ground. The computer in space, along with IEEE-488 and IEEE-802 disciplines under local control by astronauts, can provide an expansion of operations and processing capability well beyond the capability of an unmanned platform. Also, the space station using high density new memory techniques, such as Winchester disks and random access semiconductor or optical memory, can provide the unique and *new* function of providing a massive memory bank in space that can receive bursted data from space or on the ground and later transmit it back to space or to another point on the earth. This is a unique pick-it-up here, deliver it there operation.

II. Space Station Communications Heritage

The concepts of space station communications can be attributed to a long heritage of manned space platforms as viewed from the telecommunications standpoint. As shown in Table 2, a NASA low orbit space station can trace its heritage to the Apollo missions to the moon, and the tracking and data management of many military low orbit satellites that require tracking up to 90,000 passes each year. This heritage includes experience at communicating with many remote sensing satellites and has culminated with the successful operation of the shuttle. At the same time, the synergistic growth of satellite technology, computer and LAN technology, and mm-wave technology, including the technology support to the NASA ACTS satellite, now provide an excellent technology base from which to explore the communications requirements of a space station program now and in the future.

FIG. 1 shows the evolution of space station related systems with time. The contributions in technology by all these related low-orbiting spacecraft systems have provided an excellent starting point for considering the technology required for the space station. These various systems have developed long operational experience at the various frequencies listed in Table 3 and have provided the wide range of applicable technologies listed in Table 4. Actually the technology listed in Table 4 only "scratches the surface" of the total technology arsenal available to the space station designers. Other technologies include radar phase arrays and both tube and semiconductor amplifier developments for both commercial and military systems. The impact of the growth of VLS1 and new gallium arsenide monolithic microwave circuits that will be available from the DoD VHISC, NASA ACTS, and DoD MILSTAR program will further serve to make the space station telecommunications requirements realizeable.

Table 3 lists the frequencies that have been the areas of development, circa 1978, for TDRS, LANDSAT-D, etc. These frequencies (through K_u-band) represent the limit of expertise at that time in the United States - excepting the Japanese CS 30/20 GHz satellite that was built by Ford Aerospace and launched in December 1977. The frequencies above 15 GHz and in particular at 20, 30, 44, 60, and 94 GHz have now been the subject of much system development in the United States since that time. The use of the Government allocated frequency bands at 20.2 to 21.2 GHz for downlink and 29 to 30 GHz for uplink would be excellent candidates for space station earth-to-ground links that could take advantage of both the wide bandwidth (1 GHz) and the small size high-gain narrowbeam antennas that could be involved. The NASA ACTS program has made a special study of antennas in this frequency range that could be adapted to space station use.

III. Space Station Communications Requirements

Much of the NASA effort on space stations to this date has given special attention to structures, life sciences, materials processing, and the use of a space station as a staging platform and controller of both free flyer and tethered experiment platforms. Relatively little attention has been addressed to the *potential* of the space station as a communication center with the exception of its ability to support a wide range of experiments and to support a manned working environment in space.

FIG. 2 shows the basis of this system consisting of the shuttle communication system that can communicate directly to the ground based NASA GSTDN (NASCOM) system using S-band, or through the TDRS satellite to ground using S-band and K_u-band. It can communicate to EVA at VHF and is designed to also communicate with detached payloads. This is a fairly sophisticated system and once the TDRS has been incorporated into the STS system, the full shuttle communications capability represents an excellent basis for command, control, and communication for any space station - even one having many more functions.

FIG. 3 illustrates a typical overview of a more complex space station communications network that includes integration into the NASCOM and Consolidated Space Network with interconnection to various user communities and orbiting spacecraft. This overview is derived from the results of the Communications and Tracking Workshop group of the OAST-NASA Space Technology Workshop held in Williamsburg, Virginia, in March 1983. Table 5 is a summary of the results of this and other working groups in terms of communication requirements and ranges. Of interest is the requirement of 25 Mb/s uplink and 300 Mb/s downlink, but of particular interest is the requirement of ten trillion bits of onboard storage and a recording rate to 100 Mb/s that would be required for advanced earth sensing devices to be developed during the decade.

These recommendations must be considered as very perceptive and yet supportable by present technologies that can handle high data rates and store gigabits. There is no problem with 100 Mb/s since both Intelsat and EUTELSAT are standardizing on 120 Mb/s TDMA. The rapidly advancing art of Winchester disk storage makes possible the feat of storage of over 380 megabytes on 5 1/4-inch disk at this time; and a capability of 2500 megabytes on one optical disk 12 inches in diameter is well in the range of present developments.

The present shuttle data rates in the communication links may well remain adequate for command and control for even a very major space station complex in the future, but data rates in the hundreds of megabits to support both experiments, data communications, and principal space station distributed data processing functions will require modulated carriers with bandwidths of hundreds of megahertz. For such carriers up/downlinks at 30 and 20 GHz and even higher frequencies to the 100 GHz range will be required.

FIGS. 4 and 5 illustrate two modes of space station communications to earth as presently envisioned. The use of TDRS as shown in FIG. 4 makes possible data rates up to 300 Mb/s and this may be increased to the gigabit range by use of laser frequencies in the advanced TDRS now under consideration.

FIG. 5 shows a typical direct-to-ground link that can use the 2.2 to 2.3 GHz frequency range to communicate with a tracking antenna that would use a communication path by a geostationary Satcom, such as DSCS-11, to serve as a relay between a master ground station and the space station. If the NASA ACTS 30/20 GHz satellite were to be used as the relay, a tracking station and relay system, all operating in the millimeter waves, could provide a gigabit data path for several minutes and could even involve an automatic handover to another tracking earth station and relay system further along the path.

IV. Space Station Concepts

Much speculation has occurred relative to the alternative space station designs, and concepts ranging from simple enclosures to cities in space have been born in artists' minds the worldover. However, the space station concept introduces a key cost parameter, i.e., costs of around two thousand dollars per pound that are required to lift material into low earth orbit. Thus, the space station design must be a tradeoff between sheer structure and what can be considered payload in the broadest sense. This consideration is already paramount in the design of communication satellites where a ratio of around 28:100 for payload weight (antennas and transponder) to bus weight has been achieved. In other words, a large structural system could make the inclusion of computer processing equipment into the total weight budget economically impossible.

FIG. 6 illustrates a prevailing approach to space station design where a series of functional modules, pallets, and docking modules are assembled into a structural system that can provide the many functions of power, habitation, laboratories, communications, logistics, command, etc., which are required. A recommendation for the modular approach to space stations was made by Christopher C. Craft, Jr. in SIGNAL, February 1981. Such a complex is the result of the assembly of modules that can be either lifted into space in complete form, or assembled and instrumented in space; i.e., the main tank of the shuttle can be taken into low earth orbit and instrumented after its main function of propulsion has been completed.

Already a precedence for the design of such modules has been established by Skylab and Spacelab. Also, the Soviet Union with its SALUT, PROGRESS, COSMOS, and SOYEZ series has been engaged in developing module and docking techniques for several years. FIGS. 7a and 7b are illustrations of a proposed Soviet space station complex using a complex of SALUT-7's modules that were published in "Family Science" in Japan in March 1983.

One advantage of the modular approach to a space station is that it can be built up gradually from a simple system (e.g., Skylab) to a large complex providing many functions with this buildup taking place over a period of years.

FIG. 8 illustrates a candidate communications module and an external antenna and sensor pallet. These are derived directly from the present European Spacelab design and illustrate how equipment can be arranged in a cylindrical module, and the use of a pallet (there might be several) for antennas and sensors. It can be envisioned that an advanced generation of space station design would require many different antennas to not only communicate to and from earth, and to and from TDRS, but also to the shuttle and to other space stations, and to other free flyers that are supported by the space station.

When many antennas are used, and where continuous antenna motion is required to provide tracking of each of these several antennas to its particular distant transmitter, then interference must be avoided between the many types of steerable antennas including phased arrays located on the pallets or on the surfaces of the appropriate modules.

One key advantage of the module and pallet system for a space station is its serviceability in space either by EVA or by automatic submodular replacement. A major difference between the low earth orbit communication platform over a geostationary Satcom is the reliability built into the design of each subsystem. In a geostationary Satcom designed to operate for 10 years in synchronous earth orbit, all components must be designed to operate with a reliability that guarantees a *total reliability* sufficient to assure the life-goal. Until the era of Satcom repair arrives, this need for the highest Satcom reliability produces much impact on satellite design and cost.

In the space station system, reliability of individual equipments is still a very important item, but the ability to repair or replace equipments on space pallets or within the modules greatly reduces or eliminates the present geostationary Satcom reliability problem where a single point failure could result in loss of capacity or even loss of the satellite.

V. Space Station Communication System

The space station, as a communications center, must provide many functions, i.e., command, operation and controls, navigation, tracking, supporting functions of power, altitude control, thermal and vibration control, communications in support of space station operations and functions, data processing, and communications in support of people who will live on the space station. These functions can at best be performed very inefficiently by the inhabitants of the space station by themselves; however, it is the team of man and the computer that makes the vast communications and data processing capability of a space station possible.

In considering the role of communications and data processing on a space station, it must be recognized that the early architects of the space station were oriented primarily toward structures, experiment-support, solar power, or space biology and life sciences. The computer, except for its role as system controller and automatic pilot in Apollo, space shuttle, and selected satellites, has been a relative stranger in space, particularly in areas of communications support and data communications. These early uses of the computer were not only necessary, but vital to the first role of the space station to study man's adaptation to the space environment and to provide an enhanced ability for acquiring experimental data.

The handling of the data of many experiments for science and application missions, such as earth resources, has been studied by L. H. Kasulka of McDonnell Douglas Astronautics Company (NASA Contract NAS8-33592) for a "Science and Applications Platform (SASP)" in 1980 and resulted in a recommendation for an onboard command and data management system. Mr. Kasulka studied the need to have a multiplexer and demultiplexer system that could combine the different data for a wide range of experiments to produce a data stream of up to 300 Mb/s. After analyzing the data requirements a large number of experiments, primarily in the earth resources category, he concluded that a space station could acquire data up to 10×10^{13} bits per day with a need to provide a record/dump rate up to 100 Mb/s for a record capacity of 3×10^{11} bits.

At this Space Station Symposium, W. Swingle from NASA JSC and Dr. C. McKay of the University of Houston, also discussed the concept of space station information systems using a space operations distribution network and an end-to-end interconnected-network information system that would interconnect both flight systems and ground systems making use of new technology distributed data processing systems that are undergoing a worldwide technological explosion in business and new electronic office systems.

Carrying the concepts of Swingle and McKay farther, it is clear that both cable networks and local-area-networks can provide the interconnection required for the data processing requirements of a space station, both within the space station complex, between the space station and other free flying complexes, and between the space station and a ground network.

The space-ground interconnect is of particular importance since there is a global move toward integrated services data networks (ISDN) and the formation of "digital highways" that can be accessed in almost any region over which the space station will pass.

FIG. 9 and 10 illustrate various candidate space station communication systems. FIG. 9 shows a schematic of the four elements of one communication system; i.e., an earth-space link via TDRS or direct to ground, an antenna farm, a network communications system including all RF and modems required to transduce between antenna radiation and baseband, and a space station local area network (LAN).

FIG. 10 illustrates in more detail a typical local area network that interconnects computers, memories, and IEEE-488 and IEEE-802 controllers in each module to provide an overall system with a data rate capability of 10 Mb/s in 1983 and exceeding 100 Mb/s by 1990. In addition, a wideband cable system (similar to a CATV system) is used to interconnect each module to handle data rates up to 300 Mb/s but this system capability could be extended to well beyond a gigabit by 1990.

The LAN and wideband cable system is based on present technology using cable or fiber optics. Already inexpensive VLSI integrated circuit chips have become available for Ethernet and the growth in both VLSI and business communications will continue to be the source of new computers, microprocessors, graphic displays, memory, and related software for space station use. Also, experimental systems are under way to develop the techniques of interconnecting local area networks, PABX boards, and multiplexed data/voice/video systems by computers (interconnection of electronic office buildings). This growth in technology and hardware will be applicable to the space station as a distributed data and communications center in constant communications with earth and other spacecraft.

VI. Data Transport Concept

Because the space station is a moving node in space, it can only provide continuous communications to a particular point by a global TDRS or global NASCOM system using earth based tracking antennas as illustrated in FIG. 5. However, a very major function of the space station can be to receive data from the ground, store it, and transport the data to a point where it is "dumped" or transmitted to ground. As already mentioned, the capability of the space station to store and handle trillions of bits of data per day is almost overwhelming and gives insight into a potential major use of space stations.

One system of receive-store-and-dump is one that derives information either actively (sensors or buoys) from the earth, provides the required processing and storage, and then dumps to an appropriate earth user. One such system would be the use of side-looking-radar in an earth sensing service as shown in FIG. 11. Here, the data received from the SLR after a particular segment of earth has been covered is either stored in raw form and then dumped or processed and then dumped.

FIG. 12 shows a very exciting application of the storage and data processing potential in the receive-store-dump service (data transport). Here, the duty is transmitted by a tracking station up to the space station during a pass. At 100 Mb/s for 5 minutes, 30 gigabits of data could be received and stored by the station; a 56 kb/s uplink could produce almost 20 Mb in 5 minutes in the space station. These data accumulations can be handled by present day or near future storage technology.

At the dump site or sites this data could be returned to earth at an appropriate data rate. The S-band link could handle a 100 Mb/s QPSK or MSK carrier, and mm-wave frequencies that have the advantage of narrow antenna beams (30/20 GHz) could be the pathway for higher data rates into the gigabits.

The following describes five illustrative scenarios of how the data transport process could serve United States interests.

- The space station could pick up data from the U.S. State Department in Washington, D.C., and deliver it to selected embassies in the world. It could also reverse the process.
- The space station could pick up data from a U.S. Navy installation and deliver it to selected ships or bases in the world.
- The space station could pick up data from a multinational corporation office and dump it to selected foreign offices.
- The space station could receive data *upon command* from earthbound data sources and deliver this data to a main center.
- The U.S. Worldwide Military Command and Control System, WWMCCS, which is used by the president and National Commander Authority for dispatching critical commands, uses a packet switched worldwide intercomputer network. The space station could provide a unique receive-store-and-dump function between the various globally distributed computers.

Conclusion

FIG. 13 shows a projection of space station telecommunication requirements and objectives envisioned as they might occur into the 1990's. It is assumed that the first space station will follow closely shuttle and EVA communication capabilities, which are followed initially by support of experiments, manufacture, sensors, and life sciences.

However, it has been shown that the man-computer combination in space will lead to implementation of a space borne electronic "office" and data/communications center that can use the onboard now-developing local area networks and cable systems that are coupled to the digital ISDN networks on earth. It is recognized that the unique and overwhelming data storage capability of the space station can lead to very advanced applications of distributed data processing between earth and space as mentioned above. Also, unique functions of data transport and data retrieval by the space station can be used to serve worldwide communications networks in support of the interests of the United States Government.

OPERATIONAL BASIS OF SPACE STATION EVOLUTION

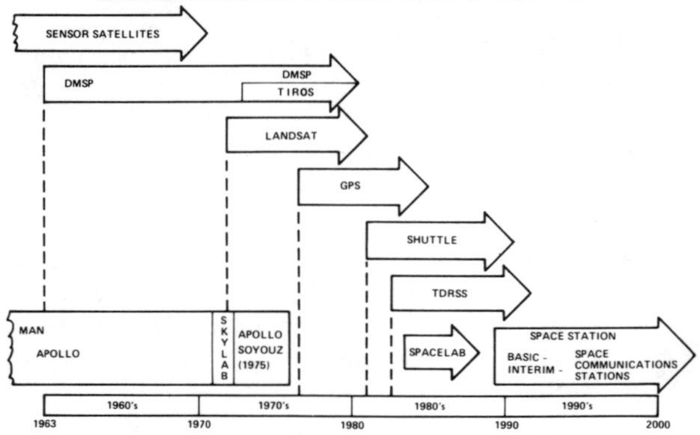

Figure 1.

SHUTTLE COMMUNICATION SYSTEM

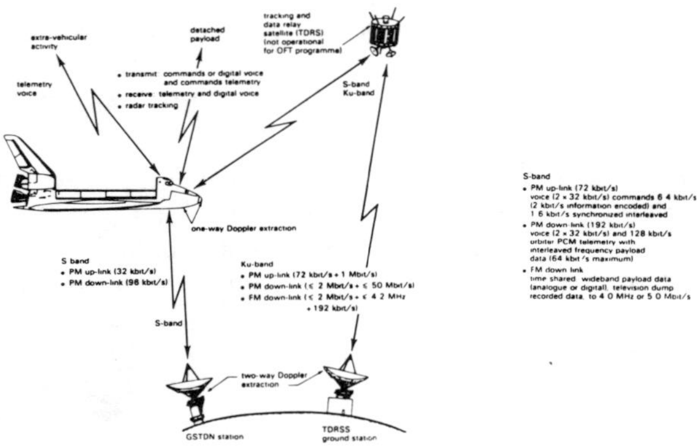

Figure 2.

SPACE STATION COMMUNICATION SYSTEM

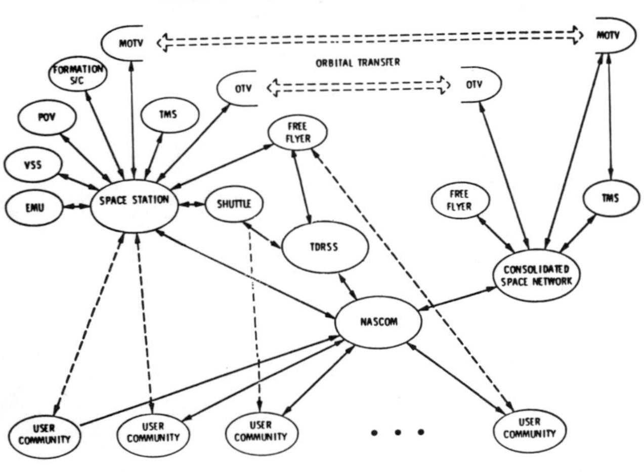

Figure 3.

SPACE STATION COMMUNICATION VIA TDRSS

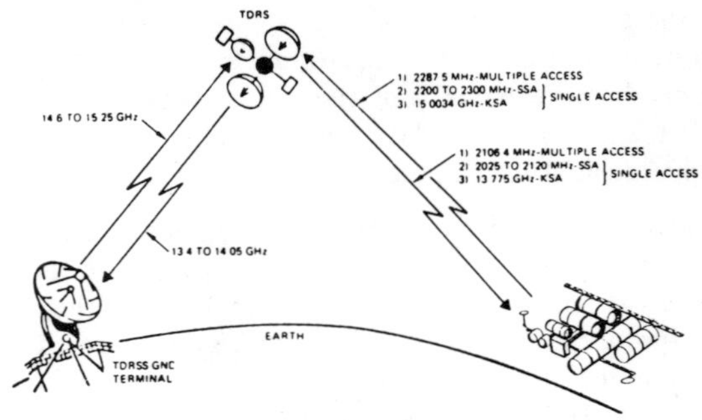

Forward and Return Link Frequency Plan

Figure 4.

SPACE STATION COMMUNICATIONS VIA TRACKING EARTH STATION

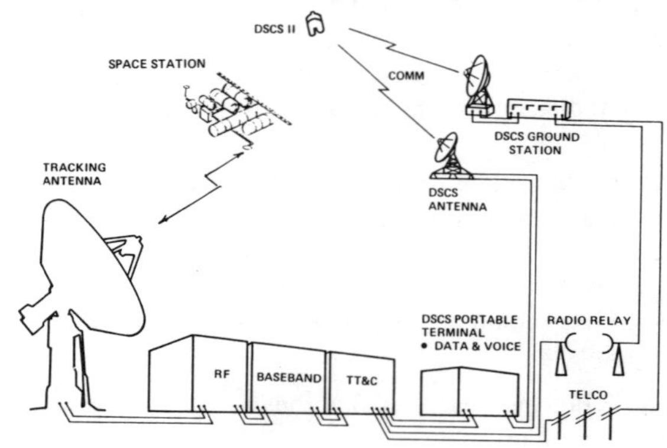

Figure 5.

CANDIDATE SPACE STATION COMPLEX WITH COMMUNICATION MODULE

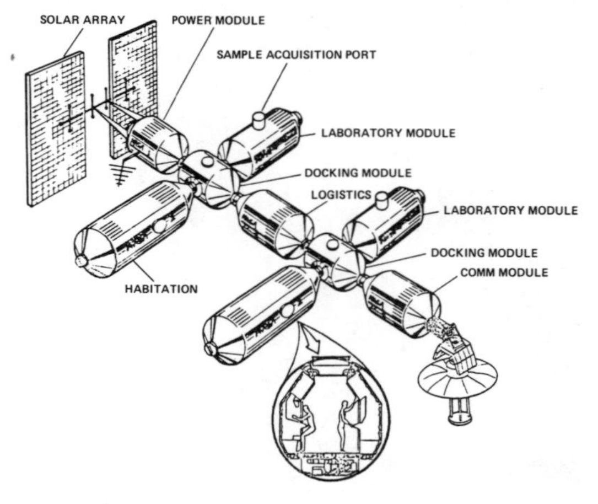

Figure 6.

SOVIET SPACE STATION - PROPOSED COMPLEX

Figure 7a.

Figure 7b.

CANDIDATE COMMUNICATION MODULE AND ANTENNA PALLET CONCEPT

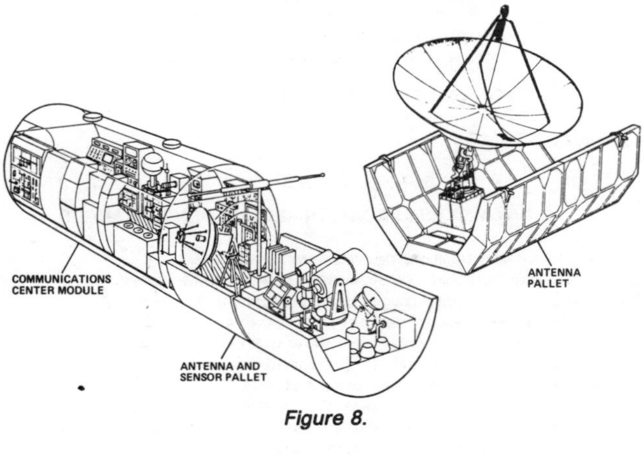

Figure 8.

SPACE STATION LOCAL AREA NETWORK AND COMMUNICATION SYSTEM

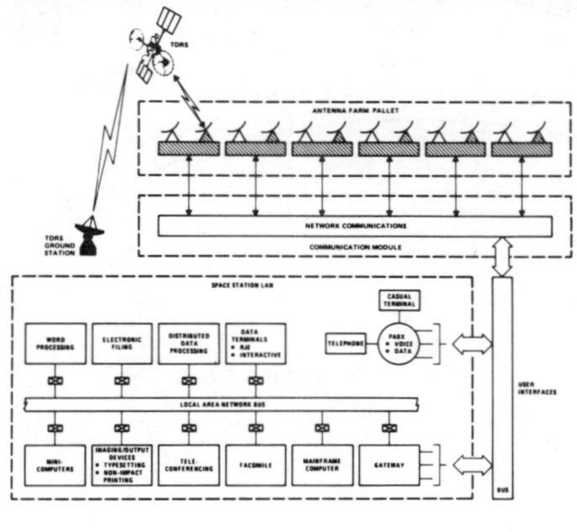

Figure 9.

CANDIDATE SPACE STATION COMMUNICATION SYSTEM

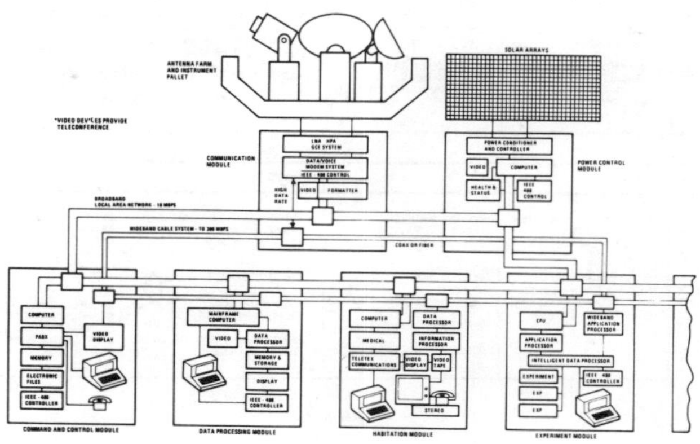

Figure 10.

SPACE STATION IN EARTH SENSING SERVICE

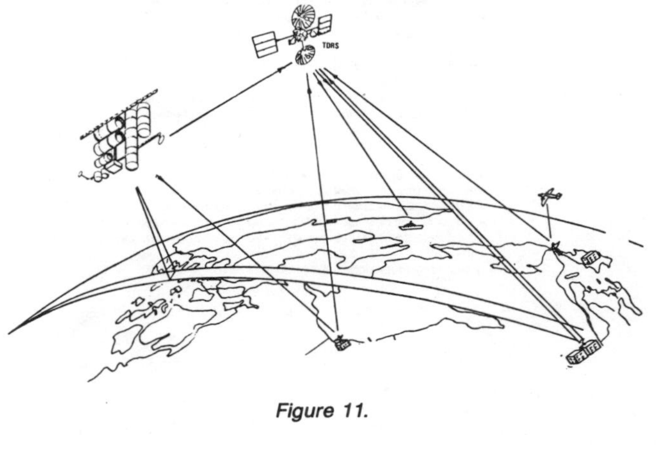

Figure 11.

SPACE STATION DATA TRANSPORT

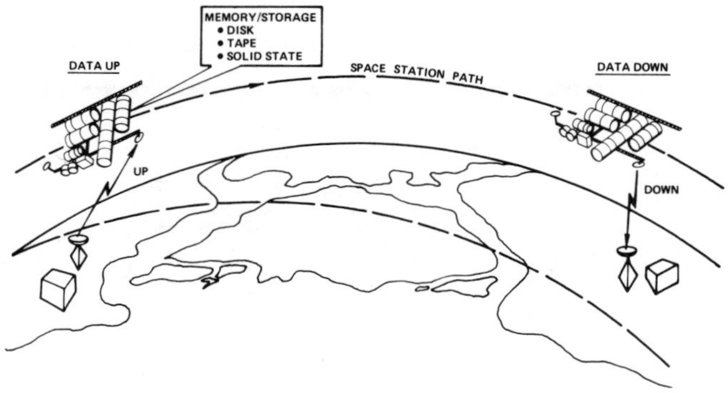

Figure 12.

EVOLVING TELECOMMUNICATION REQUIREMENTS IN SPACE STATION DEVELOPMENT

	1980'S		1990'S	
REQUIREMENTS	SIMPLE TELECOM REQUIREMENTS FOLLOWING SHUTTLE - S-BAND UPLINK PM-72 KBPS - S-BAND DOWNLINK PM-192 KBPS - S-BAND FM DOWNLINK TV/DATA TO 4 MHZ - KU DOWNLINK PM TO 50 MBPS	SAME AS SHUTTLE DATA COMMUNICATIONS UP TO 150 MBS VIA TDRS SENSOR SYSTEMS MAXIMUM DATA RATE 150 MBPS (TDRS-CAPACITY LINK)	INCREASED ON BOARD PROCESSING WITH LOCAL AREA NETWORK (LAN) AND DISTRIBUTED PROCESSING (10 MBPS LAN) INCREASED VOICE / DATA / TV REAL-TIME TWO-WAY COMMUNICATIONS BETWEEN SPACE STATION AND EARTH MAXIMUM DATA RATE 300 MBPS	SPACEBORNE FULL CAPACITY LAN PLUS WIDEBAND CABLE SYSTEM LAN ≈ 120 MBPS CABLE ≈ 1 GBPS DISTRIBUTED REAL TIME PROCESSING BETWEEN EARTH AND SPACE

	1980'S		1990'S	
OBJECTIVE & FUNCTION	DEVELOPMENT OF THE SPACE SHUTTLE AND EVA CAPABILITY DEVELOPMENT OF MAN - COMPUTER AND MAN - COMMUNICATIONS INTERFACE IN SPACE	INITIAL MANNED SPACE OPERATIONS CENTER - (5 MAN) STATUS AND HEALTH DETERMINATION SPACE EXPERIMENTS SPACE MANUFACTURE	INTERMEDIATE - MANNED SPACE OPERATIONS CENTER - (8-MAN) SPACE BIOLOGY – MANNED DATA CENTER IN SPACE SPACE-EARTH DISTRIBUTED DATA PROCESSING	FULL-UP MANNED SPACE OPERATIONS CENTER (12-MAN) MANNED SPACE COMMUNICATIONS/ COMPUTER CENTER. COMPATIBLE WITH EARTH ISDN

Figure 13.

Table 1.

GEOSTATIONARY COMMUNICATION SATELLITES VS LOW-ORBIT SPACE PLATFORMS

- PURPOSE AND OBJECTIVES OF GEOSTATIONARY COMMUNICATION SATELLITES
  - DESIGNED TO OPERATE OVER A PARTICULAR AREA OF EARTH SURFACE
  - DESIGNED TO PROVIDE MAXIMUM BANDWIDTH AND INTERCONNECTIVITY
  - OBJECTIVE OF COST EFFECTIVE COMMUNICATIONS ACCORDING TO A PRE-DETERMINED TRAFFIC / USER PATTERN
- PURPOSE AND OBJECTIVES OF LOW ORBIT SPACE PLATFORMS
  - OBJECTIVE IS TO EXPLORE THE FRONTIER OF SPACE TO DETERMINE MAXIMUM ECONOMIC / SECURITY BENEFITS FOR THE U. S. GOVERNMENT
  - PURPOSE IS TO EXPLOIT THE PRESENCE OF MAN IN SPACE
    - HUMAN FACTORS IN SPACE - VISUAL
    - MAN-CONTROLLED COMMUNICATIONS AND REMOTE SENSING
    - EXPLOITATION OF COMPUTER DATA PROCESSING IN SPACE
    - EXPLOITATION OF SPACE MANUFACTURING AND COSMIC EXPLORATION
    - EXPLOITATION OF SPACEBORNE DATA PROCESSING AND MANAGEMENT
  - PROVIDE A COMMUNICATIONS AND OPERATIONS CENTER IN SPACE

Table 2.

THE TECHNOLOGICAL BASIS OF MANNED SPACE PLATFORMS - FROM THE TELECOMMUNICATIONS STANDPOINT

- HERITAGE EXPERIENCE FOR A NASA SPACE STATION IN LOW ORBIT
  - THE APOLLO MISSIONS TO THE MOON
  - THE APOLLO-SOYEZ AND SPACELAB EXPERIENCE
  - THE TECHNOLOGY EXPERIENCE OF FIVE GENERATIONS OF COMMUNICATION SATELLITES
  - THE GOES, TIROS, NIMBUS, LANDSAT AND METGOSAT EXPERIENCE IN REMOTE SENSING
  - THE TRACKING AND DATA MANAGEMENT OF 45 AIR FORCE LOW ORBIT SATELLITES REQUIRING TRACKING OVER 90000 PASSES PER YEAR
  - THE SUCCESSFUL REALIZATION OF THE SPACE SHUTTLE AND ITS VOICE / DATA / COMMUNICATION SYSTEM SERVED BY THE NASA GSTDN, MCC, AND POCC SYSTEMS
  - THE SUCCESSFUL REALIZATION OF TDRS SATELLITE AND GTSTD BOTH OF WHICH MAKE REAL TIME SPACE STATION COMMUNICATIONS POSSIBLE
  - THE GROWTH OF S-BAND AND MM-WAVE TECHNOLOGY AND THE SYNERGISTIC GROWTH OF COMPUTER AND COMPUTER NETWORK TECHNOLOGY
  - THE SUCCESSFUL REALIZATION OF THE JAPANESE CS 30/20 GHZ SATELLITE - BUILT IN THE U. S. WHICH OPENS THE DOOR TO MM-WAVE SATELLITE COMMUNICATIONS
  - THE NASA ACTS TECHNOLOGY PROGRAM WHICH HAS DEVELOPED MANY RF BUILDING BLOCKS REQUIRED FOR FUTURE SPACE STATION COMMUNICATIONS AT DATA RATES UP TO 250 MBPS

Table 3.

PRINCIPAL CANDIDATE SPACE STATION FREQUENCIES

PRE-SPACE STATION

- DMSP/TIROS
  - VIDEO - HIGH RES. (HBPT) : 1698-1701 MHz
  - VIDEO - MED. RES (APT) : 137.5 MHz
  - COMMAND UPLINK : 148.56 MHz
- LANDSAT-D
  - TRANSMIT (TDRS) 15 GHz
  - RECEIVE (TDRS) 13.775 GHz
  - TRANSMIT (THEM.M.) 8.212 GHz
  - TRANSMIT (MULTI-SPEC) 2.287 GHz
  - RECEIVE 2.106 GHz
- TDRS
  - TO GROUND 14-15 GHz (UP) 13-14 GHz (DOWN)
  - TO USER SATELLITES RETURN AND FORWARD 2.2-2.3 GHz 13 GHz
- SPACE SHUTTLE
  - S-BAND PM (32-142 KBPS) 22875/2217.5 MHz 21864/2041 MHz
  - S-BAND VIDEO (FM) OR DATA 2250 MHz
  - TDRS COMPATIBLE KU-BAND

SPACE STATION

- TRANSMISSION DIRECTLY TO AND FROM GROUND
  - S-BAND (2.2 TO 2.3 GHz)
    - COMPATIBLE WITH GROUND SPACE TRACKING AND DATA NETWORK (GSTDN)
  - KU-BAND
    - OPERATE WITH TDRSS GROUND STATIONS - UP TO 300 MBPS DATA RATES AVAILABLE.
- TRANSMISSION TO AND FROM OTHER SPACECRAFT
  - KU-BAND
    - OPERATE WITH TDRS USER-SATELLITE LINK.
  - 23 GHz
    - USE CCIR INTERSATELLITE FREQUENCIES
  - 60 GHz
    - AVOID INTERFERENCE FROM EARTH BASED SIGNAL SOURCES

Table 4.

HERITAGE TECHNOLOGY WHICH CAN BE USED BY THE SPACE STATION

ANTENNA TECHNOLOGY - CRITICAL TO SPACE STATION COMMUNICATIONS
- TDRSS STEERABLE KU-BAND ANTENNAS
- LANDSAT-D S-BAND AND KU-BAND ANTENNAS
- THE MM-WAVE ANTENNA'S OF JAPAN CS-1, LES 8/9, MILSTAR

MM-WAVE LOW NOISE TECHNOLOGY
- FET AMPLIFIERS PROVIDING 3 dB NOISE FIGURES AT 21 GHz AND LOW NOISE AMPLIFICATION TO 60 GHz

MM-WAVE POWER AMPLIFIERS
- MAJOR TWT ADVANCES TO 20 WATTS AT 40 GHz AND 75 WATTS AT 20 GHz
- FET AND IMPATT AMPLIFIER DEVELOPMENTS (NASA) AT 20/30 GHz

COMPUTERS IN SPACE
- NASA DEVELOPMENT OF THE NSSC-1 STANDARD SPACECRAFT COMPUTER
- MICROPROCESSORS ON INSAT-A, PAYLOAD ASSIST MODULE PAM-D
- MICROPROCESSORS IN GPS AND LANDSAT-D

COMPUTER NETWORKS
- GROWTH OF BASEBAND (EX: ETHERNET) AND BROADBAND (EX: CABLENET) LOCAL AREA NETWORKS AND DEVELOPMENT OF PROTOCOLS

ADVANCED MODULATION AND DATA FORMATTING TECHNIQUES
- DEVELOPMENT OF MSK MODULATION FOR THE NASA ACTS SATELLITE
- DEVELOPMENT OF PACKET SWITCHING TECHNIQUES AND PROTOCOLS (X25)

DEVELOPMENT OF IEEE-488 NETWORK CONTROLLERS

Table 5.

DATA/COMMUNICATION REQUIREMENTS/RANGES

- MONITORING : REQUIREMENTS : EXPERIMENT DEPENDENT
- ENCRYPTION/DESCRIPTION : REQUIREMENTS : EXPERIMENT DEPENDENT
- UPLINK FREQUENCIES : REQUIREMENTS : TDRS COMPATIBLE, S-BAND
- DOWNLINK FREQUENCIES : REQUIREMENTS : TDRS COMPATIBLE, S-BAND
- UPLINK DATA RATES : RANGES : COMMAND UP TO 20 KBPS TDRS CAPABILITY 0.1 TO 25 MBPS
- DOWNLINK DATA RATES: RANGES : 0.01 TO 300 MBPS
- VOICE - HOURS/DAY : REQUIREMENTS : AS REQUIRED
- LIVE TV - HOURS/DAY : REQUIREMENTS : 0.25 HOUR/DAY TO 3 HOUR/DAY
- DATA TYPES - ANALOG : REQUIREMENTS - MOST EXPERIMENTS USE ANALOG DATA
- DATA TYPES - DIGITAL : REQUIREMENTS - TO 128 KBPS
- ON BOARD STORAGE - REQUIREMENTS - 5 TO 10 X 10^{12} MBPS
- RECORDING RATE (KBPS) : REQUIREMENTS - TO 100 M BITS
- DATA DUMP FREQUENCY PER ORBIT - REQUIREMENTS - TO ONE PER ORBIT

ENVIRONMENTAL CONTROL AND LIFE SUPPORT (ECLS) DESIGN OPTIMIZATION APPROACH

Harlan F. Brose*
United Technologies Hamilton Standard
Windsor Locks, Connecticut

Introduction

When we talk about a space station -- a manned space station -- we are really talking about putting man in orbit as a resource. It would thus seem appropriate to take very good care of this resource. The environmental control and life support (ECLS) of the space station is a key element in how comfortable and productive this particular resource will be.

When I talk to people about ECLS, especially for the space station, there seems to be an aura of mystery about it. When I talk about closed-loop life support, what I really mean is partially closed -- partially is the key word. We are presently addressing closing the oxygen and water loops. Any mystery is in closing the carbon loop, the food loop. The unknowns in this area create the mystique. Closing the oxygen and water loops is not mysterious, but it is an extremely complex subject. I plan here to go through a tutorial on the selection of an ECLS system for a space station. Let me first say that approximately 75% of the equipment that provides the ECLS system doesn't have anything to do with whether or not the equipment is labeled closed loop or open loop.

Toward the end of this presentation, I'll give a scenario on that last 25% of the equipment that does get involved with whether or not we want to close the oxygen and water loops.

Space Station Scenario Impacts ECLS Design

First we have to know what kind of space station we have, or at least define the range that the space station life support has to address.

1) Crew size vs time. Certainly, if we have a small crew, open-loop life support is not much of a penalty. If we have a large crew, more resupply is required, which might be a very large penalty.

2) The modular evolution of the space station is significant if we're going to do it all right away, similar to what was planned in Space Operations Center (SOC) concept. We must make an earlier decision about closing the life support loop. If we build it up slowly with time, the life support equipment might very well evolve with time.

3) The power concept is a key factor in integrating with the life support system if the power system is of a regenerative type. If it's a battery system, there is very little integration. If it's a fuel cell system or a regenerative system using fuel and electrolysis cells, many possibilities for integration exist. I'll talk more about that particular type of integration later, but you all know that the astronauts now drink fuel cell water. There's no question on the potability of that water. If we rely completely on solar array and battery-type systems, that particular type of integration isn't available. If we have large cryogen storage in space, the boil-off alone might be passed through fuel cells to provide adequate water for all drinking and hygiene purposes. So there can be quite an integration aspect to the type of power system that is selected and the cryogen storage system for a space station.

5) ACS orbit-keeping concept selection. If it's hydrazine, there's not much potential for integration. If hydrogen and oxygen are used, there's more potential integration. If resisto jets are used, there is a significant integration possibility. I will give two specific examples later. The dumping and vent restrictions have a lot to do with what type of life support system we ought to be using. Right now, for example, the Shuttle toilet is vented into space to freeze dry the contents and keep them biologically stable.

6) We talked about regenerative CO_2 systems that have the option of dumping the CO_2 overboard. Those type systems would be classified as venting systems. There are other ways of handling those particular functions: for example, the CO_2 can be used in resisto jets to provide orbit-keeping, at least to get two for one on that sort of thing -- sending the CO_2 away at a high velocity will probably not contaminate the orbit path significantly.

7) Extra vehicular activity (EVA) requirements. If we want to close the ECLS equipment on the space station and if we do a lot of EVA, it probably also should use closed-loop life support. Right now the present EMU suit does boil water to keep the astronaut cool, as well as lithium hydroxide, which is nonventing, but is an expendable. If there is a lot of EVA required for the space station, closing that backpack up and interfacing it with the cabin life support loop has impact on the life support equipment and modifies the backpack considerably.

ECLS Selection Process

In the tutorial process, the first step in selecting a life support system is to define the space station scenario or at least the range of scenarios. All along, the objective would be for most systems I can imagine, and especially for the life support system, will be to allow the mission requirements to dictate the scenario and not have

*Program Manager, Environmental Space Systems.

the life support system compromise the intended mission in any way.

The second step would be to get the basic requirements established, or at least a range of requirements. A number of specific requirements will fall out during the selection of space station concept. Most physiological-type requirements are pretty well known: i.e., the CO_2 partial pressure level, the amount of water the astronauts need to drink, etc. However, the proper care and feeding of the crew will result in a number of subjective-type requirements that will have an impact on productivity and we should pay quite a bit of attention to it -- it's not an easy issue. For example, the amount of water we provide for hygiene. On a daily basis, all of us use all the water we want for hygiene. It would place a restriction on the crew to severely limit them on their use of water. Possibly on an early space station, we will have to limit water. As the space station grows, we certainly would want to provide all the water that the astronauts might want.

Hygiene facilities can be very crude, as they have been really on all of the spacecraft so far, but for space station we would want home-like facilities. As a matter of fact, for all of these areas -- the subjective ones -- the best guide I can think of is to take what we normally interface with here on Earth and, if necessary, very carefully back off from that point.

A well-fed crew is a productive crew. Submarines, for example, have very good meals, and there is a very close comparison between a space station and submarines. The type of menu, whether there is a lot of water in it or not, will have an impact on the life support system, making it an important consideration in establishing some of the requirements.

The noise level in the spacecraft will be important. If the crew is highly motivated they can probably live with quite a bit of noise. As time goes on, we will probably want to provide areas away from the distraction of machinery-type noise. So, the hygiene facilities, which might tend to have some noise connected with them, probably should be isolated from the sleeping quarters.

There is -- and has been -- equipment on every spacecraft to provide all of these functions. However, for a space station, using the types of equipment on previous spacecraft, the resupply penalty would probably be quite high. So we talk about closed-loop equipment. Remember only about 25% of the total life support equipment category is involved in closing the loop. Primary concern has been the areas of atmosphere supply, atmosphere revitalization, and water processing. Equipment in the waste management and hygiene areas still needs a reasonable amount of development. Showers, for example, even though used on Skylab, still require a reasonable amount of development.

Once we have all the candidate concepts for providing the life support functions, we can conduct payback analyses. Those payback analyses are between advanced concepts, once you've made the decision to use an advanced concept, vs present flight technology. Payback analyses are a part of the story. When I say payback analyses, I really mean getting down to the dollars and cents. You have to look at what you might want to select from a dollars and cents standpoint and see if it's reasonable for a space station. All aerospace engineers have been trained to worry about weight, power, and volume. Weight and power are really the same thing. Power is weight on a space station. And for a one-time launch of a set of equipment such as the space station, it really isn't very important. As a matter of fact, it's, let's say, \$1,000/lb. And for \$1,000 in a weight scrape program, you don't get very far. And if I can add a pound to improve the safety and reliability of equipment, I certainly would do that. Reducing risk and making the equipment more maintainable would certainly be the intelligent thing to do for one-time launched equipment. However, resupply is a "scar" penalty that has to be provided continually. Here weight becomes very important, as it is in the Shuttle vehicle, which has to continually bring up the 65,000 lb with however much the Shuttle weighs -- any weight that can be removed from the Shuttle means more payload. We don't have that situation with a space station.

Any system that reduces crew time has an advantage. Remember our resource is crew time and we surely don't want the crew involved in taking care of the life support system.

Functional independence of each life support area is an extremely important aspect of selecting a system. Some of the subsystems integrate very

Table 1 ECLS selection process

1) Define objectives of space station scenario(s):
    a) Mission needs dictate scenario(s)
    b) ECLS system shall not compromise mission
2) Define ECLS requirements
    a) Specific: (most physiological needs are well defined)
    b) Subjective:
        - Hygiene water quantity
        - Hygiene facilities
        - Menu
        - Galley facilities
        - Noise level
        - Degree of automation
3) Define equipment options for ECLS functions:
    - Cabin ventilation and thermal control
    - Heat transport and rejection
    - Atmosphere supply
    - Atmosphere revitalization
    - Water processing and management
    - Waste management hygiene support
    - Galley function
4) Conduct payback analyses: advanced concepts vs flight-proven technology
5) Make final technology selections based on following factors (listed in approximate order of importance):
    - Safety/reliability
    - Risk/technology readiness
    - Mission flexibility
    - Functional independence
    - Crew time required
    - Maintainability
    - Resupply requirements
    - Orbital autonomy
    - Volume
    - Power
    - Weight

closely within the life support system. If one system goes down, everything goes down, and the backoff is to emergency systems. Other systems are independent of interfaces, and they go on doing their thing even if other subsystems are down. These are the better systems and, even if they weigh more, they still probably ought to be selected.

Orbital autonomy is very important. We don't want a ground crew having to monitor the life support equipment all the time, nor do we want the station crew monitoring the equipment. So it has to be very automatic.

Table 1 summarizes the ECLS selection process.

Space Station ETCLS Basic System Objectives

I'll now quickly go through a scenario for the life support system for a space station that I think fits most of these objectives noted in Table 1.

I think that the initial system will obviously be driven by cost. Nevertheless, any equipment we install we don't want to have to take out later just because we want to close the loop, so to speak.

I believe we will probably resupply drinking water, possibly in the form of hydrogen and oxygen processed through fuel cells. We can certainly use the condensate for the hygiene water. The Russians have been using the condensate for drinking water. I do not suggest that. I think it's not very clean water, so it would require quite a bit of watching and quite a bit of cleanup in order to drink it, but we certainly can wash with it. And I believe we should install regenerative CO_2 removal on the space station. It ought to be a concept that can grow so that at first it probably dumps the CO_2 overboard or through resisto jets and later has the capability, if desired, to be used in closing the oxygen loop.

Space Station ETCLS Basic System

Figure 1 is a top level schematic of that basic system I just described. Don't get confused by all the numbers. What we are concerned with here is that 25% of the equipment I mentioned earlier that is involved in balancing or closing the loops. I said we would resupply the drinking water (more about that below), but we must also bring up the oxygen, nitrogen, and food. These weights shown are in pounds per day for eight crew members. If you want to think about decreasing it to two crew members, just use one-quarter of the totals shown.

What you bring up, you've got to bring down. We probably will not want to dump any more than we have to in the orbit path.

In this particular scenario, I show the toilet venting overboard, a technology that might need improvement. There are techniques of avoiding the overboard dump. The Russians, for example, deorbit their waste periodically, a technique that would result in no contamination of the orbit path, would gassify the material on re-entry, and would not contaminate the Earth atmosphere. As I said, what we brought up we have to bring down. So you can

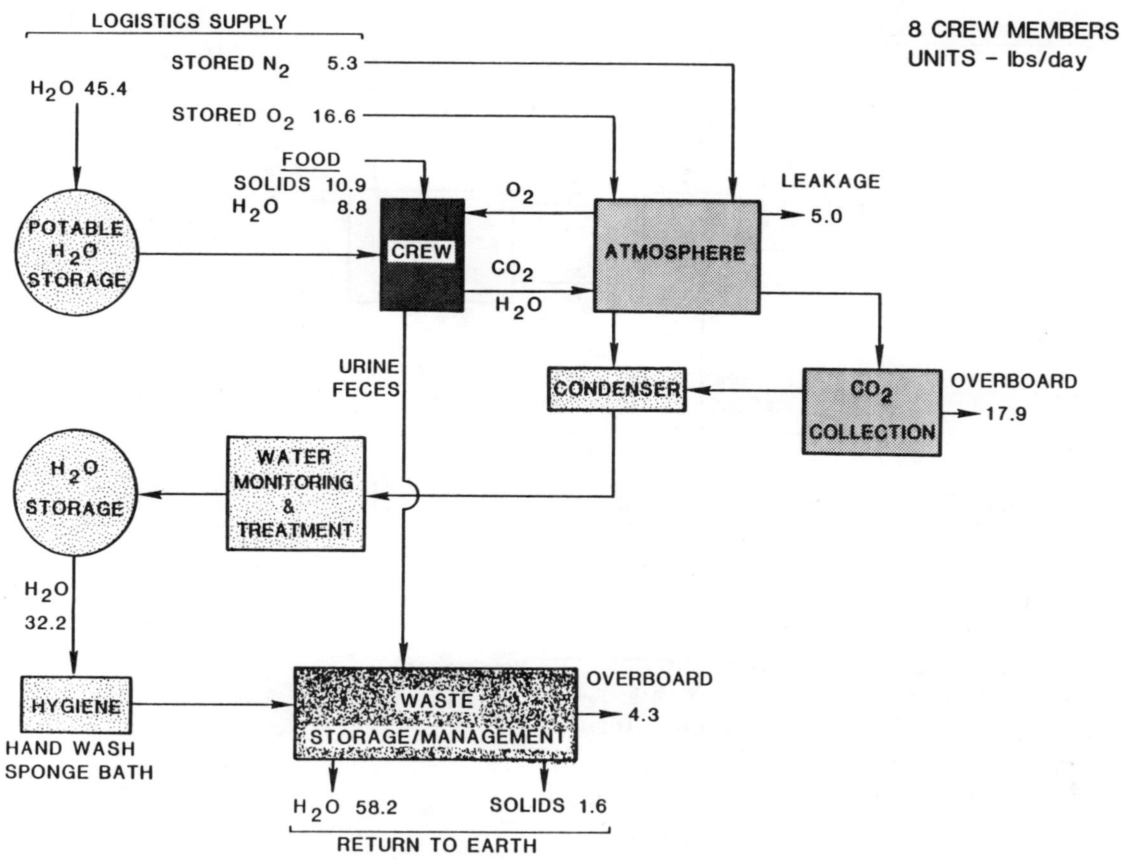

Fig. 1 Space station ETCLS basic system.

see this scenario requires a reasonably high logistics resupply requirement. However, if we think about it a little, it may not be a bad system for a long period of time. We could take the CO_2 and run it through resisto jets. We have to dump it continuously if we're going to do it that way. By varying the temperature and, therefore, the orbit-keeping propulsive effect, we can have a continuous thrust that gets the CO_2 away from the spacecraft. If CO_2 is not enough, I suggest that a better approach than using hydrogen and oxygen or hydrazine for propulsion may be to take some of the atmosphere from the spacecraft and add it to the CO_2 for additional thrust. Then, take the attitude control resupply to the station in the form of oxygen and nitrogen. One reason I suggest this is that we've always had reasonably leaky spacecraft, i.e., when we talk about trace contaminant buildup with time, we've had rather leaky spacecraft.

We're getting a lot better at making tight spacecraft. The Shuttle's very tight, the Spacelab is very tight, and we'll probably do better with time. Even though a lot of docking interfaces are shown on the space station buildup scenarios, I still think we're going to have a very tight spacecraft. If we do, we will become very concerned about trace gases building up with time. We get into unknowns. A 5 lb/day leakage is shown in Fig. 1, but if that leakage were zero, for example, it might be very desirable to use the attitude control gas in the form of oxygen and nitrogen, add it to the atmosphere, and take those trace contaminants along with the CO_2 overboard for attitude control.

I mentioned before that we are using the condensate for a very austere hygiene situation, with a Shuttle-type handwash and sponge bath. It turns out there is enough condensate water for the type of bathing they do on the Shuttle. With cryogen storage or scavenging and boil-off or cryogens that could be put through a fuel cell, there might be additional water available to add to this scenario. This basic schematic might be good for a very long time for the space station. If other loops close and we have to live with the austere hygiene situation, we could start thinking about processing wash water and closing the loop further.

Note that 45 lb of drinking water, the amount needed for a crew of eight, brought up in the form of hydrogen and oxygen is, if passed through a fuel cell first, approximately 45 kW·h of power and makes quite a dent in the dark side power requirement, and provides drinking water.

Space Station ETCLS Intermediate System Objectives

Assuming that the basic system isn't satisfactory because of vehicle evolution or because of what the cryogens are used for and so forth, the first requirement would be to improve the hygiene amenities by adding the ability to take showers and probably to wash clothes. The minimum amount of clothing is about 2½ lb per manday and it will be a penalty in volume and weight to keep resupplying clothing. Washing clothing will pay off pretty early. We can then start processing water to provide these hygiene amenities and determine whether or not we can, in fact, in the long term, get drinking water from processed water on-orbit.

Figure 2 shows an intermediate ETCLS system in which we continue the no-throwaway design concept.

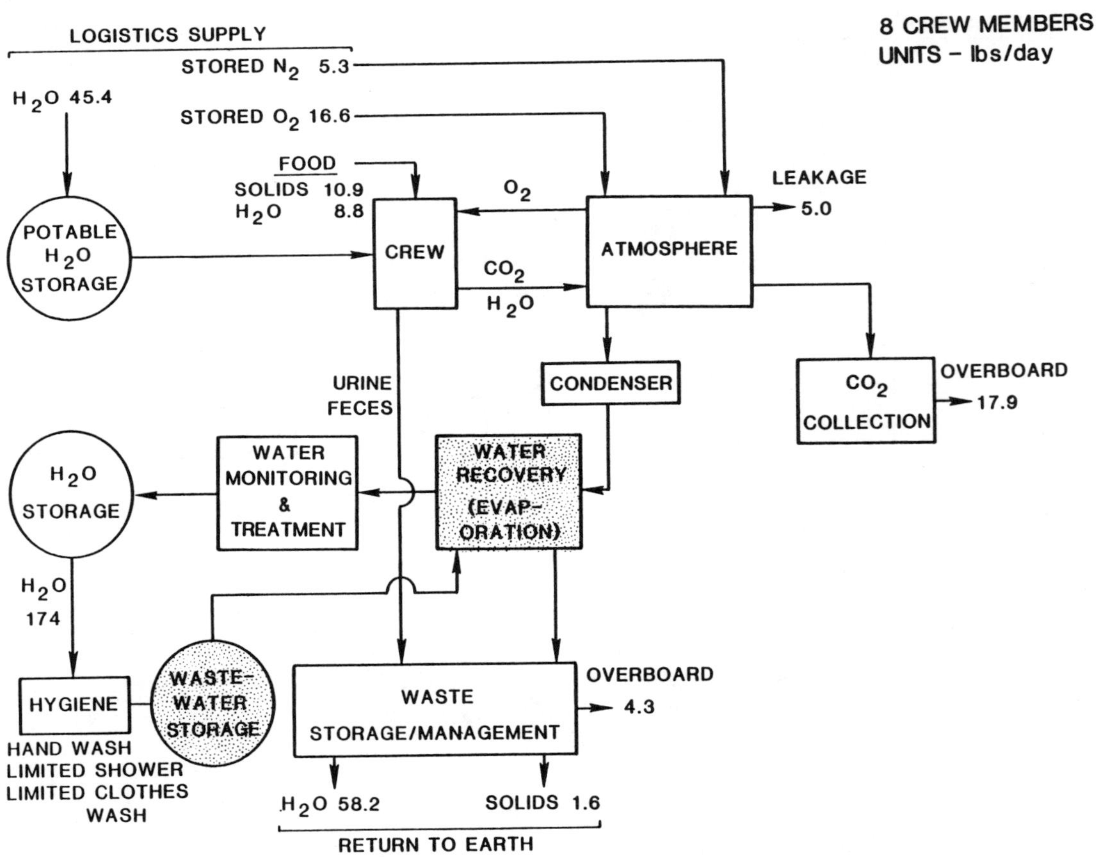

Fig. 2 Space station ETCLS intermediate system.

The white items are the same as in Fig. 1. Improved hygiene amenities have been added by processing the water. In other words, we've increased the water flow around this particular loop from the 32 lb/day shown in Fig. 1 to 174 lb/day which provides a shower every other day and limited clothes washing.

You don't have to necessarily leave a hole in the first module that comes up for the rest of this equipment. This process equipment can be included where, for example, you're going to improve the hygiene capabilities. All space station wash water can be pumped to the process equipment. It's fairly easy to pass waste water through bulkheads: certainly the drinking water would be passed through the bulkheads. So this equipment might be brought up in a later module. For this first step, we are still dumping CO_2 overboard and we're still bringing drinking water up, and everything else is the same as in the basic system. What we've really improved is the hygiene situation. Some waste water still has to be brought back down to Earth. I don't think we want to dump the water to the orbit path. We may, however, deal with it in deorbit bags or something of that order.

Space Station ETCLS Growth System

The final step in closing the life support system I've been describing -- the oxygen-water loops -- is shown in Fig. 3. Full hygiene amenities, all the water the crew might want, and getting rid of the resupply penalty is provided in the final system. And again, this equipment can be added in another module that comes up. It doesn't have to have space left for it in the first module.

In order to provide more hygiene amenities, we really just add more processing equipment and increase the water to somewhere around 316 lb/day for eight crew members. But if that number were 600, it probably wouldn't make much difference in the equipment size or weight. But 316 lb does provide showers every day and a couple of washer loads of clothing for the crew.

The CO_2 is now going through a Sabatier reaction and being reduced to water and methane. The water, made at 1000°F out of pure gases, is good for drinking, which goes directly into offsetting the drinking water requirement for the crew.

In order to close the loop, we have to get rid of the drinking water resupply penalty. A technique for doing that is to add water electrolysis if it isn't part of the power system already. Water electrolysis is necessary for the CO_2 reduction part of this scenario. The hydrogen goes into the CO_2 reduction process, is added to the CO_2, and breaks it down to water and methane. The methane must be vented overboard.

We can electrolyze more water than is required for metabolic and leakage oxygen, run it with the hydrogen through a very small fuel cell (about a 2 kW), and have fuel cell quality drinking water. Take my word for it, this system balances and we really haven't had to address the situation of having to drink processed urine, so to speak. We would have no problem in defining a specification requirement for the electrolysis unit that would take our processed water and electrolyze it. On the other

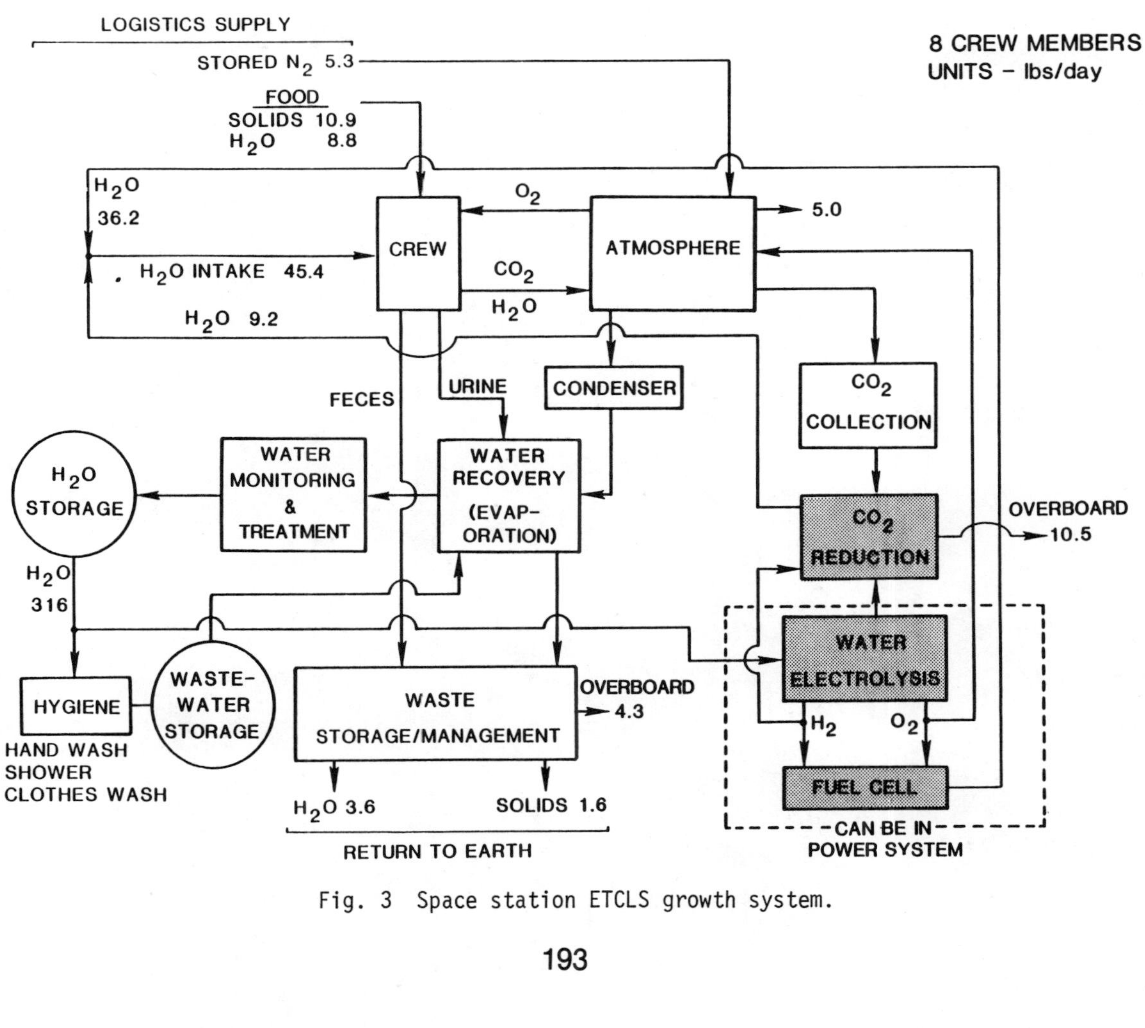

Fig. 3 Space station ETCLS growth system.

hand, I do have a problem in trying to convince the medical people that the processed water is good enough for drinking.

Thus, most of the resupply has been eliminated. There's no oxygen required, no drinking water required, and, therefore, we have to take very little back down to Earth.

Because of the metabolic process of the food combustion, we do end up with a surplus of water. And remember, that's still with 5 lb of leakage. Fig. 3 shows a very well-balanced system. This is what I would call a closed-loop environmental control and life support system. However, to the purest, we still haven't closed the food loop.

ECLS Technology Recommendations

My recommendations are that we concentrate on orienting our system technology development to support a range of space station scenarios that are going to continue right on through phase B studies, that we maintain a technology base to provide an appropriate life support set of equipment however the scenario comes out. We should develop a system technology to support any of the various space station scenarios being considered.

A major concern is to avoid getting trapped into selecting a specific life support system design too early. Cover the bets, so to speak, with flexible technology.

LUNCHEON SPEECH, JULY 20, 1983

George E. Mueller*
System Development Corporation
Camarillo, California

From a data processing point of view, the space station presents an interesting anomaly. One of its primary purposes is to gather data; huge amounts of data will continously pour into the space station. And imbedded in that data will be vital information. Therein lies the anomaly---the more potentially useful data we have to process, the more difficult it is to extract the useful information from it.

We got a taste of this problem on Skylab. That comparatively modest vehicle, limited as it was by the communications devices of a decade ago, still generated warehouses full of data that we still haven't entirely assimilated today. Yet, indisputably, the data that Skylab gathered contained extremely useful information. Just as one example, the results from eight different solar telescopes mounted on Skylab's Apollo telescope mount dramatically advanced our understanding of the sun. There were needles in that haystack.

The space station haystack, if I may belabor that simile, is more analogous to one of those huge grain elevators they have in the Midwest. The task force at NASA Langley Research Center in May 1983 identified more than 100 missions, nearly half of which would be high-data-rate space science and applications missions. The task force estimated that the data requirements of just the initial orbital configuration would be 80 million bits/s.

It will not be a simple task. As Bill Swingle has pointed out, "The space station proper will be just one element in a potentially large and complex network of distributed elements in a space operations system. Besides the space station itself, the orbiting elements will include the Space Shuttle, communications relay and navigation satellites, orbital-transfer vehicles, extravehicular crew members, and tended free flyers. Ground elements will include control centers, receiving stations, data archives, and a host of users, all demanding, in some form or another, data and some control access." **

Clearly, just sizing the processing load will require functional partitioning between space and ground segments. We need to define what makes sense in the context of mission requirements, communications requirements, physical limitations, and cost.

We must also factor another important capability into this complex formula--the crew. With a truly interactive system, the crew can do much to reduce both the processing and communications loads. With the onboard crew, we are not limited to pre-established mission profiles as we are with unmanned vehicles. They can redefine a mission as it evolves, analyzing displayed data, altering sensor coverage, and reprogramming when events so dictate.

Consider a space telescope, for instance. With an astronomer on board, it could be operated like an Earth observatory. Rather than surveying the entire universe, the telescope could be used in an adaptive fashion. When elements of scientific interest are detected, the onboard astronomer could focus on those elements. Not only would this approach vastly decrease the volume of data to be reduced and analyzed, it would probably result in information of far greater scientific importance.

The same concept could be beneficially applied to all the sensors. When warranted, a crew member could reconfigure an array, concentrating on different spectral regions. In effect, operations could be conducted much like a reconnaissance mission. Rather than operating in a survey mode, recording the full spectral coverage, an operator-attended system could select a region of interest and scan it with different sensors.

I dwell upon the importance of the astronaut as an integral part of the information processing system because it has such implications on the overall data processing requirements and architectural design. I believe that well-planned, extensive applications of the astronauts' reasoning capability as an integral part of the onboard data processing system will enable us to perform a far larger portion of the processing on the spacecraft. Indeed, coupling human capabilities with equipment based on the emerging computer technology should enable partitioning of the system between ground and space without compromise of mission effectiveness.

We must remember the enormous strides the computer industry in general and the semiconductor industry in particular have made since the days of Skylab. Today's tiny, high-speed VLSI devices, perhaps best typified by the 256K RAM chip, today's distributed architectures, today's fiber optics busses, today's densely packed magnetic storage devices--all greatly increase the processing capability we can afford to place on the spacecraft.

And by 1991, even these marvelous devices will have been outstripped by the new technological developments even now emerging from the laboratory: robotics and other advanced automation techniques, artificial intelligence, optical storage, still more advanced semiconductor devices. All these and more will still further enhance the processing capability we can afford to place on board.

*Chairman and President.

**See William L. Swingle and Charles W. McKay, "Space Station Information Systems."

Technologically, therefore, I do not see any barriers to an extraordinarily powerful onboard space station information system in this time frame. I do not mean to minimize the difficulty in designing the architecture for such a system, however. We are still facing very large data loads and a system of many diverse distributed elements.

Some of you may remember the beginning of the Shuttle program, a proposal I made for autonomous electronic modules interconnected by a simple four-wire cable (two wires for power and two for signal). These electronic modules were to be self-testing and self-diagnosing with internal microprocessors to not only identify a problem, but with enough memory so that they could detect performance changes and give an easy prediction of problems. A part of the concept included providing enough redundancy so that when a failure was detected, the defective circuit could be isolated and a new one switched in.

The idea proved to be difficult to implement at the time although today many of the advanced avionic systems are partially utilizing self-diagnostics.

I do believe that the technology on LSI circuits has advanced to the point where it is relatively easy to provide the necessary redundancy, error correction, and self-diagnostics to implement the concept on the space station with a life cycle cost substantially less than conventional systems, while at the same time providing a system architecture which is most adaptable to the changing requirements of a permanent space station.

Let me outline how this concept could be applied to a permanent space station. First, we need to know the environment in which the system is to operate. I make the following assumptions, any or all of which are challengable:

1) That to the greatest possible degree the information system should be autonomous, i.e. having a minimal requirement for ground support.

2) It should be reliable and, if necessary, repairable on board.

3) It should be intelligent to the maximum possible extent.

4) It should be modular in physical architecture, easily interconnected, and designed for additions.

5) It should be user friendly and expansion and adaptation should be transparent to the user.

6) The role of ground support should be passive, that is, provide expert assistance and advice only when called upon.

7) As the station becomes operational the experimenters for major experiments will be on board. For those experiments or operations which do not require direct human involvement the onboard system should be trained to handle all routine chores and also be capable of detecting problems and calling them to the attention of the onboard personnel. Only if the problem is beyond the capabilities of the machine and crew to repair will the ground support personnel be called in.

Second, we need to describe at least one kind of system that could meet these requirements. I believe that it will have the following parts: A local area network interconnecting application specific machines using fiber optics technology. This network will be readily extendable--on board, through the use of standard interface units and to other networks or communications nodes, either in other space modules or on Earth through the use of gateways (translating gateways, if that is necessary). Since the bandwidth of these cables is large, the local area network will also carry voice and video throughout the station and, through suitable gateways, provide televideoconferencing interconnection with the external world.

The modules with which the local area network interconnects will be:

1) Microprocessor-based "expert" machines for operating both the space stations and those manufacturing processes that have reached a stage in development where they can be automated.

2) Information storage modules for maintaining a record of results of experiments and operations.

3) Specific application modules for supervising, operating, and processing data (to obtain information) for experiments.

4) Communications modules for internal and external televideoconferencing.

5) Various monitoring, control,and display devices for the conduct of experiments and the operation of the station.

The modules will in general consist of a front-end processor for diagnostics and interfacing with the network as well as buffering data from the main microprocessor. The function of this last processor is to measure the performance parameters of the attached system, to control the system based on a set of rules and criteria developed by the expert or experts responsible for the system, to flag anomalies, to extract from the data the essential information giving the results of the system operation, and to forward the pertinent information to the central memory.As I have described them, these modules become the distributed brains and operators of both the support and "core" systems which operate the space station, as well as performing the same functions for the manufacturing and experimental systems on the space station and as delivered to the station from time to time.

The communications modules will operate on a separate carrier frequency and will, I expect, for this generation be the "conventional" combination of digital voice and analog video.

The systems monitors and the displays can be a common design of user-friendly "smart" terminal capable of fully utilizing the resources of the application specific machines on board. Then displays would serve for both data processing and televideoconferencing modes.

As you can see, this approach divides the data processing and control function into many relatively small applications specific machines.

It makes operation of the system easy, it simplifies maintenance, and it permits an orderly evolution of both experiments, manufacturing operations, and station augmentation without destroying the basic integrity of the system. It must be noted that these desirable results are possible only if these modules are vigorously designed to be standalone, independent, and self-contained.

This leads the next and greater challenge--writing the software programs which will implement the system. Here the modularity of the application specific systems are an aid. Also, insofar as a common design philosophy and a requirement for a limited set of standard piece parts are enforced, many of the software subsystems can be used on all machines. For example, a basic diagnostic package could be implemented with the ability to tailor it to specific measurements and comparisons with standard parameters for trend analysis and failure forecasting. Again, a common implementation of a knowledge-based system could support the expert in all the modules.

However, as in all such systems, it is necessary to begin the software architecture well before the final selection of hardware, as well as being sure to establish configuration specifications and controls in the beginning. Also, one should not expect perfection in the first implementation, but rather should provide for an iteration evolution of the information system for the outset.

Finally, I'd like to make two observations: properly implemented, the distributed processing proposed here will markedly reduce the amount of data transmitted by the experiments and operations, thus dramatically reducing the required bandwidth and data storage requirements; and, second, the life cycle costs will be drastically reduced, primarily through the elimination of ground support equipment and personnel and secondarily through reduction of ground processing and storage equipment.

PANEL: SPACE STATION ISSUES AND PROJECTIONS

Moderator: John D. Hodge
Director, Space Station Task Force
National Aeronautics and Space Administration
Washington, D.C.

At the end of any Symposium such as this one it is always useful to devote some time to evaluate whether the objective of the Symposium has been met. Our objective was "to present in a concise but fully documented manner to interested sectors and individuals the current thinking on the designs, technology, and potential applications of an Earth-orbiting space station." I think we have achieved that. It is also useful to have a wrap-up portion of the Symposium to provide a means of getting some feedback from those who have patiently listened to the papers and panels presented.

As Bob Freitag pointed out in the remarks he made to introduce the panel on Space Station Issues and Projections in my absence, we made an effort in this closing session to hear from representatives of the various disciplines that we feel will be served by a space station. We received such feedback from the international community, the Defense Department representative, a member of the media, a contractor, an industry participant, a member of NASA management, and a scientist. They were candid and straightforward and I'm sure that their thoughts and opinions will be carefully considered by all of us as we move ahead in the planning and other activities that will lead to a space station in orbit by 1991 or 1992.

PANEL DISCUSSION

Moderators:
John D. Hodge
Director, Space Station Task Force
National Aeronautics and Space Administration
and
Robert F. Freitag
Deputy Director, Space Station Task Force
National Aeronautics and Space Administration

Horst Anton
Deputy Head
Department of Shuttle/Spacelab Utilization
DFVLR

George E. Mueller
Chairman and President
System Development Corporation

Charles W. Cook
Deputy Assistant Secretary (Space Plans and Policy)
U.S. Department of the Air Force

Neil Sheeley
Astrophysicist
Naval Research Laboratory

Craig Covault
Space Technology Editor
Aviation Week & Space Technology

Richard G. Smith
Director
NASA Kennedy Space Center

C. James Dorrenbacher
Vice President and General Manager
McDonnell Douglas Astronautics Company

Robert F. Freitag: This panel is designed to summarize the conclusions that have emerged from this three-day symposium and to give everyone the opportunity to speak out or to ask further questions -- we even invite short speeches if they're held to a minimum.

This panel was assembled in order to air any ideas, thoughts, comments, or critiques that need to be emphasized or that have been neglected during the symposium. Our objective in this particular panel was to pick representatives from several different areas of interest in the space station that may or may not have been covered in earlier talks, so that the broad viewpoints of the technical, operational, outside, user worlds, etc., will have a chance again, "to have at us."

A NASA Task Force has been studying the proposal for a space station for about 14 months now. James Beggs of NASA commissioned the Task Force in May 1982. The purpose of this symposium has been to lay out in as much detail as possible the work of the Task Force to date and to ask -- are we on the right track?, what's missing?, what needs more emphasis?, etc.

I think we've made tremendous progress thus far. Many may think after hearing the presentations that things haven't changed much -- they're pretty much the way they were two, five, ten years ago, or whatever time you took your snapshot. But there is a lot of difference in who is understanding what's going on, and I think that's very, very important.

The very fundamental approach of taking an intense look at the mission requirements has been extremely effective. Rather than just taking a snapshot, an activity has been set up to stick with the study, continuing to iterate what the mission requirements are.

We're moving ahead on the concept development, although I know that our activities in this area are a little disappointing. Many of you wanted to see pretty pictures of what we were talking about. We think we've come up with an idea to solve that problem: a book that explains how to draw your own space station. In this volume are any type of space station components you'd like, furnished with vellum paper. Anyone who would like to can take a crack at building a space station. We'd like to see them!

Seriously, that is the task ahead of us. The Task Force wants to digest the mission requirements, digest the utilization, make sure that we've heard all suggestions, criticisms, etc., and then -- and only then -- make sure that the space station concept is driven by the requirements of the potential users. This is difficult when you have available the many studies and examinations that have gone before. But I think we can expect to see in the next few years real progress in this program.

From edited transcript.

Comments have been made by the science applications people, the international people, the Department of Defense people, and the commercial people. We're anxious to make sure that we have the continued input of the NASA centers, the aerospace industry, and other government agencies. We discussed a bit about the political aspects involved in the senior interagency group, and Jim Beggs has outlined the activities going on to understand what's key in that area.

We have tried to look at every option that we think is practical and I think we have not closed our eyes to any particular one. We will undoubtedly hear many more as the next year unfolds.

In the aggregate, we have told you what we know and think and why we're doing it. I know it revealed some gaps and some areas that we want to work on. We are certainly going to move toward refining the time-phased mission sets to make them more realistic, to integrate the international requirements that came in somewhat after the domestic requirements, to maintain a constant dialogue with the group, and -- I'll say over and over again -- to keep talking to the users, and I want them to keep talking to us.

When we talk concept development, it's simply the ring shown in Fig. 1: starting with the mission sets, which we intend to define in terms of the functional requirements, and proceeding around the ring clockwise until we come up with a "cut" at what we think the architecture will be. We've been through this ring only once. In the next year we intend to go around it at least four times.

We have established the task for the next 14 months, and we're at the halfway point of this schedule, see Fig. 2. We have shown the first cut: to define the scope and the overall size of the project as we see it. We're now in the process of the second iteration, which will yield better definition so that we can prepare the FY '85 budget, which will be submitted to the Office of Management and Budget in September.

Almost immediately after that we will start on the third iteration of the definition, which we anticipate having by the end of 1983, the output of which will be the fodder for the generation of the first set of specifications for a detailed definition.

As those specifications are being generated and as they appear, we will continue with the fourth iteration, to be completed at the time of the appearance of the FY '85 budget, when we expect to go into detailed definition. That is the planning, and if the planning is approved within the next few months -- approved in the form of approval of an '85 budget item of significant proportions -- we will be ready to proceed.

We will continue to move forward on the advanced development of sub-systems. We will continue to work out the parameters and the policies associated with international cooperation. We will continue to do everything we possibly can to bring about precursors to the space station that will be developed within the framework of the Shuttle or Shuttle operational activities, such as tether and teleoperator retrieval systems.

Meanwhile, we are defining a whole series of management options, procurement options, planning for the actual internal and external management. As the program emerges, these options will be presented to the Administrator for his choice, decision, and assignments.

We certainly hope to maintain the momentum and I think that we will have no problem in that. There's plenty of work to do and meetings such as this will greatly stimulate our efforts. There will obviously be the inevitable setbacks along the line in various and sundry areas, but we're really ready to go.

Charles W. Cook: I have indicated there were some possible areas where the space station could be used in support of DOD activities. In other words, if higher authorities decree that there should be a space station in the overall national interest, the DOD would be sure to use it in research and development with regard to how the military man in space can best contribute to DOD's mission. I also indicated that however at the present time there were no DOD plans to develop or to help develop a space station.

I thought it might be useful to add a little perspective with regard to requirements -- what they are and what they aren't. There are some things that you should understand about requirements. The first is that they tend to be a bit subjective. To get a validated requirement means that you have at least two people agreeing that it's a good thing to do.

Requirements tend to follow capabilities, rather than vice versa. Theoretically, in the purest of worlds, people know what they want, they lay it out as requirements, and then go ahead and develop it. Actually, it doesn't work quite that way. As a matter of fact, it tends to, at least

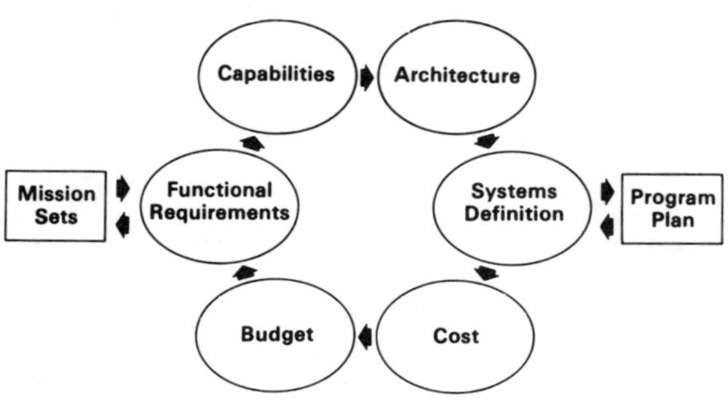

Fig. 1 Concept development.

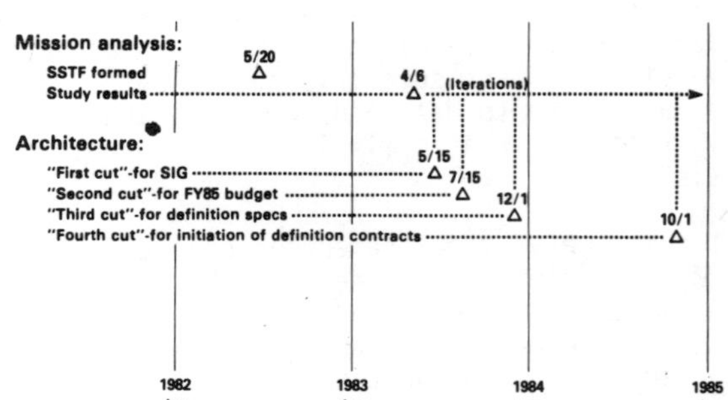

Fig. 2 Space station definition schedule.

in my experience, work the other way around -- that once a capability has been developed, you can see what it can do, you can see how to use it, and then you proceed ahead with it.

The last point is one that Dr. DeLauer[1] mentioned, that is, if you develop a solid requirement and say, "By golly, we've got to have it," you can be sure that before the words are out of your mouth, somebody's going to put a bill in your hand, which is usually fairly substantial one.

There are two other things about requirements. One is what I call bureaucratic rule No. 16: if you don't have a good reason for doing something, list 10 or more requirements. Thus, if you find somebody who has a long list of why we ought to have something, you might want to question the need. The other is that they tend to vacillate. Not only that, but they are sometimes subject to change without notice. An example here that you might find interesting is shown in Fig. 3.

John Hodge with his statement there is shown in Fig. 3a. Figure 3b shows the current DOD response. And then from the wings we may have Fig. 3c. Then Fig. 3d, which you know is....That's another way of saying we all have bosses and the boss is always right.

I would also like to point out that there's a whole spectrum of answers that can be given to any question. Every young girl soon learns very well the spectrum of answers to a proposition. Engineers generally aren't quite as fast on their feet, so it might be useful to illustrate the range of answers that can have been given to the question (Fig. 4). It goes all the way from "never, never and get lost" to "yeah, we're ready now." Putting into perspective the answers that I gave yesterday, you'll see that it's somewhere down around "maybe, give me a call next week or next year or in the next several years" or "maybe yes, but not now." So in the overall scheme of things, on a scale of 10, I give it about a 6.5 to 7. I think that might put DOD's attitude into a little better perspective.

Horst Anton: I would like to give you some of my personal views on the international issues under discussion here:

1) I want to define "umbrella agreement." I believe that an umbrella agreement on international cooperation similar to the one already existing between Germany and the United States for the utilization of the STS Shuttle should be established for the space station. This might be done by extension of the STS agreement and even without having an approved program on NASA's side.

2) I see a need to establish the ground rules for participation pretty soon. These rules should, for instance, attack such questions as constraints for cooperation, forbidden items, strategic items, policy from a procedural point of view, how would it develop in Europe, can European developments be brought into the space station program to the benefit of both partners to reduce investment costs. If European industry is to undertake joint endeavor agreements, will the same rules apply as for American industry?

3) Use of and evolution from existing hardware. We would like to see a NASA policy as to how, from a technical point of view, the smooth transition and evolution from existing Spacelab and STS hardware can be made to partially reduce new developments. Of course, the question of standardizing interfaces has to be asked here.

4) Communication and exchange of information. In the past, information exchange between the different organizations involved has not been covered by any rules. Nevertheless, so far, it has worked quite well. I clearly see the need for more intense communication both ways for all steps in the program.

5) Proprietory rights, that is, intellectual property. This question has been raised several times during the discussions at this symposium, and is indeed an important issue. In my opinion, NASA policy has to be established. This becomes even more important as a long-term direction of the space station will be in commercial and application uses.

6) Definition of operational costs. We all know from past experience that the operational costs often -- in fact, always -- have been underestimated. We would appreciate it if NASA would

Fig. 3

NEVER
HELL NO! (WITH EMPHASIS)
HELL NO (WITHOUT EMPHASIS)
NO
NO?
MAYBE (WE'LL SEE)
YES (BUT, NOT NOW)
YES (TENTATIVE)
YES (NOW!)
DO IT

Fig. 4 Spectrum of answers to the question.

1 See Richard D. DeLauer, "Military Space Activities and a Space Station."

try to solve this question pretty soon, while in the early stages of the game.

My feeling is that all of these six issues have to be solved before we start Phase B activities. In my opinion, NASA is on the right track. The list of open questions is relatively small compared to those we have already solved. The issues seem to me to be similar to the early discussions we had when we were starting the STS program. We have solved them for STS, and I can assure you that we will try all we can do from our point of view in helping to also solve these things for the space station.

Question: I'd like to ask Dr. Anton if he can characterize the current state of interest among German or European industry in participation in MPS and space station.

Horst: We have held several symposia and circulated several questionnaires in both Germany and the rest of Europe, as you have done over here in America. We had a big meeting in December 1982 just to explore this question. My personal feeling is that there is no scientific discipline really saying that they are the driver for a space station, but at the moment, all of the disciplines are saying that the space station will be there, that we will make use of it, and that we will really enjoy participating in it.

Craig Covault: It has been valuable devoting a whole week to the space station. Most of us have been so busy with Shuttle over the past several years that it has been difficult to focus on the next U.S. goal in space.

Checking back I found the first large article I did on the Space Shuttle (and by now we're up to about 300 on Shuttle) was 13 years ago in 1970. It was an interview with Buzz Aldrin who was only a year back from the moon. Buzz was assigned to the Space Shuttle Task Force in Houston and I was working under a Reader's Digest grant. The Space Shuttle was then at about the same point the station is now, a year or so away from formal program status.

The situation in the country was a lot different then compared to now, however. NASA budgets were going down, public interest in space was on the wane, and the nation's attention was focused on Vietnam. But, even with these problems NASA was successful in obtaining Shuttle program approval. Now the situation is much different. NASA budgets are holding their own, the nation's economy shows improvement, and public interest in space is up.

I cannot advise you on whether to build a space station or what type station to build. I can raise some questions that I think are important to your overall strategy, however.

Issue No. 1 is the question of leadership from the top -- support from the President. On all of the Space Shuttle missions President Reagan has taken time out to say something or do something in connection with the flight. Atending the mission 4 landing is the best example. Unfortunately, most of what he has said on the air-to-ground link with the crews has involved reminiscing about life on the ranch and how that compares to spaceflight -- there has not been a lot of substance there. This is an observation, not a criticism. The point is he has "been there" and that's an important and good sign.

I think if his advisors will let you, you can win the President on a station decision. On the topic of advisors, the President's Science Advisor George Keyworth has been less enthusiastic, but his words have carried far more substance about quite appropriate concerns. His message has been to make sure you've gotten your station act together and a plea that station activity not be the detriment of Shuttle and Spacelab utilization.

The President and Keyworth have to some extent given the impression the White House is going in two different directions -- apparent pessimism vs apparent enthusiasm. Now we are starting to hear a more upbeat approach by Dr. Keyworth with his comments on long-term space goals. I do not think you should jump to enthusiastic conclusions about this until pinning Dr. Keyworth down in more detail. I do believe, however, the overall vibes from the White House indicate a GO for some sort of station or platform development.

All of us I believe would welcome long-term U.S. space goals, but you should be careful not to let such goals dilute more near-term objectives over the next five to ten years. NASA's station strategy has to date been quiet and conservatively paced. This is a technique Administrator James Beggs has used successfully to hold the line and even increase some recent Agency budgets. In the case of the station NASA has been concerned about getting ahead of the White House. It's a question of strategy and appearance. The conservative approach toward modest new starts in the budget is one thing, a large new U.S. space goal like a station is another.

I'm beginning to believe you should be more out front and aggressive, especially in explaining the multifaceted station elements you have in mind as compared to the old Skylab/Salyut approach. NASA and the aerospace community have done extensive study to provide direct input for station justification. You have done your homework and now its time to state the case more clearly and more publicly. Also, if NASA really believes a station is vital to sustain the Agency's space development capability and that this is itself important to the nation -- then it should not be so shy about saying so. If you can state your case directly to the public I am sure they will support you. They are ready for a major new space goal with the Shuttle as a stepping stone.

Look at the evidence: up to 1 million travel to the Kennedy Space Center for Shuttle launches; several hundred thousand have driven to Edwards for landings; 780,000 called a special number to listen to the mission 6 crew and on mission 7 that number grew to 840,000. During the Orbiter Enterprise U.S./European tour 2.2 million people specifically traveled to the viewing sites, not counting many more who made some effort to view the city flyovers.

You have gotten the ball rolling, now you have to keep it going.

A view I hear often is that even with the best station user justification, the station decision will be a political decision. If you start seeing support from the White House, don't be so sure that it's because you have made a strong technical case, but rather because the White House smells the political hay that can be made from the success of the Space Shuttle and missions such as Voyager and Pioneer sailing out of the solar system.

In this regard, an interesting area to watch will be the White House/OMB interaction in the station decision process. I believe OMB is willing to approve some sort of NASA space platform development, but only a relatively low level effort. This type of development would have little or no political clout. It would be a routine NASA new start, not a national space goal the White House could tout as a significant Presidential decision to push U.S. capability in space.

While on politics let me raise another issue. If so many people think politics is going to be a significant factor in the station decision, then why is NASA so hesitant to raise the Soviet space program as an issue? We are all in agreement the U.S. should not be doing a station development simply because the Soviets are, but NASA cannot ignore the big station push by the United States' prime competitor in the world. I'm personally tired of going to hearings, or speeches, or astronaut news conferences where someone asks a legitimate question about the Russian program with the NASA response superficial at best. No one that I know of in NASA has said anything really strong about the Soviets since astronaut Tom Stafford unloaded on them in 1974 for the monkey business they were trying to pull in Apollo/Soyuz program discussions.

In the current situation, I'm not talking about criticism. I'm talking about informing the public, Congress, a sizeable portion of DOD, and the California lawyers in the White House about a Soviet space station development comparable to the U.S. Apollo program in expenditure and effort.

Let me remind you about what the Soviets have in development:

1) A booster larger than the Saturn 5 moon rocket. They have totally revamped the design since failures with a similar class booster 15 years ago. I expect this one will work.

2) Large space station elements for launch on top of this booster. Imagine the hardware they could place in orbit by flying this booster and payload only two or three times per year.

3) A heavy space shuttle with an orbiter nearly identical in size to ours.

4) A small winged spaceplane manned vehicle for launch on an expendable booster. They have already flown subscale versions.

5) Continuing expansion of existing Salyut station capability with the Progress tanker and uprated Soyuz.

6) A space station tug development that includes a large Gemini-size re-entry vehicle to return station products to Earth. The Soviet station tug now operational could transition into a Soviet teleoperator maneuvering system like that sought eventually for a U.S. station.

It is obvious the Soviets long ago established a large space station development goal and are now executing their plan. I believe they hope their station development will put them on the road for a manned Mars expedition in our lifetime.

There is also no doubt the Soviets already have leadership from the top on this station push. Soviet Party Chairman Yuri Andropov in his first address to the Plenem (an equivalent to our State of the Union Message) specifically endorsed establishment of permanently manned Soviet space facilities.

Soviet periodicals repeatedly state the Soviet goal of "Sovyetski Gorod" -- a Soviet city in space. They call it Cosmograd and with the developments just cited it's clear they have the foundation laid for such manned space facilities.

In short, it's time for NASA to pull it's head out of the sand on the Soviets.

The agency's public affairs office is totally unequipped to handle the simpliest queries on Soviet space flight -- and the agency's managers have been ducking the chance to comment. The Soviet program is an issue because their space station push is part of their military, technology, and international strategy -- not because it would make a convenient tactic for U.S. station approval.

Were NASA to raise this more visibly the Soviet program might get more media and public attention. I can offer one example. About three years ago, Omni Magazine asked me to write an article on the Soviet space program. I did not put a lot of work into it, and it was just a compilation of old Aviation Week findings, but the reader response at Omni to this thing was tremendous -- I believe the second largest reader response they had had to date. The typical comment to Omni was, "I had no idea."

I would like to say a few more words on the media because the media is very important to your space station kickoff. Just as you consider your approach to the White House on the station issue, you should also consider your approach to the media -- or make sure the White House does when it discusses the space station. That sounds simple, but I will cite two cases where the space community and the White House blew it.

The first is the Reagan space policy released July 4, 1982, in connection with the Shuttle mission 4 landing. There was an almost universal perception in the media that this was a military space statement, not the national policy it was designed to be. Media coverage of the policy failed to convey the policy's objective of maintaining U.S. space leadership, both civil and military. The civil side was ignored, as was the importance of language allowing consideration of station facilities.

Why? First, the White House was uncooperative in discussing for background the policy prior to its release. If they had not treated it like a big secret they could have laid some valuable groundwork in explaining their objectives. Second, the President's speech at Edwards was not particularly hard-hitting. Finally, the White House backgrounder held when the policy finally was released was a group grope on relatively short notice. Since all the journalists that normally cover spaceflight were out of town covering a spaceflight, the background was to a group of media who had little savvy on the space program. The results were predictable.

Another example of where the White House message got lost was President Reagan's so-called Star Wars speech. The fact it drew the Star Wars label indicated media misunderstanding of the message that was directed more at overall defensive technology than space warfare. The White House approach with the defensive technologies speech suffered from some of the same problems that affected the space policy speech and the result was wide spread nay-saying in the media.

If a space station announcement runs into similar problems, you've got trouble. So what can you do to prevent it?

First of all, I think you are always better off saying more, rather than less, early in the game. The media by and large is not very savvy on technology in general, let alone spaceflight specifically. You have to explain this stuff frequently for it to soak in, especially when it comes to the multielement space station concepts now under planning.

As for worrying about "getting ahead of the White House," how often have you heard the comment about the man in the street already being way ahead of the bureaucracy when it comes to caring for the U.S. space program.

Another thing you can do is insist when the White House has something to say on space, that the statement carry sharp substance -- not generalities. After all, you are the space experts, not the White House.

A third measure would be to strive for NASA public affairs offices with a more aggressive attitude. They need a stronger understanding of space flight hardware and operating procedures and what it takes to convey that to the media.

A final thing you could do is bring the younger participants in NASA and industry out to do more of the explaining. An example is the group of new astronauts brought into NASA in 1978 and 1980. They are as sharp and enthusiastic a crowd as you will ever find. Of course, they are already busy--but I think they could be turned loose more to explain what's going on in both the U.S. and Soviet space programs. They represent action, results, and adventure -- not bureaucracy.

Another problem is the media itself. Much reporting on technology and especially spaceflight reporting is filled with serious errors. The emphasis is on show biz and superficiality. I'm not talking about the "correction box" type of errors here. We all make those and always will. I'm talking about some really poor explanations of spaceflight being conveyed to the American public. Some examples: the Washington Times here a few months ago carried a large article about NASA having an offensive Space Shuttle capability based on a supposed astronaut-manned maneuvering unit ability to kill 12 Russian satellites in a seven day mission. The same guy that wrote that gem followed it up a few weeks later with a big article about how NASA Administrator Jim Beggs has assigned himself to fly on the Space Shuttle.

We laugh at this stuff, but the paper was serious. A lot of people probably believed that nonsense, and when the first manned maneuvering flights are made you can bet the Soviets will use the Washington Times' inaccuracy as propaganda.

Turning to Shuttle mission 7, wouldn't you agree that most of the stuff written about Sally Ride was pure drivel. Of course, it was historic, but the media's show biz hype was ridiculous. I think Ride's experience also showed a pretty shallow understanding in the media about what a Shuttle crew actually does in both preparing for a mission and then flying it. Another mission 7 example: NBC routinely called the Space Shuttle a capsule. That really is an insult.

Then there was Newsweek's large drawing supposedly showing how the mission 7 crew -- heroically led by Sally Ride -- was going to deploy the German SPAS pallet satellite. Unfortunately, Newsweek's drawing actually showed deployment of the shipping crate used on the ground to haul the SPAS around in. It looked like this giant coffin going overboard -- Newsweek did not know the difference between the crate and the actual spacecraft. It's bad enough that the writers make such errors. What's just as bad is the editors that oversee the writers don't know any better either.

In addition to these errors, I know the media has not adequately conveyed the adventure of space flight on the Shuttle and the marvel of this spaceship to the public -- remember NBC's capsule description. I know that adventure is there, I've seen it in the simulators dozens of times.

You need to convey to the public the adventure side of future space station operations as well as the technological side. A great place to start would be the extraordinarily deep space observations that could be made with large observatory structures built in orbit with space station support. The discoveries we expect from Space Telescope are just the beginning of what you could do in our lifetime with station-based astronomy and astrophysical observation.

Most members of the media and NASA public affairs believe they have to be superficial in explaining space flight lest the public not understand. I don't believe the public is given enough credit for its ability to understand space flight in general and specifically what's going on in individual programs like Shuttle. My experience has been when you give them more accurate detail and less show business, they appreciate it. It stimulates their interest.

I think if you can find a way to push and stimulate that extra accuracy and detail that's lacking, you will make the public feel more a part of the space adventure and secure an even stronger base of support for a new national goal such as a space station.

Comment: I would second most of what Craig says, because I work from a slightly different angle, being on both a daily newspaper and a monthly magazine. I also see a lot of the drivel and incompetence going on out in the field, and it irritates me because I think my profession is degraded when improper reporting like that takes place. Long before calling the Shuttle a capsule, NBC News did a thing called the little spaceship that couldn't in October 1980. Well, the little spaceship has gone on and done it over and over again, but, to my knowledge, NBC never backtracked and said, "Well, maybe we were wrong."

This goes on not only at the network level but at the local affiliates -- it's just now that some degree of expertise is beginning to be developed at the TV stations in Huntsville about the space program. I've largely, regrettably, been without competition for a long time.

I would suggest that NASA and its contractors be a little bit more aggressive in calling errors to the attention of the press. We all inadvertently screw up occasionally -- we get in a rush occasionally or we just misunderstand something. We will correct those. But there are people out there who are reporting, giving the image of being knowledgeable about things, and completely reporting it backwards. Those are the ones you have to go to and try to turn their heads around to the front. Even if they disagree with you on the philosophy of why we should do space station, or should not or whatever, go after them when they report known facts wrongly. It doesn't do my profession any good when that goes on and certainly won't do the taxpayer or the Agency or the contractors any good if they don't correct it.

Comment: I have a request that Mr. Covault seek out meetings with editors and publishers to tell them what he's told us today. A major problem that I find is that editors, especially on small dailies, don't consider space newsworthy.

Comment: Is it possible for NASA, Aviation Week & Space Technology, or anyone else who has got any clout to lean a bit harder on the U.S. Air Force, the DOD, or whoever is the keeper of the secrets to let us have a bit more of the intelligence information, the high-altitude reconnaissance inventory which is available to them on Soviet activities?

I seem to remember when Aviation Week published a planned view of the alleged Russian orbiter, the same edition carried an editorial by Mr. Gregory saying that it was a double-delta while the article showed a straight-delta. In other words, the data had been doctored.

I think the same thing occurred in the case of imagery of a different sort concerning the IUS malfunction. Would it be possible to press a little harder for release of U.S. Air Force imagery data that is relevant to the task of spreading fact rather than fiction?

Freitag: The comment is well taken. This is a subject that's under continuing review. I don't think it's a subject I can comment on here, but we will keep the problem in mind, as to both this comment and Craig's most perceptive remarks.

Covault: Relative to the Soviet orbiter picture, if I recall, that came out in Caspar Weinberger's latest discussion on Soviet military power, and I think we ought to be glad DOD turned that loose. That was an important thing that they did decide to turn loose.

Relative to the shape of the vehicle, the first thing I thought of when I saw it was the paper airplane that's on the seal of the Space Command. It looks very much like that type of planform.

We checked and these guys said, "Hang tight, that's the way it looks" so that's the way we ran it. But, within two days, people came out of the woodwork who'd seen that picture and said, "That looks just like our orbiter." It turned out that they had, in fact, airbrushed the picture to a new design.

I think there's an argument for holding back the imagery when it shows the best capabilities of the system. But when you've got a vehicle that's the size of the broad side of a barn, as all of us know that reconnaissance spacecraft is, it's silly to play games. I think a good artist's concept that just keeps the basic size and shape, but maybe gets rid of the rivets that would show how good you can be when you want to be, might be the way to approach it.

C. James Dorrenbacher: I have to agree with most of what Craig said, but it sure hurts when he says we've got to get those younger guys out there. I've been chasing this space station rabbit since about 1961; and if we don't catch it this time, I don't think I'm going to live long enough to see it. So maybe we'd better get some younger guys out there -- you're right, Craig.

I would like to reflect a bit on what we've learned since 1961. By the way, NASA didn't run that initial competition in 1959 -- it was a design concept competition sponsored by the London Daily Mail. They received two proposals, one from McDonnell Douglas and one from the Boeing Company. We won't mention who won.

But we learned a lot of things that are applicable today, most of which I've heard during this symposium and some of which seem to me to need further emphasis as we embark on the program.

Figure 5 is a brief summary of what's gone on since about 1962, in space station studies. Incidentally, something called OSSS is down there in the lower left-hand corner --that was one of the first DOD space station studies and you really ought to get it out and read it. It's not all that bad.

We've gone through, really, three eras. There was an era that took place in 1962-1969, in which the principal problem was that there was no united position within NASA. Everybody in NASA had their own idea of what the problem was and their own idea what a space station ought to look like; something like seven years was spent fighting each other on this problem. We dealt with crew sizes ranging from 4 to 50, we dealt with stations that were zero gravity and stations that had the spin to provide an artificial gravity field. Behind all that, however, there was some pretty good requirements work done, much of which is still valid today and much of which has been carried into the second era.

Incidentally, one of the things that came out of those early studies was the conviction that autonomy was essential. Onboard data processing was tougher to do in those days, but the need was obviously there. Most importantly, it was recognized that the cost driver of the space station system is really the logistics, operations, and resupply operations. Once you get by the original investment, that's where the money's going to be spent. Frankly, I haven't heard an awful lot about that in the last few days, except, "Oh, we got the Shuttle".

Well, the Shuttle is a thing. The Shuttle is not by any means a logistical operational system. I would admonish anyone estimating the costs of this program, whether it be 5, 10, 20, or 30 year projections, that they had better start boring in on this aspect of the cost. If you don't do this carefully and accurately, the program won't survive. That first space station program survived into the 1970s -- we even got into a Phase B. The big debate there was really more where should we start, on the logistics system or on the station itself.

In this competition, the Shuttle program got the nod -- I think properly so -- and the station program got the back seat. That is why we are where we are today.

Thus, a couple of years ago when General Abrahamson started to generate the plan that is now in being, he asked a number of industry representatives for our reactions. The first thing I said was, "For God's sake, don't start another Phase B." Let's not go out and draw a bunch of inboard profiles. Rather, let's get some requirements and things of that ilk -- I think we're now calling it architecture, which is a good word -- together, and let's understand what we're about. Then when we finally understand what we're doing, how much money we think we can spend, and how fast we can spend it, the technology is available to configure the station.

Of course, there are going to be debates and a lot of decisions to be made in the area of modularity. Incidentally, modularity was a fifth thing that came out of those early studies: it had to be growable, it had to be adaptable, it had to be a changeable device.

So, my first comment is that I think NASA is on the right track. I think they started at the right end and I think they're spending their money on the right things, with the possible exception that I have not yet seen a good logistics plan or the thinking behind it.

The real question is, where and when do we start, how much are we prepared to spend, and how rapidly are we prepared to spend it? In my mind, a space station program is going to be different than the Apollo and Shuttle programs. It's going to have a hard time surviving. It is not going to be the glamour girl on the street, because once you get it there, there will be a lot of hard, experimental development work. This has some appeal to the public, but not necessarily the instant peak glamour that the others did.

So, we have a two-fold problem. One is getting it started and the other is how to keep the engine going once it gets started. To supply the station and to keep it doing anything will take a continuing, fairly high level of investment; but that's where the payoff is.

I suggest that you don't try to justify the program on a cost/benefit basis. It's a case you can't prove. If you want to make an OMB cost-effectiveness case, you're not going to win it. But you can make a benefits case out of it, and those benefits, as has been pointed out, can have a great deal of political attraction, not only to the zealots, but to the man on the street and to industry -- and I'm glad to see industry getting genuinely interested in this program and what it might do for itself.

Don't expect miracles. Industry can't take chances beyond certain limits, but I think that

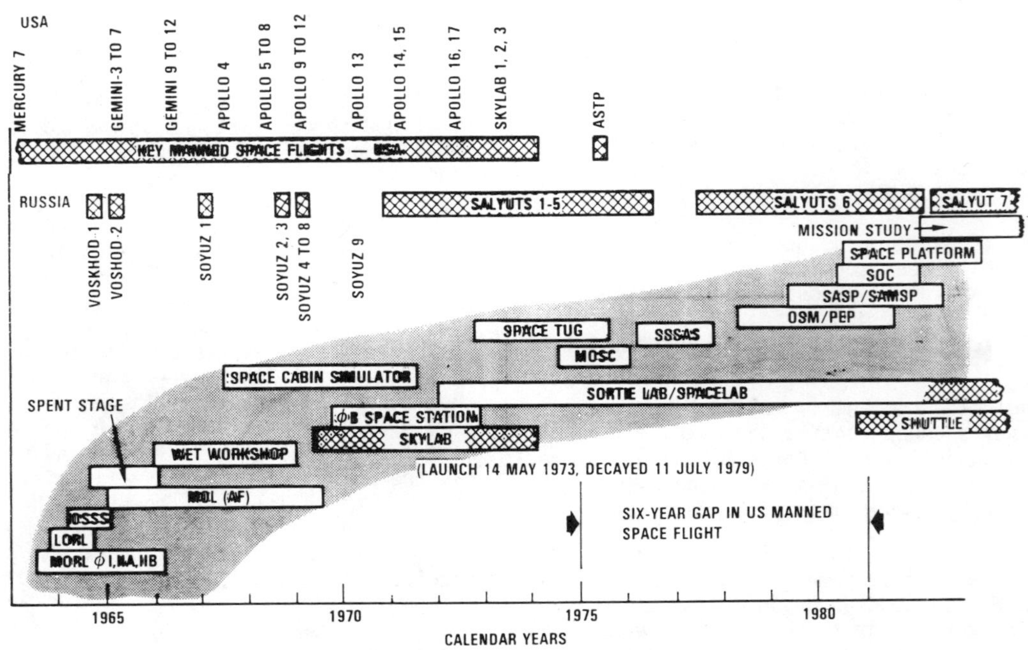

Fig. 5 Space station heritage.

when they realize that the program is real and that it's going to be around, you'll see industry stepping forward to do things in that space station.

Above all, when you construct the initial budgets for the program, make them realistic, because this program is going to have to survive on meeting its objectives. If two or three years into the program, 50% or 100% cost overruns surface, I think the program will die. It hasn't got a window or a finite one-element objective it has to meet. However, it has to survive on its performance and its utility, and that has to be a long-term thing that must be brought out in both the budgeting of the program and its initial presentation.

There are probably more things in Fig. 6 than we can afford in the beginning. We're probably going to have to make some compromises as to what comes first, how much of it we can afford, and when we can afford it. That's where the going will get tough between now and 1985 in deciding just what sort of program we're really going to embark on.

I think we now have a fairly united NASA on the program. You never can get everybody to agree, but contrary to what we had years ago, NASA is fairly united and I see much of the rest of the community coming into agreement. There will always be those who will say no, but you have to accept that they have a point and you have to argue around and over that point.

You need the concurrence of the user community, whether that be industry, science, or applications. Be sure to find out what the logistics system is really going to cost. And last, but probably not least from an industry standpoint, don't start a Phase B unless you really have the support of at least the current Administration. These false starts to industry, with the large expenditures that go with Phase Bs really wrap us around the axle. We really are looking for something of substance; then we'll carry our part as we always do.

To summarize, the technology is in hand and the time is right. I think when the station comes a reality, DOD will find that there's more value there than they can perceive now. Requirements don't generate the article, sometimes the article generates the requirements -- it's always been that way.

I'll wind up with two more visions of the future. If we do it right, we have a station like Fig. 7 in 1990, and if we're still learning something by then, we might get Fig. 8 by 2000.

Cook: The point that was made here a moment ago of work that was done 20 years ago being pretty darn good -- it occurs to me that over the last year or so there have been four or five areas in which I've been involved where we looked back at things that were generated 20 years ago, and the reactions have been "Boy, those guys were really smart," "They really did a good job back then," "They had it just about right at the time." I just want to say I hope that 20 years from now, whoever is sitting here will look back and say, "Gee, those guys did some good things."

George E. Mueller: The past sometimes is a prologue, but not often. Unfortunately, you have to repeat mistakes in order to really understand them. But there are several challenges, and I'll mention just three of them. One is maintaining the support of the public. In my definition, the public support can be spelled "OMB" and, to some extent, Congress.

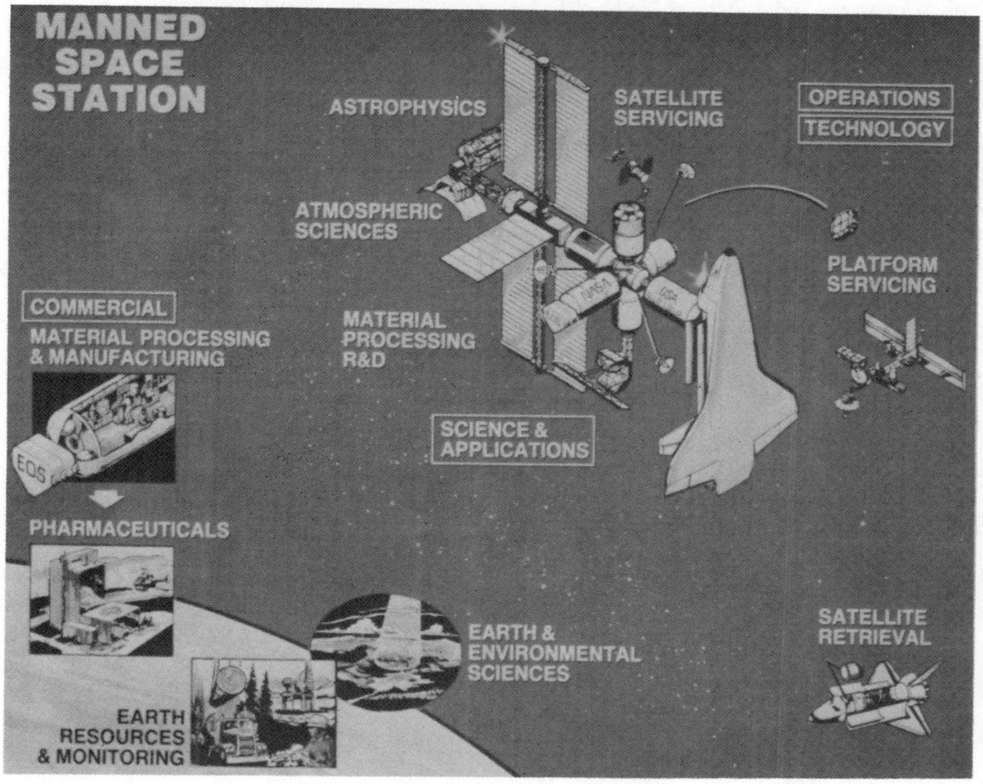

Fig. 6 Manned space station.

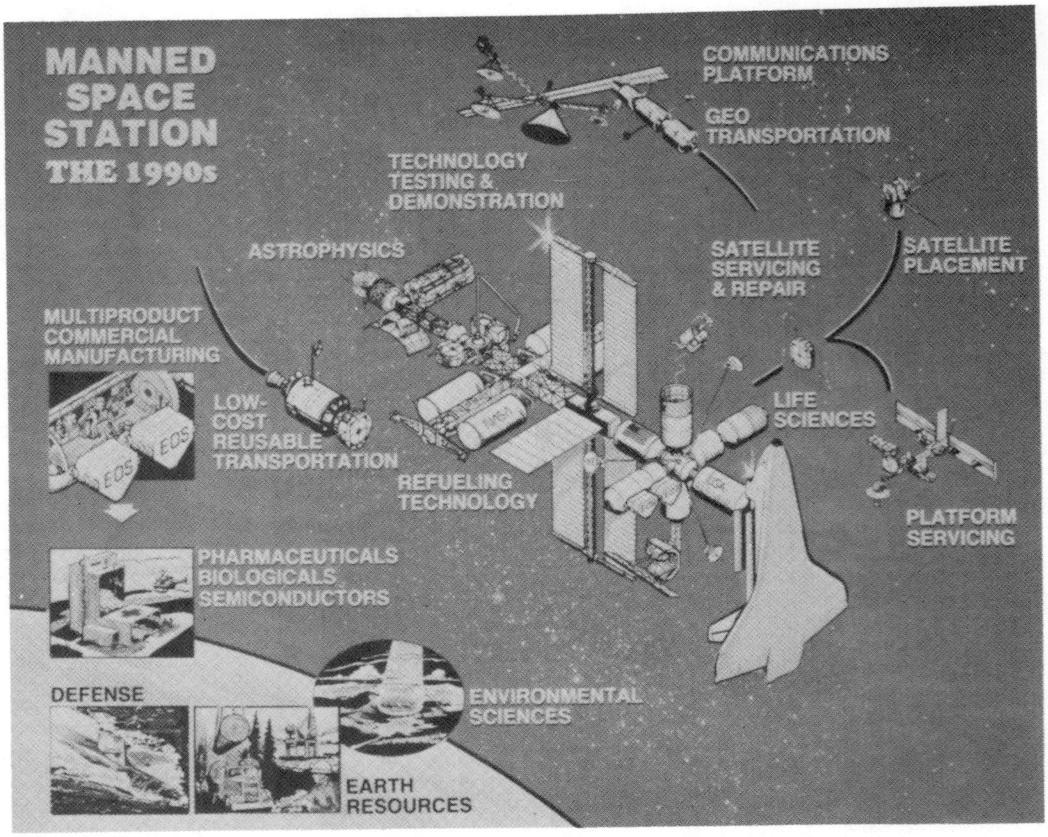

Fig. 7 Manned space station in 1990s.

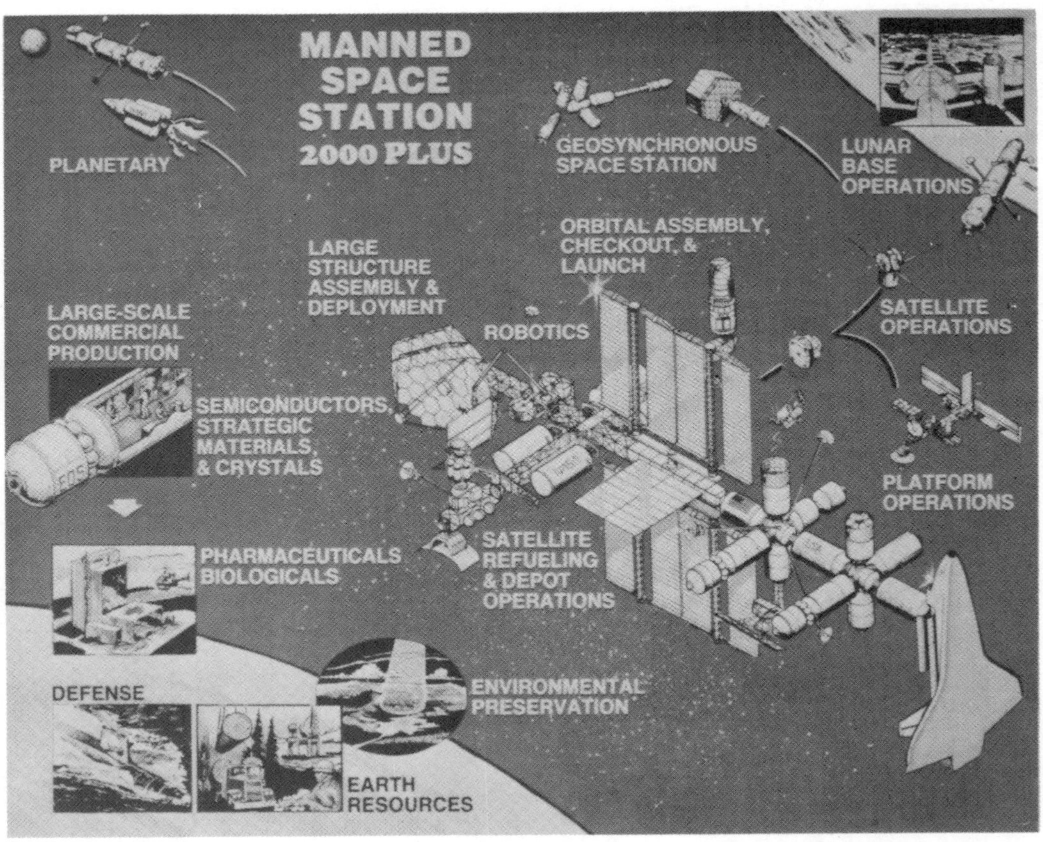

Fig. 8 Manned space station in 2000.

In order to do that, we need, I think, to be sure that we have real international participation in the space station. This means that they have meaningful things to do and that they do meaningful things. They must become a part of the operation. You can say that we ought to maintain our technological lead and all of the other things that mitigate against true international cooperation, but my observation over many years is that the only way to maintain a technological lead is to work on technology, to push it. You are not going to maintain it by trying to conserve it or hide it.

Another way of maintaining credibility is to make more efficient use of the whole Shuttle package. Going into orbit, the Shuttle has both a relatively large amount of aluminum and a considerable amount of residual propellents. Not many people recognize the fact that we could put a larger payload in orbit if we didn't deorbit the external hydrogen tank. That actually costs payload without gaining payload. Since we have the sun as a source of energy, it seems to me we ought to take the aluminum up into orbit and do something with it -- if nothing else, melt it down and cast it into shapes that are useful. Also, we ought to use the residual propellents for various activities in space. There are several thousand pounds of hydrogen and oxygen left over.

A third way to maintain public interest is to have some simple, overriding, easily understood mission objective. For example, the purpose of Skylab was to answer the question, "Can man survive and perform effectively over a very long time in zero gravity?" And it demonstrated rather conclusively that that, in fact, was possible. That finding was essential to the follow-on missions envisaged when we started the Shuttle program.

The second challenge is the cost of logistics. Somehow or another, we've gotten ourselves in a mode that says that the Shuttle we designed 15 years ago must now be the answer to all space transport needs. It is fairly clear today that we know how to do a much better job than we did when we started the Shuttle 15 years ago.

For example, the single staged orbit and return is a quite feasible concept at this point in time, and we really ought to be building the second generation Shuttle to support the fifth generation space station, at least in the study form. Or at the least, the second generation station, since I regard Sklab as being the first generation.

The third thing is that we seem to have gotten ourselves hung up on the idea that there will be a space station. And maybe that's a big enough step. But I think it's clear, if the space station is to be useful at all, there must be more than one -- more than one Western space station, let alone the Russians. I would guess that the first economically attractive space station will be a geosynchronous hub for communications and navigation -- that is, really for establishing the basic parameters on how to survive over long periods of time in interplanetary flight...because someday we're going to go to Mars!

<u>Neil Sheeley</u>: I've identified three issues for the space station: 1) the role of man, 2) the relative roles of science applications and the military, and 3) requirements for scientific success. I don't pretend to be knowledgeable about requirements for other kinds of success and, in fact, I can't even pretend very much about requirements for scientific success, but I will address these issues in terms of the experience with the first space station, Skylab.

I think the role of man is an easy issue to make projections for. In fact, it's probably so easy that it is not really an issue. In my opinion, it's essentially the same as it was with Skylab. As demonstrated by the first crew, we'll have space station repairs. Remember the parasol that was used to replace the heat shield on Skylab and thus drop the temperature, making the laboratory habitable? And then there was the release of one of the main solar panels to provide power for the laboratory. Space station repair is well documented.

There were other repairs, of course, for example, of the instruments. Many people have forgotten that many of the Apollo telescope mount (ATM) instruments had problems. The coronograph had debris removed from its occulting disk several times during EVA. The crew replaced a jammed film transport and pinned open telescope doors that had jammed shut, among other repair jobs. Four of the five ATM instruments recorded with film had to be reloaded. Then, of course, there are specialized observations: in the case of solar observations, selecting targets of opportunity such as solar flares, placing a spectrograph foot on the flare, and taking a spectrogram of the target of opportunity.

There were the new things -- tests, improvements, and types of observations. Ten years ago the SX70 was a brand new camera, and one thing that didn't work on the first manned mission (there were three manned missions during the Skylab program) was an XUV monitor. Actually, the instrument worked but the cathode ray tube image was so faint that it was not usable. An infrared viewer was taken into orbit on the second mission which enabled the crew to see the solar changes. The SX70 allowed them to record these changes so that they could put them on the Velcro mount next to their console.

What man will not do, in my opinion, is routine observations that can be done better by automation. In fact, we would prefer to make our observations from the ground by automation. We are always trying to do that because the excitement of science is not being on the mountaintop freezing to death trying to get that spectrogram. The excitement is learning, of which there are two aspects: seeing the new observational fact and, the thrill of understanding.

But the thrill of understanding is a human process that probably doesn't take place in space. Maybe someday it will; but, even on the ground, when you have a good idea on a mountaintop -- later, when you get back down to 1500 feet or whatever, you find the idea wasn't so good after all.

The second issue -- the relative roles of science applications and the military -- I think could be difficult. But maybe it's not an issue -- perhaps the military has bowed out.

Furthermore there is a strong overlap as far as science and applications are concerned. Applications, in the sense of weather forecasting, etc., fit in very nicely with science, such as solar terrestrial studies and understanding the sun/Earth system, including the heliosphere.

However, to review what went on during Skylab, there were several major activities that I can recall: the Apollo telescope mount; the solar instruments; and the medical experiments (remember the onboard bicycle, drawing of blood samples, etc.?). There was also something called the Earth resources experiment package. Some beautiful pictures were taken of the Earth during that experiment.

There was another general category called corollary, which involved a variety of smaller experiments, some of them in astrophysics. For example, one of the astronauts obtained XUV stellar spectra; two scientists from the Naval Research Laboratory obtained some very beautiful auroral images from, I believe, a hand-held camera. And, there was materials processing, of which I know little. Finally, there were some very special experiments from high school students. For example, remember the fish swimming in Skylab? The object was to find out which way they would swim -- that is, which way is up? You may remember that "up" turned out to be, at least for the second generation of those fish, toward the light. The first generation, if I remember correctly, was really mixed up.

The story of the ants on the recent Shuttle flight had much publicity -- but do you remember the spiders on Skylab? And I raise a trivia question -- do you remember the names of the prime and backup spiders on the Skylab mission?

At first, there was intense competition among these various activities. Everyone wanted time blocks for his experiments. About 24-36 hours in advance, a calendar was filled in -- actually it was filled in in the middle of the night, sometime between 11 and 7 -- a terrible shift, by the way. It seemed that nothing happened at all during the night until just before the shift ended when everything happened. We had a very competent scientist who was responsible for making the decision as to who got what time block. This intense competition disappeared during the third mission, especially when it was extended from 56 to 84 days. In fact, the question became what to do with this time. I know that at the Naval Research Laboratory we were a bit frustrated because we didn't have enough film -- we would have liked to photograph all the time of the full sun (there was only one experiment that was not photographic).

Another thing that you may not remember is an experiment we did in January 1974. In preparation for the future Shuttle, we let Ed Gibson, the science pilot, make his own observing program for the day. He has written about it in some of the post-mission literature.

The third issue, requirements for scientific success, is perhaps controversial in some respects -- but issues by definition are controversial. The answer, in my opinion, should be a facility to help us do the same kinds of things we've done in the past, except better, cheaper, more flexibly, and more permanently and continuously.

Let's look at failures of Skylab. After all, we know that Skylab, especially with the solar experiments, were a great success. However, nothing is perfect. What were the things wrong with it?

The solar observations were not completely new. The rocket observations prior to Skylab had shown coronal loops, coronal mass ejections, coronal holes, and some of the other things seen during Skylab. What Skylab did was to show these features with better spatial resolution, and it showed them more frequently and for a longer period of time -- pictures were taken at intervals of time ranging from seconds to, of course, essentially a nine-month mission and they were made at an active phase of the sunspot cycle. It led to some very important and convincing results, such as the fact that coronal holes are the source of high-speed streams in the solar wind. They cause geomagnetic activity and, in fact, produce the auroras that the experiment with a hand-held camera showed -- those beautiful auroras from the coronal holes.

In fact, newspaper headlines occasionally report a great flare to light sky. Someone at the Naval Observatory sees this flare and they assume that within a few days there will be a shockwave that reaches the Earth and produces various disturbances. But no headline ever reports a coronal hole to light sky. Yet we know that the coronal hole is over there on the east rim of the sun, that it's rotating around, and that in a few days it will be pointed toward the Earth. And about three days after that the high-speed stream will reach Earth and produce not a single incidence of geomagnetic activity, but a rattling of the magnetosphere for several days.

The point is, Skylab was a long-duration mission much better than anything we had previously, but it was not long enough. And it was soon ended.

Many people spent years in preparation for the mission, some starting in 1965, some as late as '73 -- but not one of the original principle investigators was there when the data came back for analysis. It was too long a time. This exciting new data became old data and grew progressively older until splashdown, Feb. 8, 1974. So people invested considerable amounts of their time for an observatory that lasted only nine months. That is not very rewarding.

It is now 10 years later and we are still doing exciting science. It's not the same stuff that we did in Skylab. In some respects, it's more exciting. We've continued to look at coronal holes, at other phases of the sunspot cycle down through sunspot minimum (a declining phase during Skylab). We have gone down through sunspot minimum, up through solar maximum, and we're now coming down again. We have found ways of looking at the coronal holes from the ground. And we have discovered that the association between the coronal holes and the high-speed streams breaks down at other phases of the sunspot cycle. Instead of Skylab, we have unmanned satellites observing coronal mass ejections from the sun. In this particular case, because of a unique spatial relation between Earth-orbiting satellites observing the sun and the Helios I spacecraft in orbit around the sun, an orbit which causes it to dwell for six-month periods on the east and west limbs of the sun (as seen from Earth), the Helios spacecraft can observe the shock waves at 90 deg to the sun/Earth line -- the same place that we are best able to see mass ejections from Earth. Thus, we are now beginning to understand how these great eruptions from the sun make shocks and disturb the heliosphere.

So science is marching on. I think that one of the requirements for a space station to be of value to the scientific community is that it somehow provide a living, growing research environment -- that it not be a one-shot affair.

I don't know if this is a good analogy, but maybe there should be something akin to the National Science Foundation, except that it should be within NASA -- maybe the NASA Science Foundation -- which would be a strong arm of NASA that could perpetuate and insure the continuity of vital scientific activities.

Finally, flexibility is important. Even in the nine-month Skylab mission, there was considerable flexibility, i.e., data during the first mission were used in planning the second mission and exposures were modified as discovered.

With a permanent space station, the old problems will be replaced by new ones -- for example, the study of coronal holes at other phases of the sunspot cycle. Therefore, it is important to maintain the complimentary missions to the Space Shuttle. We should continue sounding rockets, unmanned satellites, in both Earth and solar orbits (i.e., the Pioneers, the Voyagers, the International Sun-Earth Explorers, the Helioses), as well as the ground-based observations to provide the continuity so essential to scientific experiments.

Freitag: We are extremely sensitive to the comment of how long it takes at the present time to go from concept to realization on space projects. As we have continued to move forward with the Shuttle, we are seeing what I consider a fairly remarkable shortening of that time. I thought it extremely significant that only a couple of weeks ago there was a NASA/McDonnell announcement that one of the test engineers at McDonnell Douglas had been selected to fly with their commercial equipment aboard a Shuttle flight in the spring or summer of 1984.

It's interesting to note that in the last two or three years the lead time for, say, selecting a payload specialist has dwindled from three or four years down to six or so months. I certainly think that as we proceed with the Shuttle and then the space station we will have bettered that time. If we can get away from the embellishments of some of the more difficult concerns such as safety and similar considerations, perhaps we can respond to your plea on that point.

Richard G. Smith: After hearing the presentations and comments of the last few days, I would like to share with you some of my perceptions -- or what you might call my concerns.

I grew up in a NASA development center, then served a year in the pergatory of Washington, and was fortunate enough to go the Kennedy Center and be thrust into the element of operations. I found that to be a very enlightening and revealing transition, one that causes my concern about the space station.

Early in the Saturn program, while discussing with one of our stage contractors how a certain logic should be designed, I and the other Marshall people were accused of being ancestor worshippers. At the time, we were debating whether the logic should be relay or solid-state.

Now as I look at how we have developed systems, that statement has more meaning to me -- we are ancestor worshippers in a lot of ways.

Change is very slow and very difficult. The Shuttle grew out of the techniques we'd used on Apollo, Skylab, and ASTP. These techniques were extended slightly to develop the Shuttle program. That was appropriate. We probably did not give enough attention to the operational aspects of Shuttle during the early formulation of that program, but we are going through that evolution today. The element of logistics comes slowly, but it comes surely. We too often think of logistics as a sort of spare part operation -- but that's just the tip of the iceberg associated with logistics. We did not pay enough attention to that.

In a few areas in the Shuttle program, we ventured out a little boldly and were marching to a slightly different drumbeat than most of the rest of the system. Thus, some interesting things happened as we were getting ready for the first launch. We all were going through a period of taking a second look at where we stood -- we looked under all the rocks and we considered all the risks. Were we ready to launch that first bird?

The Kennedy Center -- and I can't take any credit for this, because I wasn't there -- had developed a new processing technique called Launch Processing System, a distributed computer concept, antique by today's technology. The center had gone a step further and used engineering language for all of the programming. The software was, in fact, an extension of the test procedures. Since a lot of documentation hadn't been required, the system engineer responsible for the test was responsible for the software. There was no elaborate validation program.

People came in and started looking at this process. Every group that came in looked at it as if it was a set of flight software, written in machine language, which had to be compressed to save memory in the onboard computer. They also assumed that the program had gone through an extensive amount of validation. Deliberately and premeditatively, we had taken a different approach. Half the comments we got weren't even useable because they didn't grasp the concept, which was different to start with.

Hearing the technology discussions at this symposium, I'm not worried about the techonology available for a space station. We'll have to push in a few spots and there will be some problems with it, but I'm not worried about that. What I'm worried about is: are we as a group able to change how we do things, really ready to design an evolving space station and to consider all of its operational capabilities and complexities from the beginning? Are we capable of developing the logistics required for that? We are used to designing specific missions. Skylab was three specific missions, preplanned with all the food, all the water, etc. at the beginning -- basically no resupply.

A Space Shuttle mission today is a preplanned, predetermined mission. Nothing like that's going to happen with the space station. This may not be totally foreign to us, but it is very different. We're going to have to change our thinking patterns and let our minds wander a little freer, and be a little less conservative than we are today. I think we've got a big challenge to do that. I'm not worried about the group of people that is working on the space station testing. I'm not even worried about most of the group at this meeting. Oh, I am worried a little bit, because I've known most of you for years; I'm not as young as I used to be, and I think with age we sometimes get a little inhibited by our thought process and a little more resistant to change.

But we have got to be the leadership in cultivating the attitude that this is a different animal, and we've got to design it that way. We have the tools and the capability. That's not the problem. I'm not worried about the architecture of the overall structure nor architecture of the onboard data system.

That is the concern I have about this program and I think it's one that we all must give consideration to. This shouldn't be perceived by anyone to be negative. I think the station is the right next step and I think it's the right tool. I agree with the statement that we shouldn't try to set it on a cost basis -- after all, we don't know how we're going to use this thing. It is a bit of the old chicken and egg problem. I'm not surprised at DOD's position today. About halfway through the program we're going to see them climbing aboard -- I'll bet that right now. Without a checkbook, but they'll come aboard.

The challenge is to let our minds be really free in making the adaptive system that we have talked about during this symposium. And, frankly, we must do things differently. We're not going to be able to check out and verify everything on the ground as we do today.

I recently took a tour of the new Trident submarine. I kept comparing it to a space station. We've talked about autonomy here. But, let me tell you, that Trident captain has autonomy. He's out for 70 days with almost zero communications. He doesn't transmit once. We ought to look at that and understand what some of his problems are. We could probably learn from that situation.

Question: We've heard a lot of concerns addressed during these three days: the autonomy, the data architecture, the system architecture, making sure that enough room is left on space station both in terms of budget and physical space, etc. so that science and applications can be done easily; concerns that there is not an analogous repeat of the disastrous inertial upper stage decision that was made back in the mid-70s; keeping it user-friendly, using the appropriate levels of technology while still trying to improve technology, etc.

Jim Beggs keeps on talking about being able to buy the space station by the yard. What is the minimum size of that first "bolt of cloth" that has to be bought to make this thing effective and worthwhile? If you can't get that minimum size, will NASA have the courage to say, "If we can't do it right, we won't do it? We'll stick with the sortie-mode and free-flying unmanned satellites?"

John D. Hodge: The question is one we've been wrestling with. Mr. Beggs did say you can buy it by the yard, and I happen to agree with him. But you have to buy it by the right yard. And it's important to know what the whole bolt is before you decide which yard to pick first.

As Luther Powell has said, we're investigating how to build a system that captures everything fitting the mission set we envision in the year 2000, and then backing away from that vision to see what makes sense at the beginning.[2]

What we have to do is to look at the overall concept very carefully from two points of view: 1) that the first piece we build has a reasonable, useful capability and 2) that it has the ability to go from that initial point to the point we want to be by the end of the century. That is an engineering challenge of a kind we have not faced before. This is not a dead-ended program in that respect.

I agree absolutely that the biggest challenge we have is our ability to do business a different way. It's not that we can't build this thing. I don't think we have a problem building it. We've got enough smart people around this country to do that. It is a management challenge.

With regard to the last comment asking whether we have enough chutzpah to say no if we don't have enough, I don't know the answer to that. I think we must put our story forward to the people who have to make the decisions assuming a decision is going to be made in a positive sense. We must be eloquent enough to make the story understandable. Also, we must look at the whole in order to make the parts successful. We are going to try to be eloquent and we are looking to you to help us be eloquent in that debate which will be taking place in the next three or four months.

Question: In talking over the years, to many of the investigators who are now on the Shuttle missions, I think one of the things they believed was that there would be a chance, after their first flight or two on the Shuttle, to unload their experiments, leave them in orbit for a longer time, and regain more of the investment that Neil Sheeley stressed so very correctly.

Many of them were thinking in terms of some sort of a platform that would be developed early. Yet during this meeting, I've heard people say that an early platform would reduce the focus on a space station in the 1990s and take away monies that could be better spent for the space station. How are you going to tell those people who devoted many years to developing experiments, hoping to fly them again for a longer time in the '80s, that they'll have to wait until the early '90s, by which time, their equipment will be obsolete? -- "Sorry, we led you down the primrose path!"?

Hodge: We did make promises early in the Shuttle era that, in fact, we have not been able to

---

2 See Luther Powell, "Architectural Options and Development Issues."

keep. I think it is vitally important that we be able to maintain those assets and resources in space. In many respects, that's what Frank McDonald[3] was saying yesterday -- that these things are getting so expensive and take so much of a person's time (sometimes half a career) to get going, we have got to be able to continue to accommodate them in a reasonable time frame.

There is no question in my mind that a platform, unmanned and sophisticated, is a part of the future. If you look at the development cycle of the space platforms, it has both unmanned and manned elements. There's not that much difference between the two in the time needed for development. If the two are not designed together the chances are that each will have to be redesigned every time we need another element, thus forfeiting some of the cost and time savings that would accrue from designing them together. Perhaps we could build an unmanned platform a year earlier than if both were done together. But, there seems to be no doubt in anybody's mind that in the end we've got to put the manned element onto it. And if the two configurations are done sequentially, they both will end up costing a lot more money.

My personal feeling about that is that we should not talk about manned and unmanned elements as though it was an either/or situation. We're going to have both. The question is, what is the best way to put those two pieces together in a symbiotic way so that we can get the best advantages out of both of them.

One thing absolutely definite is that we should not raise the expectations of the users about what we can expect to give them, as we have done that in the past. It is going to be difficult to keep the faith.

Comment: I view the space station in the big cosmic picture. It's a stepping stone, it's a base camp, it's the enabling technology mentioned earlier. It is something that will lead to all sorts of stuff that I predict, partly from intuition, will lead to an explosive renaissance of mankind in space around the year 2000.

I suppose I'll continue to always be an advocate of the farther future; thus, one of my concerns is that we make sure that the nearer future leads to the farther future and that options are not cut off. An example of this is the possibility that the volume will be quite limited. It seems to my intuitive sense that you want to fill the cargo bays as much as possible with modules that are as big as possible. If the total program costs of the initial space station version are 2-3% more because you have modules that are longer than might be currently considered, at least you have more space to spread around in.

I have a feeling that the space station of the 1990s is going to be much different from what many engineers envision as being a sort of antiseptic, battleship grey facility with lots of consoles and Star Wars/Star Trek science fiction kind of fantasy. I see it more as a relatively nice accommodation, something that would maybe more resemble a hotel in ambience and comfort because, in the 1990s you're going to be dealing with so many different crew members. It won't be like a submarine crew with a single-minded mission and purpose; it's going to be a very diverse group of people with many different missions.

My own guess is that, instead of having lots of consoles and bicycle exercising equipment, there will be space made available for people, for example, to psychologically withdraw from the rest of the group and sit and look out a Jules Verne type of oval window amid rocking chairs and potted palms -- just to stretch the images. Maybe it sounds frivolous when first broached, and it certainly sounds frivolous within an engineering/government community, but what I'm suggesting is that, as the 1990s approach, I think that every bit of volume is going to be wanted, needed, and used.

If people want to counter and say Parkinson's Law will hold and it's going to cost more than 2-3% more to have the larger volume -- let's say to double the volume -- then my answer to that would be, do nothing with that volume. Just cordone it off and discipline the program so you pay only 2-3% extra. Then further downstream, the space station can grow internally, and you actually save money instead of having to build new modules earlier.

That is just an example of the kind of concern I have with the program. Otherwise, I think you're right on target and wish you the very best.

Hodge: When we thought about putting this symposium together it was September 1982 or so. It seemed that somewhere around the anniversary of the setting up of the Space Station Task Force was a good idea, because it would be time for us to take stock in what we have done, to decide what we needed to do, and when we should do it.

What we wanted to do was to get everybody involved -- to really address the issue of why do we want to build a space station, as well as what it was we wanted to do.

I want to thank everyone who has participated with us in this last year. It's been very exciting, it's been fun, and it's been somewhat frustrating. But I think we have opened a communication path to everybody -- the users, the aerospace industry, the potential commercial people and to potential international partners. The dialogue has been started.

In this first year we're beginning to understand what it is that we're going to have to do in the year or so ahead before we start to get down to the serious design. Figure 1 is a chart that says we're somewhere on a carousel, which we've got to keep going around until we really know what it is we're doing. We're going to go through at least four times between now and the time that we bring the design contractors aboard.

We have some challenges. We have some cynics. We have some users who don't believe we're going to follow through. We have some people who are not sure what they're going to use it for yet. But the dialogue must continue.

---

3 See Frank B. McDonald and Kenneth J. Frost, "Space Station Architectural Issues: The View from Science."

We have some technology challenges. I think we will be able to meet them -- we have to be ready to meet them by the time design decisions are necessary. Those must be thought through very, very carefully. However, we must be sure we don't try to tackle problems that aren't really there.

We've got this real big challenge: can we build something that in 10, 15, or 20 years will still be useful?, will be bigger than it was when it started?, will be more capable than it was than when it started?, and will have different technology in it by the time we're through?

We've got to know how best to use the Shuttle between now and the time that we get a space station up there. I think it's absolutely essential that we use the Shuttle to its ultimate as concerns testing, doing experiments, and furthering our understanding of how to operate in zero gravity. That's particularly true with regard to the development of the human capability itself -- how can we best use man to increase productivity?

I thank all of the speakers for taking the trouble to carefully put something thoughtful together. Also, NASA needs your help. Perhaps we should do this again next year to present what happened on the second time around and the third time around and the fourth time around that carousel. Then maybe we will begin to understand what it is we're going to do and how we're going to go about making it happen.

☆ U.S. GOVERNMENT PRINTING OFFICE : 1983 O - 424-649